T0293669

Handbook of Polyhydroxyalkanoates
(Bioengineering Essentials)

Handbook of Polyhydroxyalkanoates (Bioengineering Essentials)

Edited by Barbara Moore

SYRAWOOD
PUBLISHING HOUSE
New York

Published by Syrawood Publishing House,
750 Third Avenue, 9th Floor,
New York, NY 10017, USA
www.syrawoodpublishinghouse.com

Handbook of Polyhydroxyalkanoates (Bioengineering Essentials)
Edited by Barbara Moore

International Standard Book Number: 978-1-64740-398-0 (Hardback)

Cataloging-in-publication Data

Handbook of polyhydroxyalkanoates (bioengineering essentials) / edited by Barbara Moore.
   p. cm.
Includes bibliographical references and index.
ISBN 978-1-64740-398-0
1. Poly-beta-hydroxyalkanoates. 2. Biopolymers. 3. Bioengineering. I. Moore, Barbara.
QR92.P58 H36 2023
660.63--dc23

TABLE OF CONTENTS

Permissions

List of Contributors

Index

PREFACE

Polyhydroxyalkanoates (PHAs) are bio-based and biodegradable polyesters produced naturally by numerous microorganisms such as certain bacterial species. Some of the microorganisms which are used for the production of PHAs are Ralstonia eutropha, Aspergillus eutrophus, and Cupriavidus necator. The properties of PHAs, including their biocompatibility and biodegradability, make them appropriate for a number of applications in various industries. In the biomedical sector, some major applications of PHAs are in the areas of drug delivery, surgery, tissue engineering, bio-implant patches, and wound dressing. PHAs are green plastics having positive social and environmental impact in contrast to conventional plastics, in regards to their production and recycling. There are no chronic or acute health effects of PHA, even if they are utilized in-vivo. This book provides comprehensive insights on the biomedical applications of polyhydroxyalkanoates. It presents researches and studies performed by experts across the globe. This book aims to equip students and experts with the advanced topics and upcoming concepts in this area of study.

The researches compiled throughout the book are authentic and of high quality, combining several disciplines and from very diverse regions from around the world. Drawing on the contributions of many researchers from diverse countries, the book's objective is to provide the readers with the latest achievements in the area of research. This book will surely be a source of knowledge to all interested and researching the field.

In the end, I would like to express my deep sense of gratitude to all the authors for meeting the set deadlines in completing and submitting their research chapters. I would also like to thank the publisher for the support offered to us throughout the course of the book. Finally, I extend my sincere thanks to my family for being a constant source of inspiration and encouragement.

Editor

PHA Production and PHA Synthases of the Halophilic Bacterium *Halomonas* sp. SF2003

Tatiana Thomas [1,2,*], **Kumar Sudesh** [3], **Alexis Bazire** [4], **Anne Elain** [2], **Hua Tiang Tan** [3], **Hui Lim** [3] and **Stéphane Bruzaud** [1]

1   Institut de Recherche Dupuy de Lôme (IRDL), Université de Bretagne Sud (UBS), EA 3884 Lorient, France; stephane.bruzaud@univ-ubs.fr
2   Institut de Recherche Dupuy de Lôme (IRDL), Université de Bretagne Sud (UBS), 56300 Pontivy, France; anne.elain@univ-ubs.fr
3   School of Biological Sciences, Universiti Sains Malaysia (USM), Penang 11800, Malaysia; ksudesh@usm.my (K.S.); tiang93@hotmail.com (H.T.T.); lhlim1993@gmail.com (H.L.)
4   Laboratoire de Biotechnologie et Chimie Marines (LBCM), IUEM, Université de Bretagne-Sud (UBS), EA 3884 Lorient, France; alexis.bazire@univ-ubs.fr
*   Correspondence: tatiana.thomas@univ-ubs.fr

Abstract: Among the different tools which can be studied and managed to tailor-make polyhydroxyalkanoates (PHAs) and enhance their production, bacterial strain and carbon substrates are essential. The assimilation of carbon sources is dependent on bacterial strain's metabolism and consequently cannot be dissociated. Both must wisely be studied and well selected to ensure the highest production yield of PHAs. *Halomonas* sp. SF2003 is a marine bacterium already identified as a PHA-producing strain and especially of poly-3-hydroxybutyrate (P-3HB) and poly-3-hydroxybutyrate-*co*-3-hydroxyvalerate (P-3HB-*co*-3HV). Previous studies have identified different genes potentially involved in PHA production by *Halomonas* sp. SF2003, including two *phaC* genes with atypical characteristics, *phaC1* and *phaC2*. At the same time, an interesting adaptability of the strain in front of various growth conditions was highlighted, making it a good candidate for biotechnological applications. To continue the characterization of *Halomonas* sp. SF2003, the screening of carbon substrates exploitable for PHA production was performed as well as production tests. Additionally, the functionality of both PHA synthases PhaC1 and PhaC2 was investigated, with an *in silico* study and the production of transformant strains, in order to confirm and to understand the role of each one on PHA production. The results of this study confirm the adaptability of the strain and its ability to exploit various carbon substrates, in pure or mixed form, for PHA production. Individual expression of PhaC1 and PhaC2 synthases in a non-PHA-producing strain, *Cupriavidus necator* H16 PHB⁻4 (DSM 541), allows obtaining PHA production, demonstrating at the same time, functionality and differences between both PHA synthases. All the results of this study confirm the biotechnological interest in *Halomonas* sp. SF2003.

Keywords: halophilic bacteria; polyhydroxyalkanoates (PHAs); PHA synthases

1. Introduction

Polyhydroxyalkanoates (PHAs) are valuable bio-based and biodegradable polymers produced by numerous bacterial species [1,2]. Their properties are close to those of conventional petroleum-based plastics; therefore, in addition to their biocompatibility, they are considered to be materials with high potential [3]. Actually, they can be used in various fields ranging from packaging [2] to biomedical applications [1,4], but one of the main obstacles to their commercialization and exploitation is the overall cost of production. Currently, several tools can be managed to reduce the final cost of PHA production, including the characterization of selected microorganisms coupled with the optimal selection of carbon substrates [5,6]. Indeed, a better understanding of strain metabolisms and response in front of different growth and/or production conditions participate to tailor-make PHA and enhance production yield. To date, there is an important diversity of carbon sources (monomers) that can be exploited for PHA synthesis, and as a result, a wide range of PHAs which can be synthesized [6]. PHA properties are closely linked to their bacterial producer strain, carbon substrates, and production mode [6,7], meaning that an accurate study of each parameter is required. On the other hand, the PHA production cost is still limiting their more widespread use. Over the last decade, research has notably focused on the use of low-value substrates like industrial co-products (from agri-food, waste treatment, or biodiesel industry) [6,8,9] as they can represent up to 50% of the production cost. With these carbon substrates, perfect control of PHA's structure, molecular weight and properties could be difficult. Therefore, complete studies of carbon substrates utilization and PHA synthesis are required to soundly select the most adapted carbon sources, whether it is pure carbohydrates or co-products. Another way to reduce production costs is to study the strain genome using bioinformatics and genetic engineering. These tools are also exploiting for expression of PHA synthesis operon in non-producing strains exposing less restrictive growth and production conditions [5,10].

Halomonas sp. SF2003 is a halophilic bacterium identified as a PHA-producing strain [11]. Previous studies have shown its capacity to produce polymer up to 78 wt% of cell dry weight (CDW), using conventional carbon sources but also carbonaceous by-products from food wastes [12]. Bioinformatics and phenotypic studies of *Halomonas* sp. SF2003 have demonstrated its versatility under various atypical growth conditions making it an adaptable bacterium. Additionally, genomic annotation also allows identifying various metabolic pathways directly involved, or not, in the synthesis of PHA, which can be studied for a stronger understanding of *Halomonas* sp. SF2003 PHA metabolism. Our previous study highlighted atypical characteristics and organization of PHA biosynthesis genes (*phaA*, *phaB*, *phaC1*, *phaC2*, and *phaR*) [13]. Regarding its original properties, *Halomonas* sp. SF2003 is an excellent candidate for the innovative development of biotechnological production of PHA.

The objectives of this work were to go further into the unraveling/understanding of PHA biosynthesis capability and metabolism of *Halomonas* sp. SF2003 and to identify potential carbon substrates, and in later stage, potential industrial co-products, which can be exploited for PHA production. Our work will also contribute to better understand the functionality of both PHA synthases of *Halomonas* sp. SF2003 in order to later optimize its PHA production.

2. Materials and Methods

2.1. Bacterial Strains and Media

2.1.1. Bacterial Strains

All the bacterial strains used in this study have been furnished by Research Institute Dupuy de Lôme (RIDL, University of South Brittany) collections or have been purchased to the Deutsche Sammlung von Mikroorganismen und Zellkulturen (DSMZ) collection (Table 1).

Table 1. List of plasmids and bacterial strains used in this study. DSMZ: Leibiz Institute DSMZ-German Collection of Microorganisms and Cell Cultures. Km^R: Resistance to kanamycin, RIDL: Research Institute Dupuy de Lôme, UBS: Université de Bretagne Sud.

Strain or Plasmid	Characteristics	Origin
Halomonas **strain**		
Halomonas sp. SF2003 ID CNCM-I-4786	Wild type PHA-producing strain	Sea of Iroise (France), RIDL Collection, UBS
Cupriavidus necator **strains**		
H16 (DSM 428)	Wild type PHA-producing strain	DSMZ Collection
PHB⁻4 (DSM 541)	Mutant non-PHA-producing strain	DSMZ Collection
Escherichia coli **strains**		
E. cloni® 10G	Competent cells	Lucigen
S17-1	Strain for conjugative transfer of plasmid to *C. necator* PHB⁻4	Simon et al. 1983 [14]
Plasmids		
pBBR1-Pro$_{Cn}$	pBBR1MCS-2 derivatives with *phaC1* promoter from *C. necator*. Km^R.	Foong et al. 2014 [15]
pBBR1-Pro$_{Cn}$-*phaC1*	pBBR1MCS-2 derivatives with *phaC1* promoter from *C. necator* and *phaC1* of *Halomonas* sp. SF2003. Km^R.	This study
pBBR1-Pro$_{Cn}$-*phaC2*	pBBR1MCS-2 derivatives with *phaC1* promoter from *C. necator* and *phaC2* of *Halomonas* sp. SF2003. Km^R.	This study
Transformants		
PHB⁻4/pBBR1-Pro$_{Cn}$-*phaC1*	Transformant strain with pBBR1MCS-2 plasmid expressing *phaC1* of *Halomonas* sp. SF2003. Km^R.	This study
PHB⁻4/pBBR1-Pro$_{Cn}$-*phaC2*	Transformant strain with pBBR1MCS-2 plasmid expressing *phaC2* of *Halomonas* sp. SF2003. Km^R.	This study

2.1.2. Growth Media

Halomonas sp. SF2003 is cultivated in Zobell medium (Bacto Tryptone, (Difco, BD, Göteborg, Sweden) 4 g/L, Yeast Extract (Fisher BioReagents, Pittsburgh, PA, USA) 1 g/L, sea salts (Aquarium systems, Instant Ocean, Blacksburg, VA, USA) 30 g/L, pH 7.5), with an orbital agitation of 200 rpm, at 30 °C. The medium is complemented with glucose (Labogros), at 10 g/L, for pre-cultures dedicated to PHA productions.

Cupriavidus necator H16, *C. necator* PHB⁻4, PHB⁻4/pBBR1-Pro$_{Cn}$-*phaC1* and PHB⁻4/pBBR1-Pro$_{Cn}$-*phaC2* are cultivated in nutrient-rich medium (NR medium) (Meat extract (Biokar Diagnostics, Allonne, France) 10 g/L, Yeast extract (Fisher BioReagents, USA) 2 g/L, Peptone from gelatin, enzymatic digest (Sigma-Aldrich, St. Louis, MO, USA) 10 g/L, pH (7)), with an orbital agitation of 200 rpm, at 30 °C.

The transformant strains PHB 4/pBBR1-Pro$_{Cn}$-*phaC1* and PHB 4/pBBR1-Pro$_{Cn}$-*phaC2* were selected on Simmons citrate agar plates (Thermo Scientific™, Illkirch–Graffenstaden, France), prepared following the manufacturer instructions. For the transformant strains PHB⁻4/pBBR1-Pro$_{Cn}$-*phaC1* and PHB⁻4/pBBR1-Pro$_{Cn}$-*phaC2*, the media were complemented with kanamycin (Km, Gibco, Waltham, MA, USA), at 50 µg/mL.

2.1.3. Production Media

The PHA productions were performed using a two-steps protocol. Biomass accumulation was performed in Reference 1 medium (carbon source 10 g/L, Bacto Tryptone (Difco, BD, Sweden) 1 g/L, Yeast extract (Fisher BioReagents, USA) 0.5 g/L, sea salts (Aquarium systems, Instant Ocean, USA) 11 g/L, pH (7.5) and C/N ratio 24.6), and PHA production was performed in Reference 2 medium (carbon source 20 g/L, Yeast extract (Fisher BioReagents, USA) 0.4 g/L, sea salts (Aquarium systems, Instant Ocean, USA) 11 g/L, pH 7.0 and C/N ratio 187.2). Reference 2 medium was also employed for the screening of carbon sources usable for PHA production. Nile Red agar plates have been prepared by adding agar powder (15 g/L, Fisher BioReagents, USA) and filtered Nile Red (0.5% (w/v),

Sigma-Aldrich, USA) to Reference 2 medium. For the transformant strains PHB⁻4/pBBR1-Pro$_{Cn}$-*phaC1* and PHB⁻4/pBBR1-Pro$_{Cn}$-*phaC2*, the media were complemented with kanamycin (Km, Gibco, USA), at 50 µg/mL.

2.2. In Silico Study of PHA Synthase of Halomonas sp. SF2003 PhaC1 and PhaC2

In silico analysis of PhaC1 and PhaC2 synthases of *Halomonas* sp. SF2003 was performed by confronting amino acids sequences of PHA synthases from several PHA-producing strains (all the sequences tested are available on the National Center for Biotechnology Information (NCBI) database, and accession numbers are available in the Supplementary data S1: Accession numbers of PhaC amino acids sequence). The identification of lipase box-like sequences was conducted using BioEdit software using the ClustalW Multiple alignment tool.

2.3. Cloning of Halomonas sp. SF2003 phaC1 and phaC2 Genes

To evaluate the activity of PHA synthases PhaC1 and PhaC2 of *Halomonas* sp. SF2003, *phaC1*, and *phaC2* genes were cloned. Genomic DNA of *Halomonas* sp. SF2003 overnight cultures in Zobell medium was extracted using QIAamp DNA Mini Kit (Qiagen©, Hilden, Germany). The primers PhaC1-F (5'-AGTAAGCTTAGGAGGAGGCGCATGCAGTCGCCAGCCCA-3'), PhaC1-R (5'-AGTAGCATTTAAATTCAG-GTTTGCTTCACGTAGGTG-3'), PhaC2-F (5'-AGTAAGCTTAGGA GGAGGCGCATGGACTCAGCCCAGCA-3') and PhaC2-R (5'-AGTAGCATTTAAATTCAACTCTTGT CGCTATCCTTGG-3') were designed based on the nucleotide sequence of *Halomonas* sp. SF2003 and using A plasmid Editor software (ApE). The PCR reactions were performed using KAPA HiFi HotStart Ready Mix PCR Kit (Kapa Biosystems, Wilmington, MA, USA) and an MJ Mini Thermal Cycler (BioRad, Hercules, CA, USA) according to the manufacturer's instructions and applying the following parameters: initial denaturation for 3 min at 98 °C, denaturation for 20 s at 98 °C, primer annealing for 15 s at 54 °C, elongation for 40 s at 72 °C, and final extension for 2 min at 72 °C. The denaturation, primer annealing and extension steps were repeated 30 times. Each amplicon has been digested using *Swa*I and *Hind*III enzymes before to be ligated with the pBBR1-Pro$_{Cn}$ plasmid using DNA ligation kit (TaKaRa Bio Inc., Kyoto, Japan), with an insert/vector ratio of 3:1 and following the manufacturer's instructions, to obtain pBBR1-Pro$_{Cn}$-*phaC1* and pBBR1-Pro$_{Cn}$-*phaC2* plasmids (Figures 1 and 2). The resulting plasmids were then used to be cloned into *E. cloni*® 10G cells (Lucigen Corporation, Middleton, WI, USA) by thermal shock of 45 s at 42 °C. Transformant cells were selected on Luria Bertani (LB) agar plates complemented with kanamycin (50 µg/mL) and after control PCR. Plasmid extraction was performed using the GeneJET Plasmid Miniprep Kit (Thermo Scientific™, France) and following the manufacturer's instructions. The extracted plasmids were used to transform *E. coli* S17-1 competent cells by a thermal shock of 45 s at 42 °C; transformant cells were selected on LB agar plates complemented with kanamycin (50 µg/mL). A bacterial glycerol stock of the plasmids was with a final glycerol concentration of 25% and stocked at −80 °C. Then, transconjugation between *E. coli* S17-1 cells, harboring pBBR1-Pro$_{Cn}$-*phaC1* or pBBR1-Pro$_{Cn}$-*phaC2* plasmids, and *C. necator* PHB⁻4 was performed by mixing liquid cultures of each strain before to inoculate NR agar plate and incubate it for 8 h at 30 °C. After the incubation time, colonies were picked up and used to prepare an NR medium suspension. Simmons citrate agar plates complemented with kanamycin were inoculated and incubated for two days at 30 °C. Blue colonies were picked up and used to perform control PCR using EconoTaq Master Mix (Lucigen Corporation, USA) and following the manufacturer's instructions, as the same time than subculture on LB agar plates complemented with kanamycin (50 µg/mL). Plasmid extraction was performed using the GeneJET Plasmid Miniprep Kit (Thermo Scientific™, France) and following manufacturer instructions. DNA sequencing was done by 1st BASE Sdn. Bhd. (Malaysia).

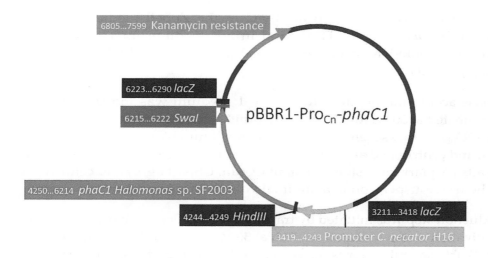

Figure 1. pBBR1-Pro$_{Cn}$-*phaC1* plasmid map.

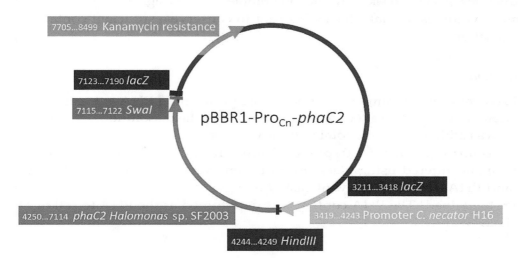

Figure 2. pBBR1-Pro$_{Cn}$-*phaC2* plasmid map.

2.4. Screening for Carbon Sources

A total of eight pure carbohydrates: fructose, galactose, glucose, maltose, mannose, melibiose, rhamnose and sucrose; and seven organic acids: dodecanoic acid, heptanoic acid, hexanoic acid, levulinic acid, malic acid, palmitic acid, and trans-2-pentenoic acid in mixture with fructose, galactose or glucose; were tested for PHA accumulation. These carbon sources were selected depending on their origin and were tested for PHA accumulation using Nile Red agar plates technique [11]. All the reagents were purchased from Sigma-Aldrich (St. Louis, MO, USA) or Thermo Fisher Scientific (Illkirch-Graffenstaden, France).

Overnight pre-cultures of *Halomonas* sp. SF2003, *C. necator* H16, *C. necator* PHB⁻4, PHB⁻4/pBBR1-Pro$_{Cn}$-*phaC1*, and PHB⁻4/pBBR1-Pro$_{Cn}$-*phaC2* were used to screen carbon sources assimilation on Nile Red agar plates. After three days of incubation at 30 °C, bacterial growth and fluorescence were checked under white and UV light.

2.5. PHA Production

The PHA productions were composed of three steps: pre-culture, biomass accumulation, and polymer production.

Pre-cultures were performed for 7 h at 30 °C in the Zobell medium complemented with 10 g/L of glucose for *Halomonas* sp. SF2003 and in NR medium for *C. necator* H16 (DSM 428) or NR medium complemented with kanamycin (50 μg/mL) for PHB⁻4/pBBR1-Pro$_{Cn}$-*phaC1* and PHB⁻4/pBBR1-Pro$_{Cn}$-*phaC2*.

For the biomass accumulation step, Reference 1 medium was inoculated at 10% (v/v) with pre-cultures, and incubated at 30 °C for 17 h with an orbital shaking of 200 rpm. Biomass accumulation was monitored by OD$_{600nm}$ measurement. Once the maximum biomass was reached, cultures were stopped, harvested, and centrifuged at 7500 rpm for 10 min at 4 °C. Then, cells were washed twice with saline water (sea salts (Aquarium systems, Instant Ocean, USA) 11 g/L) and centrifuged at 7500 rpm for 10 min at 4 °C before resuspension in a minimal volume of saline water.

The PHA production step was initiated by transfer of cell pellet of the previous step into Reference 2 medium. The cultivation was performed for 72 h at 30 °C with an orbital shaking of 200 rpm. At the end of the step, the bacterial culture was harvested and centrifuged. Cell pellets were washed with distilled water before freezing at −80 °C and freeze-drying for 48 h.

For productions in shake flasks, production volumes were designed to only use a fifth or a quarter of the maximum volumes of shake flasks in order to conserve a sufficient contact surface with air, allowing oxygenation.

2.6. PHA Extraction

Lyophilized cells were manually ground before performing PHA extraction using chloroform (25 mL of solvent per g of CDW) at 60 °C for 15–16 h. After the dissolution of the PHA in chloroform, distilled water was added, 1/3 of the total volume of chloroform, and the suspension was vigorously agitated before centrifugation at 5000 rpm for 7 min. Then, the organic phase was recovered using a sterile syringe and filtered using glass cotton to remove cellular debris. Then, the solvent was evaporated, and PHA films were solvent-casted in a glass Petri dish, at room temperature, until a constant weight obtained. The PHA content was determined as the PHA to cell dry weight (CDW) percent ratio [12,16].

3. Results

3.1. *In Silico Study of PHA Synthases PhaC1 and PhaC2 of Halomonas sp. SF2003*

In the previous work, the whole genome of *Halomonas* sp. SF2003 was sequenced and annotated, leading to the identification of two genes potentially encoding two distinct PHA synthase proteins PhaC1 and PhaC2; belonging to class I (based on gene organization and biosynthesized PHA) [13]. To further characterize this first analysis, the consensus lipase box-like sequence of both PHA synthases have been studied.

Amino acid sequences of PHA synthases, PhaC1 and PhaC2 of *Halomonas* sp. SF2003 have been analyzed and allowed the identification of two distinct lipase box-like patterns in both enzymes, beginning at position 384 for PhaC1 and position 343 for PhaC2. In the PhaC1 sequence, the pattern is composed of Glycine-Tyrosine-Cysteine-Leucine-Glycine (G-Y-C-L-G), and pattern in PhaC2 is Serine-Tyrosine-Cysteine-Isoleucine-Glycine (S-Y-C-I-G) (Figure 3). Results obtained for PHA synthases of *Halomonas* sp. SF2003 still demonstrated the distinction of both enzymes, additionally to their size and location in the genome [13]. Indeed, two different patterns have been reported: G-Y-C-L-G for PhaC1 and S-Y-C-I-G for PhaC2. The existence of different PHA synthase enzymes in the same bacterial strain has already been observed, as well as several lipase box-like sequences, like for *Halomonas boliviensis* LC1 (DSM 15516). This strain has several PHA synthases in which different lipase box-like pattern have been detected (Figure 3). Additionally, to this difference of pattern in PhaC box consensus sequences, analysis of amino acid sequences framing these active sites suggests a difference in the final structure of proteins.

Figure 3. Multiple alignments of partial amino acid sequences of PHA synthase exposing lipase box-like patterns from different bacterial species. All the sequences are available on the National Center for Biotechnology Information (NCBI) database. Highlighted sequences correspond to PHA synthases PhaC1 and PhaC2 of *Halomonas* sp. SF2003.

3.2. Screening of Carbon Substrates for PHA Production by Halomonas sp. SF2003

Visual examinations of Nile Red agar plates allow to detect colonies and bacterial growth and PHA production have been screened by detection of Nile Red fluorescence under UV-lights. *Halomonas* sp. SF2003 was able to use a majority of the tested carbohydrates as substrates for both bacterial growth and PHA accumulation: (D)-Glucose (Figure 4a), (D)-Fructose (Figure 4b), (D)-Galactose (Figure 4c), (D)-Mannose (Figure 4d), (D)-Maltose (Figure 4e) and (D)-Sucrose (Figure 4h), only (L)-Rhamnose and (D)-Melibiose were not used (Figure 4f,g and Table 2). On Nile Red agar plates, the number of colony-forming units (CFU), as well as the size of the colonies, vary from one carbohydrate to another. Qualitatively, growth of *Halomonas* sp. SF2003 seems to be more important on (D)-Glucose, (D)-Mannose, and (D)-Maltose, but comparatively, PHA production seems to be more efficient on (D)-Glucose, (D)-Galactose and (D)-Maltose, based on fluorescence intensity.

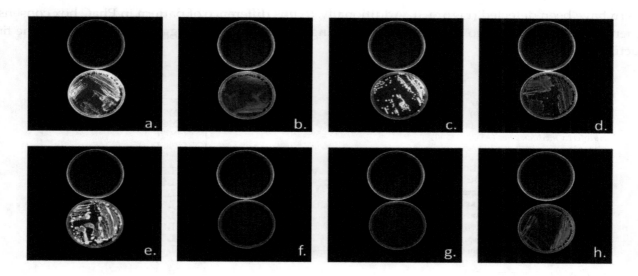

Figure 4. Nile Red agar plates screening with *Halomonas* sp. SF2003 using 2% (w/v) of different carbon substrates. Bacterial growth was evaluated under white light by the presence, or no, of colonies. PHA production was evaluated under Ultra-Violet light (UV-light) by fluorescence emission from colonies; positive results appear as "white" colonies showing their fluorescence. The positive control (medium without addition of carbon substrates) is the upper plate in (**a**) to Figure. (**a**). (D)-Glucose, (**b**). (D)-Fructose, (**c**). (D)-Galactose, (**d**). (D)-Mannose, (**e**). (D)-Maltose, (**f**). (D)-Melibiose, (**g**). (L)-Rhamnose, and (**h**). (D)-Sucrose. Observations under UV-lights performed with transillumination.

Table 2. Growth and PHA accumulation in *Halomonas* sp. SF2003 using different carbon sources.

Carbon Source	Growth	PHA Accumulation
(D)-Glucose	+	+
(D)-Fructose	+	+
(D)-Galactose	+	+
(D)-Mannose	+	+
(D)-Maltose	+	+
(D)-Melibiose	-	-
(L)-Rhamnose	-	-
(D)-Sucrose	+	+

As described previously and illustrated in Table 3, *Halomonas* sp. SF2003 is able to grow in medium with (D)-Glucose, (D)-Fructose, (D)-Galactose, (D)-Mannose, (D)-Maltose and (D)-Sucrose. The strain also seems able to use these carbohydrates for PHA production in accordance with genomic analysis [13]. Indeed, the study of *Halomonas* sp. SF2003 genome highlighted the presence of various genes coding for enzymes responsible for carbohydrates assimilation such as fructose or sucrose. However, some of the tested carbohydrates have also been used by *Halomonas* sp. SF2003, despite the preliminary study of its genome only identified a part of genes required for their total assimilation (Table 3). These results suggest the interest of performing a re-examination and annotation of *Halomonas* sp. SF2003 genome but also open the door for new studies/productions using these pure carbohydrates, which can easily be found in various food or agri-food (co-)products.

Table 3. Listing of common carbohydrates used for PHA production.

Carbohydrates	Origin	Identification of Pathway for Assimilation
Fructose *	Fruits, Honey	Total
Galactose *	Milk, Honey, Red algae	Partial
Glucose *	Food, Metabolism of living organisms	Partial
Lactose	Dairy products	Total
Maltose *	Starch degradation (barley)	n.i
Mannose *	Fruits, Plants, Mannitol	n.i
Melibiose *	Plants, Fruits	Total
Ribose	RNA	Partial
Rhamnose *	Plants	Partial
Sucrose *	Plants	Total
Xylose	Plants	Partial

* Tested in this study for PHA accumulation, n.i: not identified in the *Halomonas* sp. SF2003 genome yet.

The use of these carbohydrates for PHA production has already been reported in different bacterial species, including, or not, *Halomonas* species (Table 4). All these carbohydrates allow to produce P-3HB, and sometimes P-3HB-*co*-3HV, with a yield of production ranging from 0.11 g/L to 64.0 g/L of PHA.

Table 4. Listing (not exhaustive) of various bacterial strains using the different tested carbohydrates for PHA production.

Carbohydrates	Bacterial Strains/Species	References
Melibiose	*Burkholderia sacchari* sp. nov.	[17]
Rhamnose	*C. necator, P. oleovorans*	[17]
Glucose	*Bacillus cereus* UW85, **Halomonas sp. TD01**, *Halomonas profundus*, *Halomonas* sp. SF2003	[18–20]
Fructose	*Bacillus aryabhattai* PHB10, *C. necator, Halomonas* TD08, *Halomonas* sp. SF2003, *H. halophila*, **H. organivorans**, *H. salina*	[21–23]
Sucrose	*Azotobacter vinelandii*, **Burkholderia sacchari DSM 17165**, *C. necator, Natrinema* sp. 5TL6	[6,24–26]
Galactose	**Halomonas halophila**, *H. salina, Halomonas* sp. SF2003	[23]
Mannose	*Halomonas halophila*, **H. organivorans**, *H. salina*	[23,27]
Maltose	*B. aryabhattai* PHB10, **Halomonas sp. TD08**, *H. boliviensis* LC1 and *H. campisalis*	[17,21,28–30]

Based on data of Verlinden et al. 2007, and completed with data from other studies. Strains in bold expose the highest PHA concentrations.

Results of production show that the employed bioprocess (meaning strain, carbon sources, and production systems) significantly impacts production yields and composition of the polymer. There are plenty of systems that can be used; therefore, it is difficult to designate which one is the most effective. However, data described previously and in Table 4 show the importance of a deep study and judicious choice of the employed bioprocess. Data also demonstrate capacity of *Halomonas* species to use a wide variety of carbohydrates for PHA production in accordance with results obtained with *Halomonas* sp. SF2003, and are sometimes more efficient than non-halophilic strains. To complete data about *Halomonas* sp. SF2003 carbohydrates metabolisms, additional tests have been conducted on one simple sugar: fructose, galactose, and glucose or mixed with one fatty or organic acids, in the proportion 95:5% (mol/mol). Such acids have already been reported as a precursor for the biosynthesis of copolymers when simple sugars were used as the main substrate. The following acids, which are components of plants, fruits, or different industrial effluents (agri-food, chemical, cosmetic, pharmaceutical), were tested: dodecanoic, heptanoic, hexanoic, levulinic, malic, palmitic, and trans-2-pentenoic. In the same way than for screening tests with pure carbohydrates, bacterial growth has been evaluated by visual examination and PHA production by detection of fluorescence under UV-lights.

Halomonas sp. SF2003 can grow on majority mixtures composed of glucose or galactose and organic acids except the following: glucose-dodecanoic acid and galactose-dodecanoic/heptanoic/hexanoic acids. A mixture of fructose and acids cannot be used for bacterial growth nor PHA production, whatever the acid (Table 5). This finding suggests an inhibitory effect of acids depending on the sugar used as co-substrate. Among the mixture allowing growth, only five exhibit fluorescence under UV-lights, suggesting PHA production: glucose-malic acid (Figure 5a), glucose-levulinic acid (Figure 5b), glucose-palmitic acid (Figure 5c), galactose-malic acid (Figure 5d) and galactose-palmitic acid (Figure 5f and Table 5).

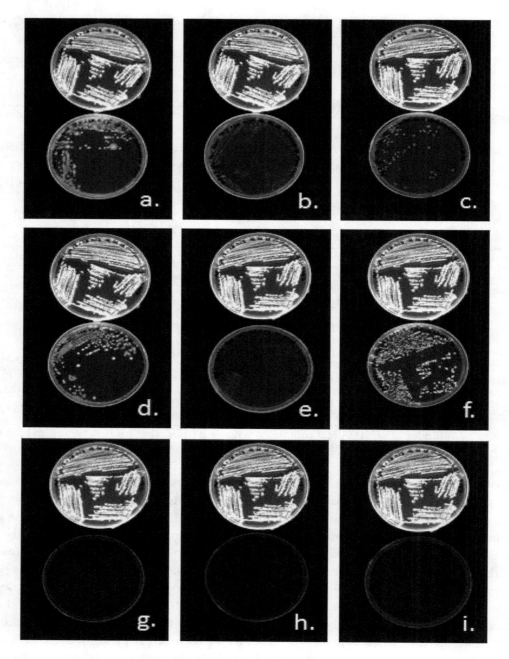

Figure 5. Nile Red agar plates screening with *Halomonas* sp. SF2003 using carbohydrates/acid mix with molar ratio 95/5%, final concentration 2% (w/v). Bacterial growth was evaluated under white light by the presence, or no, of colonies. PHA production was evaluated under UV-light by fluorescence emission from colonies, positive results appear as "white" colonies showing their fluorescence. Positive results appear as "white" colonies showing their fluorescence. The positive control (Glucose only) is the upper plate in (a–i). (a). Mix Glucose-Malic acid, (b). Mix Glucose-Levulinic acid, (c). Mix Glucose-Palmitic acid, (d). Mix Galactose-Malic acid, (e). Mix Galactose-Levulinic acid, (f). Mix Galactose-Palmitic acid, (g). Mix Fructose-Malic acid, (h). Mix Fructose-Levulinic acid and (i). Mix Fructose-Palmitic. Observations under UV-lights performed with transillumination.

Table 5. Growth and PHA accumulation in *Halomonas* sp. SF2003 using a different mixture of carbohydrates and acids.

	Carbon Source	Growth	PHA Accumulation
Glucose	Dodecanoic acid	-	-
	Heptanoic acid	+	-
	Hexanoic acid	+	-
	Levulinic acid	+	±
	Malic acid	+	+
	Palmitic acid	+	-
	Trans-2-pentenoic acid	+	-
Galactose	Dodecanoic acid	-	-
	Heptanoic acid	-	-
	Hexanoic acid	-	-
	Levulinic acid	+	-
	Malic acid	+	+
	Palmitic acid	+	±
	Trans-2-pentenoic acid	+	-
Fructose	Dodecanoic acid	±	±
	Heptanoic acid	-	-
	Hexanoic acid	-	-
	Levulinic acid	-	-
	Malic acid	-	-
	Palmitic acid	-	-
	Trans-2-pentenoic acid	-	-

According to data reported here and in the literature, it appears that numerous bacterial species, including *Halomonas* sp. SF2003, can use several pure carbohydrates for growth and also for PHA production.

This ability to exploit various carbon substrates, in addition to its capacity to grow in front of atypical/stressful conditions, make *Halomonas* sp. SF2003 a versatile strain with a high potential for biotechnological application/use [13,16]. The results of this study identify several potential carbon substrates allowing PHA production and open the door for future tests studying the exploitation of each one.

3.3. Study of PHA Synthases

PHA biosynthesis activity of *Halomonas* sp. SF2003 is due to the presence of genes coding for enzymes linked to PHA metabolism (i.e., *phaA*, *phaB*, *phaC1*, *phaC2*, and *phaR*). Interestingly, genes coding for acetyl-CoA acetyltransferase (also known as β-ketothiolase) (*phaA*), acetoacetyl-CoA reductase (*phaB*), and PHA synthases (*phaC1* and *phaC2*) are not organized in one operon but are distant from each other on *Halomonas* sp. SF2003 genome sequence. Moreover, *phaC1* and *phaC2* genes expose atypical sizes (1965 bp and 2865 bp, respectively), and conserved domain, which led to further study of both genes.

3.3.1. Cloning of PHA Synthases phaC1 and phaC2 of *Halomonas* sp. SF2003

Gene *phaC1* has been amplified using PhaC1-F and PhaC1-R primers and *phaC2* gene using PhaC2-F and PhaC2-R. PCR allowed amplicons production of approximatively 2000 and 3000 bp, respectively, corresponding to *phaC1* and *phaC2* size (1965 pb and 2865 pb, respectively).

3.3.2. Characterization of PHA Production by Transformant Strains PHB$^-$4/pBBR1-Pro$_{Cn}$-phaC1 and PHB$^-$4/pBBR1-Pro$_{Cn}$-phaC2

To evaluate the functionality of PHA synthases PhaC1 and PhaC2 of *Halomonas* sp. SF2003, screening for bacterial growth and PHA production have been performed. Likewise, with wild type *Halomonas* sp. SF2003, a total of eight carbohydrates and twenty-one mixtures, have been tested. Bacterial growth and PHA production were qualitatively checked using Nile Red agar plates technique with white light and UV-light evaluation (Supplementary data S2: Nile Red agar plates screening with PHB$^-$4/pBBR1-Pro$_{Cn}$-*phaC1* using 2% (w/v) of different carbon substrates, Picture a–h; Supplementary data S2: Nile Red agar plates screening with PHB$^-$4/pBBR1-Pro$_{Cn}$-*phaC2* using 2% (w/v) of different carbon substrates, Picture a–h and Table 6).

PHB$^-$4/pBBR1-Pro$_{Cn}$-*phaC1* was able to exploit all pure carbohydrates and a majority of mixtures of carbohydrates/acids tested for bacterial growth (Supplementary data S2: Nile Red agar plates screening with PHB$^-$4/pBBR1-Pro$_{Cn}$-*phaC1* using 2% (w/v) of different carbon substrates, Picture a–h and Table 6) except the following mixtures: glucose – heptanoic/hexanoic/trans-2-pentenoic acids, galactose – heptanoic/hexanoic/palmitic/trans-2-pentenoic (data not shown). In comparison, results obtained with PHB$^-$4/pBBR1-Pro$_{Cn}$-*phaC2* are similar, for pure carbohydrates and mixtures, except for a mixture of galactose-palmitic acid for which growth was recorded (Supplementary data S3: Nile Red agar plates screening with PHB$^-$4/pBBR1-Pro$_{Cn}$-*phaC2* using 2% (w/v) of different carbon substrates, Picture a–h and Table 6). Some results did not appear clearly positive and have been denoted as "±" making interpretation of substrates used and PHA production difficult.

PHA production has been detected with both transformant strains, in comparison with the mutant *C. necator* PHB$^-$4 (DSM 541) (data not shown), demonstrating the success of cloning experiments and functionalities of both PHA synthase genes, *phaC1* and *phaC2*. Screening tests have allowed the confirmation of correct annotation of *phaC1* and *phaC2* genes and attest to the existence of the difference between both PHA synthases of *Halomonas* sp. SF2003. Indeed, qualitative analysis of PHA accumulation, by detection of fluorescence under UV-light highlighted several differences between both transformant strains. Indeed, among all carbon substrates tested, only three seem to allow PHA accumulation in PHB$^-$4/pBBR1-Pro$_{Cn}$-*phaC1* (Fructose, Glucose-dodecanoic/palmitic acid) and nine for PHB$^-$4/pBBR1-Pro$_{Cn}$-*phaC2* (Fructose, Mannose, Sucrose, Glucose-dodecanoic/palmitic acids, Glucose-levulinic/malic acids, Galactose-levulic/malic acids).

These results suggested that synthase PhaC1 was less active or more selective than PhaC2. Actually, qualitatively, there were more carbon substrates (pure or in the mixture) that generated a fluorescence under UV-light when PHB$^-$4/pBBR1-Pro$_{Cn}$-phaC2 was used than PHB$^-$4/pBBR1-Pro$_{Cn}$-phaC1. A previous study of *Halomonas* sp. SF2003 genome and metabolisms demonstrated several differences between both synthases [13]. PhaC1 and PhaC2 had an identity of 60–70% and 65–96%, respectively, with different PHA synthases [13]. Results of both studies are in agreement with each other. Even if PhaC2 exposes some atypical characteristics (size and structure of conserved domains), it seems to be the main PHA synthase responsible for PHA biosynthesis.

Table 6. Growth and PHA accumulation in transformant strains PHB⁻4/pBBR1-Pro$_{Cn}$-*phaC1* and PHB⁻4/pBBR1-Pro$_{Cn}$-*phaC2* using different carbon sources. Legend for growth and PHA accumulation +/−: Positive/Negative.

Carbon Source		PHB⁻4/ pBBR1-Pro$_{Cn}$-*phaC1*		PHB⁻4/ pBBR1-Pro$_{Cn}$-*phaC2*	
		Growth	PHA Accumulation	Growth	PHA Accumulation
	(D)-Fructose	+	±	+	+
	(D)-Galactose	+	−	+	−
	(D)-Glucose	+	−	+	−
	(D)-Maltose	+	−	+	−
	(D)-Mannose	+	−	+	±
	(D)-Melibiose	+	−	+	−
	(L)-Rhamnose	+	−	+	−
	(D)-Sucrose	+	−	+	±
Glucose +	Dodecanoic acid	+	±	+	+
	Heptanoic acid	−	−	−	−
	Hexanoic acid	−	−	−	−
	Levulinic acid	+	−	+	±
	Malic acid	+	−	+	±
	Palmitic acid	+	±	+	+
	Trans-2-pentenoic acid	−	−	−	−
Galactose +	Dodecanoic acid	+	−	+	−
	Heptanoic acid	−	−	−	−
	Hexanoic acid	−	−	−	−
	Levulinic acid	±	−	+	±
	Malic acid	+	−	+	±
	Palmitic acid	−	−	+	−
	Trans-2-pentenoic acid	−	−	−	−
Fructose +	Dodecanoic acid	−	−	−	−
	Heptanoic acid	−	−	−	−
	Hexanoic acid	−	−	−	−
	Levulinic acid	−	−	−	−
	Malic acid	−	−	−	−
	Palmitic acid	−	−	−	−
	Trans-2-pentenoic acid	−	−	−	−

3.3.3. Polyhydroxyalkanoates Production in Shake Flasks

Results of screening tests demonstrated that several carbohydrates could be used for PHA production by the transformant strains. To evaluate the transformant strains, PHB⁻4/pBBR1-Pro$_{Cn}$-*phaC1*, and PHB⁻4/pBBR1-Pro$_{Cn}$-*phaC2*, the production of PHA was compared with *Halomonas* sp. SF2003 and *C. necator* H16 in glucose, fructose, and galactose. The functionality of PhaC1 and PhaC2 of *Halomonas* sp. SF2003 also can be compared based on the results of PHA production.

For *Halomonas* sp. SF2003, glucose was the favorite carbohydrate to produce PHA production (2.25 g/L) followed by galactose (1.23 g/L) and then fructose (1.02 g/L) (Table 7). Comparatively, *C. necator* H16 produces more PHA when fructose is used as the main carbon source in medium (2.25 g/L) rather than glucose (2.05 g/L). Production using galactose cannot be estimated due to low cell dry weight obtained.

Expression of pBBR1-Pro$_{Cn}$-*phaC1* and pBBR1-Pro$_{Cn}$-*phaC2* plasmids allow PHA accumulation in *C. necator* mutant strain PHB⁻4 (a non-PHA-producing strain), further confirm the functionality of PhaC1 and PhaC2. Similar to *C. necator* H16, PHB⁻4/pBBR1-Pro$_{Cn}$-*phaC2* uses more efficiently fructose for PHA production (1.38 g/L) than glucose. Likewise, with *C. necator* H16, PHA production tests using galactose did not allow to determine production yield for both PHB⁻4/pBBR1-Pro$_{Cn}$-*phaC1* and PHB⁻4/pBBR1-Pro$_{Cn}$-*phaC2*.

It was also highlighted that galactose is more adapted for the growth of *Halomonas* sp. SF2003 than for PHA synthesis since only 39 wt% of PHA content was estimated, whereas 86 wt% of PHA content was accumulated when glucose was used (Table 7). With *C. necator* H16, fructose seems to be more exploited for bacterial growth than for PHA production, PHA content was the same as in glucose condition (71 wt%). PHB⁻4/pBBR1-Pro$_{Cn}$-*phaC1* showed a stronger growth with galactose, very close to those obtained with glucose, rather than with fructose. However, PHB⁻4/pBBR1-Pro$_{Cn}$-*phaC1* didn't show PHA accumulation, while PHA content was quite similar using fructose or glucose as the carbon source: 33% and 30%, respectively (Table 7). Finally, PHB⁻4/pBBR1-Pro$_{Cn}$-*phaC2* uses more efficiency fructose for bacterial growth and PHA production than glucose, even if PHA contents are again quite similar (54% for fructose and 52% for glucose).

Table 7. Comparative PHA productions in shake flasks with glucose, fructose or galactose. Productions conducted in triplicate in 250 mL shake flasks containing 50 mL of medium with 2% (w/v) of carbon sources.

Strain	Carbon Source	Dry cell Weight (g/L)	PHA (g/L)	PHA Content (wt.%)
Halomonas sp. SF2003		2.63	2.25	86
C. necator H16	Glucose	2.89	2.05	71
PHB⁻4/pBBR1-Pro$_{Cn}$-*phaC1*		1.05	0.32	30
PHB⁻4/pBBR1-Pro$_{Cn}$-*phaC2*		2.63	1.38	52
Halomonas sp. SF2003		2.63	1.02	39
C. necator H16	Fructose	3.16	2.25	71
PHB⁻4/pBBR1-Pro$_{Cn}$-*phaC1*		0.79	0.26	33
PHB⁻4/pBBR1-Pro$_{Cn}$-*phaC2*		3.42	1.83	54
Halomonas sp. SF2003		3.16	1.23	39
C. necator H16	Galactose	0.79	N.D	N.D
PHB⁻4/pBBR1-Pro$_{Cn}$-*phaC1*		1.06	N.D	N.D
PHB⁻4/pBBR1-Pro$_{Cn}$-*phaC2*		0.79	N.D	N.D

N.D: Not determined.

4. Discussion

Lipase box-like sequences are highly conserved domains which have been identified as active sites of the enzymes [31–33] and play a crucial role in elongation of the polymer [34] in several PHA-producing species. These domains expose similarities with those of lipase, but the difference is in the replacement of the essential active site of lipase, a serine, by a cysteine in the lipase box-like domain of PHA synthase [35], leading to renaming these sequences as PhaC box consensus sequences [34]. In this pattern, similar to lipase, Cysteine (Cys or C) represents the catalytic amino acid and is involved in a catalytic triad (C-H-D) participating, supposedly, in the elongation step of the PHA polymer [34]. The most common described pattern is Glycine-X-Cysteine-X-Glycine (G-X-C-X-G), including Glycine-Tyrosine-Cysteine-Methionine-Glycine sequence (G-Y-C-M-G) detected in *Bacillus cereus* (ATCC 14579) or *Haloferax mediterranei* (ATCC 33500), Glycine-Tyrosine-Cysteine-Leucine-Glycine sequence (G-Y-C-L-G) found in *Cupriavidus metallidurans* strain CH34 or *Halomonas boliviensis* LC1 (DSM15516), or Glycine-Alanine-Cysteine-Serine-Glycine sequence (G-A-C-S-G) in *Cupriavidus necator* strain N-1 or *Pseudomonas fulva* strain 12-X [31,34–36]. However, variations in amino acid composition have also been described in various bacterial species like *H. elongata* (DSM 2581) or *Halomonas* sp. KM-1 for which sequences have been described (Figure 3) [31].

In silico study of PHA synthases allowed to identify two different lipase box-like sequences: Glycine-Tyrosine-Cysteine-Leucine-Glycine (G-Y-C-L-G) for PhaC1 and Serine-Tyrosine-

Cysteine-Isoleucine-Glycine (S-Y-C-I-G) for PhaC2. The G-Y-C-L-G pattern detected in PhaC1 amino acids sequence has already been reported in different halotolerant/halophiles (or not) PHA/PHB-producing strains as *C. metallidurans* strain CH34 [37] or *H. boliviensis* LC1 (DSM 15516) [38]. The second pattern, S-Y-C-I-G, founds in the PhaC2 amino acid sequence has also been reported in sequence of other halotolerant/halophiles PHA/PHB-producing strains as *Chromohalobacter salexigens* (DSM 3043) and *Halomonas* sp. KM-1 [31,39,40]. Both lipase box-like sequences of *Halomonas* sp. SF2003 PHA synthases have a tyrosine, a cysteine, and a glycine (Y-S-G) suggesting that these residues can potentially have a crucial role in the catalytic activity of the enzymes.

These results confirm the distinction of both enzymes, additionally to their size and location in the genome [13]. The differences between both PHA synthases of *Halomonas* sp. SF2003 could generate a difference in catalytic activity and potentially, in the end, impact yield of polymer production. Further research must be performed to elucidate impact of each pattern on enzymes substrates specificity and selectivity and also to validate the identification of catalytic core of both. To complete data, structural study of enzymes exploiting X-ray crystallography and/or molecular biology could be performed to confirm, or not, that the identified lipase box-like sequences play a key role in the synthesis of PHA by *Halomonas* sp. SF2003.

Alterations/modifications of PHA synthase sequences, using molecular biology, will lead to change proteins tertiary structures and potentially synthesis activity. Indeed, other studies have already been performed to elucidate the tertiary structure of different PHA synthases and to identify active sites. For example, Ilham et al. (2014) have studied PHA synthases of *Halomonas* sp. O-1 by performing site-directed mutagenesis on different residues and studying the production of the strain. They determined that appropriated changes can, positively or negatively, affect synthesis activity, bacterial growth, or molecular weight of polymers. Th substitution of alanine for Cys329 or Cys331 in *Halomonas* sp. O-1 or *H. elongata* DSM 2581 PHA synthase sequence leads to a total inhibition of PHA synthesis while substituting glycine for serine impacts polymer molecular weight. These results allowed identification of catalytic sites in enzymes and to imagine modifications in strain genes to enhance production [31]. Same kind of experiments would be one of the prospects to deeply characterize PHA synthases, PhaC1 and PhaC2, of *Halomonas* sp. SF2003. Futhermore, studies exploiting X-ray crystallography will allow to apprehend structure of the catalytic site and to confirm the role of each residue [35]. Similar studies have already been performed and reported with different species, such as *Chromobacterium* sp. USM2 [41], *C. necator* [42] or *Pseudomonas* sp. 61–3 [43].

To investigate the ability of *Halomonas* sp. SF2003 to produce different PHA, various carbon substrates and mix have been screened for growth and biopolymer accumulation using the Nile Red agar plates technique. Nile Red is a fluorescent stain of intracellular lipids, and hydrophobic domain, frequently used to detect PHA [44]. Indeed, the Nile Red represents an easy and fast detection tool for PHA biosynthesis using various technic as agar plates or epifluorescence microscopy [11,45–47]. Work has been mainly focused on "pure" carbon substrates, like carbohydrates, for a better understanding of PHA synthase activity and specificity. Eight pure carbohydrates, including five monosaccharides (glucose, fructose, galactose, rhamnose, mannose) and three disaccharides (maltose, melibiose, sucrose) found in food or natural (co-) products, including fruits, vegetables, milk or red algae, have been tested based on data available in the literature and results of previous studies on *Halomonas* sp. SF2003 [12,13,16,31,48]. The results obtained with *Halomonas* sp. SF2003 confirmed the substrate versatility of this species for both growth and PHA production. Among the tested carbohydrates, positive results have been recorded with glucose, fructose, galactose, mannose, maltose, and sucrose.

Only one carbohydrate in a (L) configuration has been tested: (L)-Rhamnose, and it does not allow both bacterial growth and PHA accumulation by *Halomonas* sp. SF2003. The inability to use (L)-Rhamnose for bacterial growth or PHA production by *Halomonas* sp. SF2003 suggests a lack in part or in totality, of required metabolic tools. This hypothesis is in accordance with the results of a previous study, which showed the presence of only a few genes responsible for rhamnose (II) degradation in the *Halomonas* sp. SF2003 genome (Table 3) [13]. At this time and to our knowledge, there are not yet reports of PHA production with (L)-rhamnose using *Halomonas* species. In contrast, ability to use this sugar as carbon source and substrate for PHA production is variable since *Halomonas* species as *H. cupida*, *H. elongata*, or *H. maura* [49,50] or other species as *C. necator* or *Pseudomonas oleovorans* [51] are capable of doing so while *Halomonas* species as *H. aquamarina*, *H. hamiltonii* (DSM 21196T) or *H. subterranea* (JCM 14608T) are not, in accordance to results obtained with *Halomonas* sp. SF2003. (D)-Melibiose was the second one carbohydrate, which does not allow bacterial growth. The results of screening tests performed suggest that *Halomonas* sp. SF2003 does not or only possesses a part of enzymes required for (D)-Melibiose degradation contrary to results suggested by our previous in silico study [13]. A limited number of studies deal with use of (D)-Melibiose for bacterial growth and only sometimes for PHA production as *Burkholderia sacchari* sp. nov. [17]. *H. cupida* is also able to use (D)-Melibiose for its growth [50] like *Bacillus* sp. (Strain SKM11) [52], *Bacillus subtilis* (Strain PHA 012), *Aeromonas* sp. (Strain PHA 046) or *Alcaligenes* sp. (Strain PHA 047) [53]. Comparatively, and similarly to *Halomonas* sp. SF2003, numerous PHA-producing species have also been reported for their disability to exploit (D)-Melibiose for growth as *Pandoraea* sp. (Strain MA 03) [54], or *Bacillus cereus* (Strain FC11) [55].

In the case of *Halomonas* sp. SF2003, and based on the results of the screening tests, it makes more sense to use (D)-Glucose, (D)-Galactose, and (D)-Maltose, which show to qualitatively allow a stronger PHA production. To complete these results, it is necessary to test a mixture of these sugars with different ratios of each of them in order to evaluate if PHA production is stronger when exploiting them alone or combined. Moreover, the evaluation of PHA production with a mixture of carbohydrates will allow the identification of potential co-products usable with *Halomonas* sp. SF2003. Indeed, bacterial growth and PHA production of various strains using pure carbohydrates is frequently tested and well reported [17,23]. However, because of their high cost, their use at the industrial scale cannot be reasonably considered and exploitation of by-products is privileged [6]. For example, the production of PHA is rarely tested with "pure" galactose but rather using products constituted by itself such as lactose sources (lactose or cheese whey and milk) or in its polymeric form such as agar in red algae in order to promote the use of various co-products. Indeed, PHA productions have successfully been performed with *Haloferax mediterranei* and *Pseudomonas hydrogenovora* on whey lactose [56–58], or with *Bacillus megaterium* [59] using acid-treated red algae. Actually, there is an important number of studies using this group of carbon substrates. However, without more precise analyzes, it is difficult to know which carbohydrates are preferentially used for PHA accumulation. Moreover, the use of these products in their original forms by bacterial strains is difficult, and consequently, some pre-treatments (hydrolysis) are required, sometimes leading to an increase in the cost and time of production [5]. Similar experiments of PHA production have also been performed using (pre-treated) co-products composed of mannose, such as spent coffee ground [60], sugar maple hemicellulosic hydrolysate [61] or ensiled grass press juice [62]. A majority of pure carbohydrates tested for bacterial growth and PHA production in this study can be found in industrial or natural products (Table 3), allowing to test the assimilation/exploitation of different (co-)products by *Halomonas* sp. SF2003.

Additional tests have been conducted on one simple sugar: fructose, galactose, and glucose or mixed with one fatty or organic acids to complete data about *Halomonas* sp. SF2003 carbohydrates metabolisms and to identify new potential carbon sources. Based on the results of the screening tests performed with *Halomonas* sp. SF2003, and using a mix of carbohydrates and acids, production tests might be achieved. In fact, a mix exposing positive results is composed of carbohydrates and acids, which are easily found in various natural products or by-products [63,64].

Levulinic, malic, and palmitic acids can easily be found in plant co-products and have already been tested in a mix with different carbon substrates for PHA production by different bacterial strains. Levulinic acid has been employed in a mix with xylose to perform PHA production with *Burkholderia cepacia* [65] or combined to glucose/fructose with *C. necator* [66]. Quantities of acid employed vary to those tested here and lead to the production of P-3HB-*co*-3HV up to 2.40 g/L with *B. cepacia* [65] and P-3HB synthesis up to 2.41 g/L for *C. necator* [66]. Alongside, previous studies for assimilation of levulinic acid has also been evaluated with *Halomonas hydrothermalis* using seaweed-derived crude levulinic acid and lead to accumulation of P-3HB-*co*-3-HV up to 1.07 g/L [63]. The second acid, malic acid, has been used as co-substrates for PHA production with different bacterial species, such as *B. sacchari*, which accumulates P-3HB up to 2.80 g/L from mix of glucose and malic acid [67]. By-products composed of malic acid from fruit pomace have successfully been exploited by *Pseudomonas resinovorans* for poly-3-hydroxyhexanoates-*co*-3-hydroxyoctanoate-*co*-3-hydroxydecanoate-*co*-3-hydroxydodecanoate-*co*-3-hydroxytetradecenoate (P-3HHx-*co*-3HO-*co*-3HD-*co*-3HDD-*co*-3HTD) production reaching 1.27 g/L [68]. Other papers reported that the addition of malic acid in the production medium of *Methylobacterium trichosporium* can promote the production of P-3HB up to 1.94 g/L [69]. Finally, palmitic acid is also frequently exploited for PHA production and with various bacterial species. Cruz et al. have tested several by-products and wastes as carbon substrates for PHA production, including olive oil, cooking oil, or biodiesel fatty acids by-products. All these products contain more or less important quantities of various fatty acids including palmitic acid. PHA production has been estimated with different species as *Pseudomonas citronellolis, P. oleovorans, P. resinovorans, C. necator* H16 and *C. necator* NRRL B-4383 and demonstrated viability of used wastes and by-products [70]. Another study exploited oil of spent coffee ground, which contains palmitic acid, for P-3HB production with *C. necator* H16 and led to productions reaching up 10 g/L [71]. Additionally, to *Pseudomonas* and *Cupriavidus* species, tests have been conducted on *Burkholderia* sp. USM (JCM15050) to evaluate the exploitation of representative quantities of palmitic acid, alone or in different by-products. The results of this study demonstrated a higher production of P-3HB, up to 1.25 g/L, using palm oil products rather than pure palmitic acid (0.14 g/L of P-3HB) [72].

Consequently, previous tests could also be completed using more or less important different carbohydrates/acids ratio, as performed in different studies. However, the Nile Red agar plates' tests are only used as screening tool and must be completed with production tests to evaluate the impact of each mix or pure carbon substrates on production yield and polymer composition.

Furthermore, additional tests exploiting different by-products derivatives of dairy, waste treatment or agri-food industries might be performed in order to evaluate viability of exploiting these co-products and to optimize PHA production by *Halomonas* sp. SF2003. This kind of production has already been done with different bacterial species, including *Halomonas* species, as well as other ones, as described previously. Indeed, Pernicova et al. 2019, have studied the viability of several *Halomonas* strains to produce PHA from waste cooking oil. They demonstrated that *Halomonas hydrothermalis* exposes the highest production yield (0.38 g/L) but also the influence of NaCl concentration on production [73]. These results demonstrate that the production medium must be wisely studied and elaborated.

The efficiency of two transformed strains harboring *phaC1* or *phaC2* genes was estimated and compared to those of the wide bacteria through lab-scale production. Data of this study confirm their functionality and existence of differences between them, including their size, sequences, location on genome, and capability of production of PHA. Indeed, PhaC2 exhibited greater PHA production capability as PHB$^-$4/pBBR1-Pro$_{Cn}$-*phaC2* accumulated more PHA in comparison to that of PHB$^-$4/pBBR1-Pro$_{Cn}$-*phaC1* under the exact same host, plasmid, and culture conditions. This is in accordance with previous results demonstrating a higher percent of the identity of PhaC2 with PHA synthases of other bacterial species than PhaC1 [13]. To confirm this result, PHA production tests must be conducted, and new genetic constructions could be tested. These tests will allow the definite evaluation of the functionality of PHA synthases. Moreover, in this study, a pBBR1MCS-2 plasmid with *C. necator* H16 promoter was used. This construction, could be responsible, in part, for the weak activity of PhaC1. Indeed, in *Halomonas* sp. SF2003 genome PHA biosynthesis genes expose an atypical distribution [13], so PHA synthases expression of PhaC1 and PhaC2 was tested separately. However, it could be possible that PhaC1 specifically requires proteins (PhaC2, PhaA, PhaB) or promoter of *Halomonas* sp. SF2003 metabolisms to ensure polymer synthesis despite that it has been identified to belong to class I of synthase (meaning that PhaC are constituted of only one subunit and does not require any additional protein to be active). To confirm this hypothesis, several different constructions might be designed using *Halomonas* sp. SF2003 promoter and PHA biosynthesis genes simultaneously, and evaluated for PHA production [35,74–76]. Among all the different constructions which could be tested, plasmid harboring both *phaC1* and *phaC2* genes, together, must be designed. This construction will allow the control of the influence of each one on the other and to check if PHA synthase PhaC1 requires PhaC2 to be active. Testing different constructions will allow a better understanding of genes activity and to identify the best combination to optimize production.

Finally, both transformant strains showed lower efficiency of PHA production than the wild type strain *C. necator* H16. Lower PHA production could be due to the weaker acetyl-CoA C-acyltransferase and β-ketothiolase activities, as already described by Mifune et al. 2008 [77]. PhaCs of *Halomonas* sp. SF2003 can also expose a lower activity than PhaC of *C. necator* H16.

The composition of production medium and production parameters used for these tests can also be responsible in part of the low production yields. Indeed, the same medium and parameters have been used for all the strains. However, *Halomonas* sp. SF2003 and *C. necator*'s wild and transformant strains do not exhibit the same origin and metabolisms. Consequently, new production medium and different production parameters must be tested. Following production of each strain in different conditions will allow more precise understanding of the activity of each strain/PHA synthase and to adjust more precisely the production step.

5. Conclusions

This study has demonstrated the functionality of both PHA synthases, PhaC1, and PhaC2, confirming annotation of *Halomonas* sp. SF2003 genome performed in our in-silico study. Performed screening tests allowed the identification of several carbon substrates, pure carbohydrates or a mix of sugars and acids, potentially usable for PHA production by *Halomonas* sp. SF2003. Substrate versatility of this bacterium opens the door for new tests in order to optimize production and also confirm its high biotechnological potential. Preliminary biosynthesis tests expose a better PHA production using glucose with *Halomonas* sp. SF2003 while *C. necator* wild type and transformant strain preferably exploit fructose. Results also highlighted higher PHA biosynthesis ability of PHA synthase PhaC2 as compared to PhaC1. These results open the door to future research in order to overexpress the *phaC2* gene of *Halomonas* sp. SF2003 to increase the yield of production of the strain. Additional research, such as kinetics of bacterial growth and PHA production, should optimize production step.

Supplementary Materials:
Supplementary data S1: Accession numbers of PhaC amino acids sequences on National Center for Biotechnology Information (NCBI) database. Supplementary data S2: Nile Red agar plates screening with PHB⁻4/pBBR1-Pro$_{Cn}$-phaC1 using 2% (w/v) of different carbon substrates. Supplementary data S3: Nile Red agar plates screening with PHB⁻4/pBBR1-ProCn-phaC2 using 2% (w/v) of different carbon substrates.

Author Contributions: Investigation, T.T.; Methodology, T.T., H.T.T. and H.L.; Validation, T.T., K.S., A.B., A.E. and S.B.; Writing—review & editing, T.T., K.S., H.T.T. and H.L. All authors have read and agreed to the published version of the manuscript.

Acknowledgments: This study was funded by Research University Grant (RUI) from Universiti Sains Malaysia (USM) (1001/PBIOLOGI/8011060 to S.K.) and Research University Grant "Homme, Mer et Littoral 2018" from University of South Brittany.

References

1.   Rehm, B.H.A. Bacterial polymers: Biosynthesis, modifications and applications. *Nat. Rev. Microbiol.* **2010**, *8*, 578–592. [CrossRef] [PubMed]

2.　Możejko-Ciesielska, J.; Kiewisz, R. Bacterial polyhydroxyalkanoates: Still fabulous? *Microbiol. Res.* **2016**, *192*, 271–282. [CrossRef] [PubMed]

3.　Keshavarz, T.; Roy, I. Polyhydroxyalkanoates: Bioplastics with a green agenda. *Curr. Opin. Microbiol.* **2010**, *13*, 321–326. [CrossRef]

4.　Valappil, S.P.; Misra, S.K.; Boccaccini, A.R.; Roy, I. Biomedical applications of polyhydroxyalkanoates, an overview of animal testing and in vivo responses. *Expert Rev. Med. Devices* **2006**, *3*, 853–868. [CrossRef]

5.　Williams, H.; Patricia, K. *Polyhydroxyalkanoates. Biosynthesis, Chemical Structures and Applications. Materials Science and Technologies*; NOVA: Annandale, VA, USA, 2018; p. 332.

6.　Jiang, G.; Hill, D.; Kowalczuk, M.; Johnston, B.; Adamus, G.; Irorere, V.; Radecka, I. Carbon Sources for Polyhydroxyalkanoates and an Integrated Biorefinery. *Int. J. Mol. Sci.* **2016**, *17*, 1157. [CrossRef]

7.　Gumel, A.M.; Annuar, M.S.M.; Heidelberg, T. Effects of carbon substrates on biodegradable polymer composition and stability produced by *Delftia tsuruhatensis* Bet002 isolated from palm oil mill effluent. *Polym. Degrad. Stab.* **2012**, *97*, 1224–1231. [CrossRef]

8.　Reddy, C.S.K.; Ghai, R.; Kalia, V. Polyhydroxyalkanoates: An overview. *Bioresour. Technol.* **2003**, *87*, 137–146. [CrossRef]

9.　Koller, M.; Maršálek, L.; de Sousa Dias, M.M.; Braunegg, G. Producing microbial polyhydroxyalkanoate (PHA) biopolyesters in a sustainable manner. *New Biotechnol.* **2017**, *37*, 4–38. [CrossRef]

10.　Balakrishna Pillai, A.; Kumarapillai, H.K. Bacterial Polyhydroxyalkanoates: Recent Trends in Production and Applications. In *Recent Advances in Applied, Microbiology*; Shukla, P., Ed.; Springer: Singapore, 2017; pp. 19–53. Available online: http://link.springer.com/10.1007/978-981-10-5275-0_2 (accessed on 15 July 2019).

11.　Elain, A.; Le Fellic, M.; Corre, Y.-M.; Le Grand, A.; Le Tilly, V.; Audic, J.-L.; Bruzaud, S. Rapid and qualitative fluorescence-based method for the assessment of PHA production in marine bacteria during batch culture. *World J. Microbiol. Biotechnol.* **2015**, *31*, 1555–1563. [CrossRef]

12.　Elain, A.; Le Grand, A.; Corre, Y.-M.; Le Fellic, M.; Hachet, N.; Le Tilly, V.; Loulergue, P.; Audic, J.-L.; Bruzaud, S. Valorisation of local agro-industrial processing waters as growth media for polyhydroxyalkanoates (PHA) production. *Ind. Crops Prod.* **2016**, *80*, 1–5. [CrossRef]

13.　Thomas, T.; Elain, A.; Bazire, A.; Bruzaud, S. Complete genome sequence of the halophilic PHA-producing bacterium *Halomonas* sp. SF2003: Insights into its biotechnological potential. *World J. Microbiol. Biotechnol.* **2019**, *35*, 50. [CrossRef] [PubMed]

14.　Simon, R.; Priefer, U.; Pühler, A. A Broad host range mobilization system for in vivo genetic engineering transposon mutagenesis in Gram negative bacteria. *Nat. Biotechnol.* **1983**, *1*, 784–791. [CrossRef]

15.　Foong, C.P.; Lau, N.S.; Deguchi, S.; Toyofuku, T.; Taylor, T.; Sudesh, K.; Matsui, M. Whole genome amplification approach reveals novel polyhydroxyalkanoate synthases (PhaCs) from Japan Trench and Nankai Trough seawater. *BMC Microbiol.* **2014**, *14*, 318. [CrossRef]

16.　Lemechko, P.; Le Fellic, M.; Bruzaud, S. Production of poly(3-hydroxybutyrate-*co*-3-hydroxyvalerate) using agro-industrial effluents with tunable proportion of 3-hydroxyvalerate monomer units. *Int. J. Biol. Macromol.* **2019**, *128*, 429–434. [CrossRef] [PubMed]

17.　Verlinden, R.A.J.; Hill, D.J.; Kenward, M.A.; Williams, C.D.; Radecka, I. Bacterial synthesis of biodegradable polyhydroxyalkanoates. *J. Appl. Microbiol.* **2007**, *102*, 1437–1449. [CrossRef]

18.　Labuzek, S.; Radecka, I. Biosynthesis of PHB tercopolymer by *Bacillus cereus* UW85. *J. Appl. Microbiol.* **2001**, *90*, 353–357.

19.　Tan, D.; Xue, Y.-S.; Aibaidula, G.; Chen, G.-Q. Unsterile and continuous production of polyhydroxybutyrate by *Halomonas* TD01. *Bioresour. Technol.* **2011**, *102*, 8130–8136. [CrossRef]

20.　Simon-Colin, C.; Raguénès, G.; Cozien, J.; Guezennec, J.G. *Halomonas profundus* sp. nov., a new PHA-producing bacterium isolated from a deep-sea hydrothermal vent shrimp. *J. Appl. Microbiol.* **2008**, *104*, 1425–1432. [CrossRef]

21.　Balakrishna Pillai, A.; Jaya Kumar, A.; Thulasi, K.; Kumarapillai, H. Evaluation of short-chain-length polyhydroxyalkanoate accumulation in *Bacillus aryabhattai*. *Braz. J. Microbiol.* **2017**, *48*, 451–460. [CrossRef]

22.　Tan, D.; Wu, Q.; Chen, J.-C.; Chen, G.-Q. Engineering *Halomonas* TD01 for the low-cost production of polyhydroxyalkanoates. *Metab. Eng.* **2014**, *26*, 34–47. [CrossRef]

23.　Pernicova, I.; Kucera, D.; Novackova, I.; Vodicka, J.; Kovalcik, A.; Obruca, S. Extremophiles platform strains for sustainable production of polyhydroxyalkanoates. *Mater. Sci. Forum* **2019**, *955*, 74–79. [CrossRef]

24. Page, W.J. Production of poly-β-hydroxybutyrate by *Azotobacter vinelandii* UWD in media containing sugars and complex nitrogen sources. *Appl. Microbiol. Biotechnol.* **1992**, *38*, 117–121. [CrossRef]

25. Miranda De Sousa Dias, M.; Koller, M.; Puppi, D.; Morelli, A.; Chiellini, F.; Braunegg, G. Fed-batch synthesis of poly(3-hydroxybutyrate) and poly(3-hydroxybutyrate-*co*-4-hydroxybutyrate) from sucrose and 4-hydroxybutyrate precursors by *Burkholderia sacchari* Strain DSM 17165. *Bioengineering* **2017**, *4*, 36. [CrossRef] [PubMed]

26. Daniş, Ö.; Ogan, A.; Tatlican, P.; Attar, A.; Çakmakçi, E.; Mertoglu, B.; Birbir, M.; Cakmakcı, E. Preparation of poly(3-hydroxybutyrate-*co*-hydroxyvalerate) films from halophilic archaea and their potential use in drug delivery. *Extremophiles* **2015**, *19*, 515–524.

27. Kucera, D.; Pernicová, I.; Kovalcik, A.; Koller, M.; Mullerova, L.; Sedlacek, P.; Mravec, F.; Nebesarova, J.; Kalina, M.; Márova, I.; et al. Characterization of the promising poly(3-hydroxybutyrate) producing halophilic bacterium *Halomonas halophila*. *Bioresour. Technol.* **2018**, *256*, 552–556. [CrossRef]

28. Giedraitytė, G.; Kalėdienė, L. Purification and characterization of polyhydroxybutyrate produced from thermophilic *Geobacillus* sp. AY 946034 strain. *CHEMIJA* **2015**, *26*, 38–45.

29. Quillaguamán, J.; Hashim, S.; Bento, F.; Mattiasson, B.; Hatti-Kaul, R. Poly(β-hydroxybutyrate) production by a moderate halophile, *Halomonas boliviensis* LC1 using starch hydrolysate as substrate. *J. Appl. Microbiol.* **2005**, *99*, 151–157. [CrossRef]

30. Joshi, A.A.; Kanekar, P.P.; Kelkar, A.S.; Sarnaik, S.S.; Shouche, Y.; Wani, A. Moderately halophilic, alkalitolerant *Halomonas campisalis* MCM B-365 from Lonar Lake, India. *J. Basic Microbiol.* **2007**, *47*, 213–221. [CrossRef]

31. Ilham, M.; Nakanomori, S.; Kihara, T.; Hokamura, A.; Matsusaki, H.; Tsuge, T.; Mizuno, K. Characterization of polyhydroxyalkanoate synthases from *Halomonas* sp. O-1 and *Halomonas elongata* DSM2581: Site-directed mutagenesis and recombinant expression. *Polym. Degrad. Stab.* **2014**, *109*, 416–423. [CrossRef]

32. Lu, Q.; Han, J.; Zhou, L.; Zhou, J.; Xiang, H. Genetic and biochemical characterization of the poly(3-hydroxybutyrate-*co*-3-hydroxyvalerate) synthase in *Haloferax mediterranei*. *J. Bacteriol.* **2008**, *190*, 4173–4180. [CrossRef]

33. Ueda, S.; Yabutani, T.; Maehara, A.; Yamane, T. Molecular analysis of the poly(3-hydroxyalkanoate) synthase gene from a methylotrophic bacterium, *Paracoccus denitrificans*. *J. Bacteriol.* **1996**, *178*, 774–779. [CrossRef] [PubMed]

34. Mezzolla, V.; D'Urso, O.; Poltronieri, P. Role of PhaC Type I and Type II Enzymes during PHA Biosynthesis. *Polymers* **2018**, *10*, 910. [CrossRef] [PubMed]

35. Rehm, B.H. Polyester synthases: Natural catalysts for plastics. *Biochem. J.* **2003**, *376*, 15–33. [CrossRef] [PubMed]

36. Tsuge, T.; Hyakutake, M.; Mizuno, K. Class IV polyhydroxyalkanoate (PHA) synthases and PHA-producing *Bacillus*. *Appl. Microbiol. Biotechnol.* **2015**, *99*, 6231–6240. [CrossRef] [PubMed]

37. Ramachandran, H.; Shafie, N.A.H.; Sudesh, K.; Azizan, M.N.; Majid, M.I.A.; Amirul, A.-A.A. *Cupriavidus malaysiensis* sp. nov., a novel poly(3-hydroxybutyrate-*co*-4-hydroxybutyrate) accumulating bacterium isolated from the Malaysian environment. *Anton Leeuw. Int. J. G.* **2018**, *111*, 361–372. [CrossRef] [PubMed]

38. García-Torreiro, M.; López-Abelairas, M.; Lu-Chau, T.A.; Lema, J.M. Production of poly(3-hydroxybutyrate) by simultaneous saccharification and fermentation of cereal mash using *Halomonas boliviensis*. *Biochem. Eng. J.* **2016**, *114*, 140–146. [CrossRef]

39. Ates, Ö.; Oner, E.; Arga, K.Y. Genome-scale reconstruction of metabolic network for a halophilic extremophile, *Chromohalobacter salexigens* DSM 3043. *BMC Syst. Biol.* **2011**, *5*, 12. [CrossRef]

40. Kawata, Y.; Aiba, S. Poly(3-hydroxybutyrate) Production by isolated *Halomonas* sp. KM-1 using waste glycerol. *Biosci. Biotechnol. Biochem.* **2010**, *74*, 175–177. [CrossRef]

41. Chek, M.F.; Kim, S.-Y.; Mori, T.; Arsad, H.; Samian, M.R.; Sudesh, K.; Hakoshima, T. Structure of polyhydroxyalkanoate (PHA) synthase PhaC from *Chromobacterium* sp. USM2, producing biodegradable plastics. *Sci. Rep.* **2017**, *7*, 5312. [CrossRef]

42. Wittenborn, E.C.; Jost, M.; Wei, Y.; Stubbe, J.; Drennan, C.L. Structure of the Catalytic Domain of the Class I Polyhydroxybutyrate Synthase from *Cupriavidus necator*. *J. Biol. Chem.* **2016**, *291*, 25264–25277. [CrossRef]

43. Chek, M.F.; Hiroe, A.; Hakoshima, T.; Sudesh, K.; Taguchi, S. PHA synthase (PhaC): Interpreting the functions of bioplastic-producing enzyme from a structural perspective. *Appl. Microbiol. Biotechnol.* **2019**, *103*, 1131–1141. [CrossRef] [PubMed]

44. Spiekermann, P.; Rehm, B.H.A.; Kalscheuer, R.; Baumeister, D.; Steinbüchel, A. A sensitive, viable-colony staining method using Nile red for direct screening of bacteria that accumulate polyhydroxyalkanoic acids and other lipid storage compounds. *Arch. Microbiol.* **1999**, *171*, 73–80. [CrossRef] [PubMed]

45. Peters, V.; Rehm, B.H.A. In vivo monitoring of PHA granule formation using GFP-labeled PHA synthases. *FEMS Microbiol. Lett.* **2005**, *248*, 93–100. [CrossRef] [PubMed]

46. Cavalheiro, J.M.B.T.; de Almeida, M.C.M.D.; da Fonseca, M.M.R.; de Carvalho, C.C.C.R. Adaptation of *Cupriavidus necator* to conditions favoring polyhydroxyalkanoate production. *J. Biotechnol.* **2013**, *164*, 309–317. [CrossRef]

47. Shrivastav, A.; Mishra, S.K.; Shethia, B.; Pancha, I.; Jain, D.; Mishra, S. Isolation of promising bacterial strains from soil and marine environment for polyhydroxyalkanoates (PHAs) production utilizing Jatropha biodiesel byproduct. *Int. J. Biol. Macromol.* **2010**, *47*, 283–287. [CrossRef]

48. Bramer, C.O.; Vandamme, P.; da Silva, L.F.; Gomez, J.; Steinbüchel, A. *Burkholderia sacchari* sp. nov., a polyhydroxyalkanoate-accumulating bacterium isolated from soil of a sugar-cane plantation in Brazil. *Int. J. Syst. Evol. Microbiol.* **2001**, *51*, 1709–1713. [CrossRef]

49. Mata, J.A.; Martínez-Cánovas, J.; Quesada, E.; Béjar, V. A detailed phenotypic characterisation of the type strains of *Halomonas* species. *Syst. Appl. Microbiol.* **2002**, *25*, 360–375. [CrossRef]

50. Vreeland, R.H. Halomonas. In *Bergey's Manual of Systematics of Archaea and, Bacteria*; Whitman, W.B., Rainey, F., Kämpfer, P., Trujillo, M., Chun, J., DeVos, P., Eds.; John Wiley & Sons, Ltd.: Chichester, UK, 2015; pp. 1–19.

51. Füchtenbusch, B.; Wullbrandt, D.; Steinbüchel, A. Production of polyhydroxyalkanoic acids by *Ralstonia eutropha* and *Pseudomonas oleovorans* from an oil remaining from biotechnological rhamnose production. *Appl. Microbiol. Biotechnol.* **2000**, *53*, 167–172. [CrossRef]

52. Chaitanya, K.; Nagamani, P.; Mahmood, S.K.; Sunil Kumar, N. Polyhydroxyalkanoate producing novel *Bacillus* sp., SKM11 isolated from polluted pond water. *Int. J. Curr. Microbiol. Appl. Sci.* **2015**, *4*, 1159–1165.

53. Sangkharak, K.; Prasertsan, P. Screening and identification of polyhydroxyalkanoates producing bacteria and biochemical characterization of their possible application. *J. Gen. Appl. Microbiol.* **2012**, *58*, 173–182. [CrossRef]

54. de Paula, F.C.; Kakazu, S.; de Paula, C.B.C.; Gomez, J.G.C.; Contiero, J. Polyhydroxyalkanoate production from crude glycerol by newly isolated *Pandoraea* sp. *J. King Saud Univ. Sci.* **2017**, *29*, 166–173. [CrossRef]

55. Masood, F.; Yasin, T.; Hameed, A. Production and characterization of tailor-made polyhydroxyalkanoates by *Bacillus cereus* FC1. *Pak. J. Zool.* **2015**, *47*, 491–503.

56. Koller, M.; Hesse, P.; Bona, R.; Kutschera, C.; Atlić, A.; Braunegg, G. Biosynthesis of high quality polyhydroxyalkanoate co- and terpolyesters for potential medical application by the Archaeon *Haloferax mediterranei*. *Macromol. Symp.* **2007**, *253*, 3–9. [CrossRef]

57. Koller, M.; Atlić, A.; Gonzalez-Garcia, Y.; Kutschera, C.; Braunegg, G. Polyhydroxyalkanoate (PHA) Biosynthesis from Whey Lactose. *Macromol. Symp.* **2008**, *272*, 87–92. [CrossRef]

58. Koller, M.; Bona, R.; Chiellini, E.; Fernandes, E.G.; Horvat, P.; Kutschera, C.; Hesse, P.; Braunegg, G. Polyhydroxyalkanoate production from whey by *Pseudomonas hydrogenovora*. *Bioresour. Technol.* **2008**, *99*, 4854–4863. [CrossRef]

59. Alkotaini, B.; Koo, H.; Kim, B.S. Production of polyhydroxyalkanoates by batch and fed-batch cultivations of *Bacillus megaterium* from acid-treated red algae. *Korean J. Chem. Eng.* **2016**, *33*, 1669–1673. [CrossRef]

60. Obruca, S.; Benesova, P.; Petrik, S.; Oborna, J.; Prikryl, R.; Marova, I. Production of polyhydroxyalkanoates using hydrolysate of spent coffee grounds. *Process Biochem.* **2014**, *49*, 1409–1414. [CrossRef]

61. Pan, W.; Perrotta, J.A.; Stipanovic, A.J.; Nomura, C.T.; Nakas, J.P. Production of polyhydroxyalkanoates by *Burkholderia cepacia* ATCC 17759 using a detoxified sugar maple hemicellulosic hydrolysate. *J. Ind. Microbiol. Biotechnol.* **2012**, *39*, 459–469. [CrossRef]

62. Cerrone, F.; Davis, R.; Kenny, S.T.; Woods, T.; O'Donovan, A.; Gupta, V.K.; Tuohy, M.; Babu, R.P.; O'Kiely, P.; O'Connor, K. Use of a mannitol rich ensiled grass press juice (EGPJ) as a sole carbon source for polyhydroxyalkanoates (PHAs) production through high cell density cultivation. *Bioresour. Technol.* **2015**, *191*, 45–52. [CrossRef]

63. Bera, A.; Dubey, S.; Bhayani, K.; Mondal, D.; Mishra, S.; Ghosh, P.K. Microbial synthesis of polyhydroxyalkanoate using seaweed-derived crude levulinic acid as co-nutrient. *Int. J. Biol. Macromol.* **2015**, *72*, 487–494. [CrossRef]

64. Ashby, R.D.; Solaiman, D.K.Y.; Strahan, G.D.; Zhu, C.; Tappel, R.C.; Nomura, C.T. Glycerine and levulinic acid: Renewable co-substrates for the fermentative synthesis of short-chain poly(hydroxyalkanoate) biopolymers. *Bioresour. Technol.* **2012**, *118*, 272–280. [CrossRef] [PubMed]

65. Keenan, T.M.; Tanenbaum, S.W.; Stipanovic, A.J.; Nakas, J.P. Production and characterization of poly-β-hydroxyalkanoate copolymers from *Burkholderia cepacia* utilizing xylose and levulinic acid. *Biotechnol. Prog.* **2004**, *20*, 1697–1704. [CrossRef] [PubMed]

66. Jaremko, M.; Yu, J. The initial metabolic conversion of levulinic acid in *Cupriavidus necator*. *J. Biotechnol.* **2011**, *155*, 293–298. [CrossRef] [PubMed]

67. Mendonça, T.; Gomez, J.G.C.; Buffoni, E.; Rodriguez, R.S.; Schripsema, J.; Lopes, M.; Silva, L. Exploring the potential of *Burkholderia sacchari* to produce polyhydroxyalkanoates. *J. Appl. Microbiol.* **2013**, *116*, 815–829. [CrossRef] [PubMed]

68. Follonier, S.; Goyder, M.S.; Silvestri, A.-C.; Crelier, S.; Kálmán, F.; Riesen, R.; Zinn, M. Fruit pomace and waste frying oil as sustainable resources for the bioproduction of medium-chain-length polyhydroxyalkanoates. *Int. J. Biol. Macromol.* **2014**, *71*, 42–52. [CrossRef] [PubMed]

69. Khosravi-Darani, K.; Mokhtari, Z.-B.; Amai, T.; Tanaka, K. Microbial production of poly(hydroxybutyrate) from C1 carbon sources. *Appl. Microbiol. Biotechnol.* **2013**, *97*, 1407–1424. [CrossRef]

70. Cruz, M.V.; Freitas, F.; Paiva, A.; Mano, F.; Dionísio, M.; Ramos, A.M.; Reis, M.A.; Andrade, M.M.D.; Mano, M.F.M. Valorization of fatty acids-containing wastes and byproducts into short- and medium-chain length polyhydroxyalkanoates. *New Biotechnol.* **2016**, *33*, 206–215. [CrossRef]

71. Obruca, S.; Petrik, S.; Benesova, P.; Svoboda, Z.; Eremka, L.; Marova, I. Utilization of oil extracted from spent coffee grounds for sustainable production of polyhydroxyalkanoates. *Appl. Microbiol. Biotechnol.* **2014**, *98*, 5883–5890. [CrossRef]

72. Chee, J.-Y.; Tan, Y.; Samian, M.-R.; Sudesh, K. Isolation and characterization of a *Burkholderia* sp. USM (JCM15050) capable of producing polyhydroxyalkanoate (PHA) from triglycerides, fatty acids and glycerols. *J. Polym. Environ.* **2010**, *18*, 584–592. [CrossRef]

73. Pernicova, I.; Kucera, D.; Nebesarova, J.; Kalina, M.; Novackova, I.; Koller, M.; Obruca, S. Production of polyhydroxyalkanoates on waste frying oil employing selected Halomonas strains. *Bioresour. Technol.* **2019**, *292*, 122028. [CrossRef]

74. Luengo, J.M.; García, B.; Sandoval, A.; Naharro, G.; Olivera, E.R. Bioplastics from microorganisms. *Curr. Opin. Microbiol.* **2003**, *6*, 251–260. [CrossRef]

75. Argandoña, M.; Vargas, C.; Reina-Bueno, M.; Rodríguez-Moya, J.; Salvador, M.; Nieto, J.J. *An Extended Suite of Genetic Tools for Use in Bacteria of the Halomonadaceae: An Overview*; Recombinant Gene, Expression; Lorence, A., Ed.; Humana Press: Totowa, NJ, USA, 2012; pp. 167–201. Available online: http://link.springer.com/10.1007/978-1-61779-433-9_9 (accessed on 8 November 2017).

76. Jia, K.; Cao, R.; Hua, D.H.; Li, P. Study of Class I and Class III Polyhydroxyalkanoate (PHA) synthases with substrates containing a modified side chain. *Biomacromolecules* **2016**, *17*, 1477–1485. [CrossRef] [PubMed]

77. Mifune, J.; Nakamura, S.; Fukui, T. Targeted engineering of *Cupriavidus necator* chromosome for biosynthesis of poly(3-hydroxybutyrate-*co*-3-hydroxyhexanoate) from vegetable oil. *Can. J. Chem.* **2008**, *86*, 621–627. [CrossRef]

2

Biosynthesis of Polyhydroxyalkanoates (PHAs) by the Valorization of Biomass and Synthetic Waste

Hadiqa Javaid [1,†], Ali Nawaz [1,†], Naveeda Riaz [2], Hamid Mukhtar [1], Ikram-Ul-Haq [1], Kanita Ahmed Shah [1], Hooria Khan [3], Syeda Michelle Naqvi [1], Sheeba Shakoor [1], Aamir Rasool [4], Kaleem Ullah [5], Robina Manzoor [6], Imdad Kaleem [3,*] and Ghulam Murtaza [7]

[1] Institute of Industrial Biotechnology (IIB), Government College University, Lahore 54000, Pakistan; javaid.hadiqa94@gmail.com (H.J.); ali.nawaz@gcu.edu.pk (A.N.); hamidmukhtar@gcu.edu.pk (H.M.); dr.ikramulhaq@gcu.edu.pk (I.-U.-H.); kanita.a.shah@gmail.com (K.A.S.); mnaqvi678@gmail.com (S.M.N.); sheebashakoor03@gmail.com (S.S.)

[2] Department of Biological Sciences, International Islamic University, Islamabad 45550, Pakistan; naveeda.riaz@iiu.edu.pk

[3] Department of Biosciences, COMSATS University Islamabad (CUI), Islamabad 45550, Pakistan; hooriakhan.pk@gmail.com

[4] Institute of Biochemistry, University of Balochistan, Quetta 87300, Pakistan; rasool.amir@gmail.com

[5] Department of Microbiology, University of Balochistan, Quetta 87300, Pakistan; drkaleemullah@gmail.com

[6] Faculty of Marine Sciences, Lasbella University of Agriculture, Water and Marine Sciences, Balochistan 90150, Pakistan; 3820150009@bit.edu.cn

[7] Department of Zoology, University of Gujrat, Gujrat 50700, Pakistan; gmurtazay@yahoo.com

*   Correspondence: kaleem.imdad@comsats.edu.pk

†   These authors contributed equally to this work.

Abstract: Synthetic pollutants are a looming threat to the entire ecosystem, including wildlife, the environment, and human health. Polyhydroxyalkanoates (PHAs) are natural biodegradable microbial polymers with a promising potential to replace synthetic plastics. This research is focused on devising a sustainable approach to produce PHAs by a new microbial strain using untreated synthetic plastics and lignocellulosic biomass. For experiments, 47 soil samples and 18 effluent samples were collected from various areas of Punjab, Pakistan. The samples were primarily screened for PHA detection on agar medium containing Nile blue A stain. The PHA positive bacterial isolates showed prominent orange–yellow fluorescence on irradiation with UV light. They were further screened for PHA estimation by submerged fermentation in the culture broth. Bacterial isolate 16a produced maximum PHA and was identified by 16S rRNA sequencing. It was identified as *Stenotrophomonas maltophilia* HA-16 (MN240936), reported first time for PHA production. Basic fermentation parameters, such as incubation time, temperature, and pH were optimized for PHA production. Wood chips, cardboard cutouts, plastic bottle cutouts, shredded polystyrene cups, and plastic bags were optimized as alternative sustainable carbon sources for the production of PHAs. A vital finding of this study was the yield obtained by using plastic bags, i.e., 68.24 ± 0.27%. The effective use of plastic and lignocellulosic waste in the cultivation medium for the microbial production of PHA by a novel bacterial strain is discussed in the current study.

Keywords: biomass valorization; biodegradation; biopolymer; biomaterials; bioplastic; plastic bag; biological materials; microbial polymers; bacterial bioplastic; eco-friendly materials

1. Introduction

Plastics and pollution are two deeply connected terms. The plastic industry is an ever-growing entity expected to reach a value of nearly $2184.26 billion by 2022 [1]. Pakistan's plastic industry is growing at an annual growth rate of 15% with 624,200 million tons/per annum estimated production rate. A value investment of about $260 billion was attracted by this industry among which half of the investments were foreign direct investments (FDI). These foreign investments contribute to 35% of the outstanding export growth of plastics in Pakistan [2]. Plastics have invaded almost all areas of our daily lives; from transportation, electronics, medical industry, textiles to packaging, etc. [3]. Depending on the plastic product, it can take approximately 600 years to degrade [4]. Plastic overuse has resulted in depleting fossil fuels, severely endangered wildlife, impacts on human health, global warming, solid waste accumulation, hazardous air, and water pollution. Plastic pollution has become such a widespread phenomenon that our environment is now home to 51 trillion pieces of plastics. This figure is 500 times more than the number of stars in our galaxy [5]. In the next 20 years, plastic pollution is expected to double putting our current waste management and recycling practices to a shame [6]. The latest research suggests that the presence of macro and micro-plastics in the environment is such immense that it is factually "raining plastics" [7–9].

For the past two decades, "Bioplastics" or "Bio-based polymers" have emerged as prominent keywords. Renewable materials, such as starch, cellulose, plant oils, chitin, pectin, soy protein, whey protein, collagen, and gelatin are important sources for bioplastic production. Polylactic acid (PLA) and polyhydroxyalkanoates (PHAs) represent the two most promising and the most important bioplastics of the modern age produced by plants, algae, and bacteria using renewable sources [10]. Over 300 heterotrophic gram-negative and gram-positive bacterial species capable of synthesizing PHAs have been isolated and identified e.g., *Methylobacterium* sp., *Cupriavidus necator*, *Bacillus* sp., *Pseudomonas* sp., *Enterobacter* sp., *Citrobacter* sp., *Escherichia* sp., *Klebsiella* sp., *Azotobacter beijerinckii*, *Rhizobium* sp., *A. vinelandii*, *A. macrocytogenes*, *C. necator*, *P. oleovorans*, and *Protomonas extorquens*, etc. [11]. The search for novel bacterial species carrying enhanced potential of PHA production has resulted in the discovery of such potential bacterial strains that produce PHAs efficiently. Some recent novel strains of *Stenotrophomonas* sp., *Xanthomonas* sp., *Staphylococcus* sp. and Haloarchaea bacteria [12] are a few more examples of such bacteria [13].

PHAs are made up of various hydroxycarboxylic acid polyesters, which are produced by a large number of bacteria. They are hydrophobic inclusions formed in bacterial cells in the excess of carbon and limitation of other nutrients, such as N, O, P, S or Mg. They are used as reserves of carbon and energy [14]. They are eco-friendly, 100% biodegradable, recyclable, non-toxic, biocompatible, and bioresorbable [15]. When buried in soil, PHAs decompose completely into carbon dioxide and water within seven months [16]. However, the cost of inoculum preparation, downstream processing, and raw materials makes its commercial preparation 5–10 times expensive than normal plastics [17]. Carbon sources cost more than 50% of the process. Therefore, the trend is shifting towards the use of cheap, waste, and sustainable substrates, which are easily processed or need no processing at all before use [18]. Most experts in the field are of the view that each step in the industrialization of PHAs still requires extensive research, optimization, and understanding to make it more sustainable in practice [19].

Since plastics are mostly based on carbon atoms, they can be an excellent carbon source to produce PHAs [20]. Plastics have many types but the most common types are polyethylene terephthalate (PET), polyvinyl chloride (PVC), foamed polystyrene and polymethylmethacrylate (Plexiglas). On the basis of chemical composition, plastics are of two types. One type is entirely made of linear (aliphatic) carbon atoms in the backbone. Other category includes heterochains, such as N, O, or S atoms in the structure other than C [21]. There are many concerns involved with using plastics in these processes. Nevertheless, with the appropriate alliance in value chains, this approach can prove fruitful in plastic waste management.

Considering the current research on sustainable carbon sources to produce bioplastics, the bioplastic industry will potentially boom in the near future. The global bioplastic production capacity is set to increase from around 2.11 million tons in 2018 to approximately 2.62 million tons in 2023, with PHAs being the main drivers of the market [22]. This research focused on the isolation of a new PHA producing bacterial strain from soil and effluent samples to be used in developing a cost-effective method for the synthesis of PHAs by synthetic waste and lignocellulosic biomass as substrates.

2. Results and Discussion

2.1. Primary Screening

A total of 65 samples, including 47 soil and 18 effluent samples were collected from various areas of Punjab, Pakistan. These samples were subjected to primary screening for the isolation of PHA producing bacteria. A total of 127 bacterial isolates were screened from these samples. Moreover, 67 isolates revealed signs of PHA accumulation by showing orange–yellow fluorescence under UV light as mentioned in Table 1. Consequently, these 67 isolates were further selected for the estimation of PHA production. Figure 1A shows an example of a PHA positive sample (sample no: 8).

Table 1. Geographical distribution of samples used for the isolation of polyhydroxyalkanoates (PHAs) producing bacteria and the number of positive and negative isolates per sample.

Label [1]	Sample Type	Area in Punjab, Pakistan	Global Positioning System Location [2]	No. of Isolates	PHA Accumulation [3]
1	Compost	Kakrali, Gujrat	32°51′8.9352″ N 74°4′25.8132″ E	2	+ in all
2	Soil	Crop field, Mandi-bahauddin	32°32′48.1″ N 73°28′00.4″ E	3	+ in isolates 2a, 2c
3	Effluent	Akzo Nobel Pakistan, paint industry, Lahore	31°28′38.9″ N 74°20′32.1″ E	1	+ in all
4	Soil	Flying Paper industry, Lahore	31°31′47.4″ N 74°21′57.3″ E	1	+ in all
5	Soil	Toyo Nasic Plastic industry, Kot Abdullah, Gujranwala	32.1112° N 74.1967° E	2	+ in 5a, − in 5b
6	Effluent	Royal industries, Pvt. Ltd. Gujranwala	32°08′03.5″ N 74°10′01.9″ E	2	− In all
7	Soil	A field in Narowal city	32°06′28.0″ N 74°52′39.2″ E	3	+ in 7b, 7c
8	Soil	Asia poultry farm, Phool Nagar, Dina Nath	31°12′30.7″ N 73°59′43.2″ E	3	+ in 8b
9	Effluent	Universal Food Industries, Lahore	31°34′58.2″ N 74°18′18.8″ E	3	+in 9c
10	Soil	Corn field, Sargodha	32°02′58.5″ N 72°38′52.5″ E	1	− in all
11	Soil	Shahtaj Sugar Mills, Mandi-Bahauddin	32°35′19.8″ N 73°27′17.1″ E	1	+ in all
12	Soil	Gabol Chowk, Seetpur, Muzaffargarh	29°14′50.3″ N 70°50′25.0″ E	3	+ in all
13	Compost	Kot Shahan, Gujranwala	32°13′51.2″ N 74°10′45.5″ E	2	+ in 13c
14	Soil	Plastic bag landfill site in Mandi-Bahauddin	32°34′41.2″ N 73°29′26.4″ E	2	+ in all
15	Soil	Indus Rice Mills, Mandii-Bahauddin	32°35′51.3″ N 73°29′27.7″ E	2	− in all

Table 1. *Cont.*

Label [1]	Sample Type	Area in Punjab, Pakistan	Global Positioning System Location [2]	No. of Isolates	PHA Accumulation [3]
16	Soil	Engi Plastic Industries, Lahore	31°19′21.8″ N 74°13′42.4″ E	2	+ in all
17	Effluent	Toyo Nasic Plastic industry, Kot Abdullah, Gujranwala	32.1112° N 74.1967° E	1	+ in all
18	Soil	Incineration plant, Wah cantt, Rawalpindi	33°46′47.8″ N 72°41′05.1″ E	3	+ in 18a, 18b
19	Effluent	Sewage stream in Gujranwala	32°11′58.4″ N 74°10′38.3″ E	2	+ in all
20	Soil	Jadeed plastic industry, Pirwadhai, Rawalpindi	33°38′08.8″ N 73°02′05.2″ E	2	+ in 20a, − in 20b
21	Soil	Fazal Paper Mills (Pvt.) Ltd., Okara	30°52′29.8″ N 73°25′09.6″ E	2	+ in all
22	Effluent	Griffon Plastic Industries (Pvt) Ltd., Kot Lakhpat, Lahore	31°27′02.7″ N 74°20′49.8″ E	1	− in all
23	Soil	Lokhadair Landfill, Lahore	31°37′37.2″ N 74°25′09.0″ E	3	+ in all
24	Soil	Wheat crops in Phalia, Mandi-Bahauddin	32°25′44.7″ N 73°35′27.5″ E	2	− in all
25	Effluent	Z.A. Food Industries, Faisalabad	31°26′25.1″ N 73°01′31.0″ E	3	+ in 25a
26	Effluent	Campbell Flour Mills, Attock	33°45′30.7″ N 72°21′43.6″ E	3	− in all
27	Soil	Gujrat Road, Phalia, Mandi-Bahauddin	32°26′08.8″ N 73°36′59.3″ E	2	+ in all
28	Soil	Agbro Poultry Breeding Farms, Jajja-Dhok, Rawalpindi	33°19′42.2″ N 73°02′14.0″ E	1	− in all
29	Compost	Near Narowal railway station, Narowal	32°06′02.2″ N 74°51′38.3″ E	1	+ in all
30	Soil	Margalla Packages Industry Plastic Bags Manufacturers, Rawalpindi	33°37′00.1″ N 73°03′09.2″ E	3	+ in all
31	Effluent	Century Paper and Board Mills Ltd., Lahore	31°30′10.3″ N 74°19′30.6″ E	3	+ in 31a, 31c
32	Effluent	Sewage stream, Narowal	32°06′29.9″ N 74°51′43.1″ E	3	− in all
33	Soil	Prime Tanning Industries (Pvt.) Ltd., Sheikhupura	31°46′53.5″ N 74°15′31.7″ E	2	+ in 33a
34	Compost	Alipur, Sheikhupura	31°43′36.0″ N 74°13′38.8″ E	3	+ in 34a, 34c
35	Soil	MashaAllah Poultry Farm, Chak Beli Khan, Rawalpindi	33°16′21.1″ N 72°54′34.7″ E	2	− in all
36	Effluent	Quick Food Industries Mon Salwa, Lahore	31°17′04.0″ N 74°10′42.2″ E	3	+ in 36a, 36b
37	Effluent	Hassan marbles industry, Jehlum	32°55′30.9″ N 73°43′09.5″ E	2	− in all
38	Soil	Lucky Plastic Industries (Pvt) Ltd., Raiwind road, Lahore	31°16′26.3″ N 74°05′47.2″ E	3	+ in all

Table 1. *Cont.*

Label [1]	Sample Type	Area in Punjab, Pakistan	Global Positioning System Location [2]	No. of Isolates	PHA Accumulation [3]
39	Soil	Noon Sugar Mills Ltd., Bhalwal Road, Sargodha	32°16′44.4″ N 72°55′02.1″ E	1	+ in all
40	Soil	Sewage stream in Kasur	31°06′41.1″ N 74°28′26.4″ E	1	+ in all
41	Compost	Sargodha	32°03′55.6″ N 72°37′58.4″ E	2	+ in all
42	Effluent	Askari Cement Ltd., Wah	33°49′01.3″ N 72°43′25.5″ E	2	− in all
43	Effluent	Jauharabad Sugar Mills Limited, Khushab	32°18′21.7″ N 72°16′28.5″ E	1	+ in all
44	Soil	Wheat field, Sargodha	32°05′38.2″ N 72°42′49.3″ E	1	− in all
45	Soil	Andrew Paints, Islamabad	33°32′35.6″ N 73°08′21.3″ E	2	+ in 45a, − in 45b
46	Soil	Arsam Pulp and Paper Industries Limited, Sheikhupura	31°41′15.8″ N 74°02′26.1″ E	3	+ in 46b, 46c
47	Soil	Amir Food Industry, Faisalabad	31°26′53.7″ N 73°05′50.8″ E	2	+ in all
48	Soil	Hero Plastic Industries, Lahore	31°36′27.1″ N 74°21′14.7″ E	1	+ in all
49	Compost	Sheikhupura	31°43′46.5″ N 74°00′33.7″ E	1	+ in all
50	Soil	Capital Chemical Industries, Rawalpindi	33°30′47.9″ N 73°13′52.7″ E	2	+ in 50a, − in 50b
51	Effluent	Pakistan Paint Factory, Rawalpindi	33°38′50.0″ N 73°03′37.1″ E	2	+ in all
52	Soil	Dandot Cement Factory, Jehlum	32°38′39.4″ N 72°58′34.7″ E	3	+ in 52b
53	Effluent	Dump site, Sheikhupura	31°43′47.5″ N 74°00′02.0″ E	2	+ in all
54	Soil	Kamalia Sugar Mills, Toba Tek Singh	30°45′59.7″ N 72°34′49.5″ E	3	+ in 54a
55	Soil	Oval Ground, Government College University, Lahore	31°34′22.2″ N 74°18′24.8″ E	1	− in all
56	Effluent	Anarkali Bazar, Lahore	31°34′04.7″ N 74°18′16.5″ E	2	− in all
57	Soil	GCU Girls Hostel, Lahore	31°34′08.1″ N 74°18′16.6″ E	1	+ in all
58	Soil	Tariq Plastic Industry, Gujranwala	32°12′12.7″ N 74°10′22.9″ E	1	- in all
59	Effluent	Expert Advertising and Packaging Ltd., Lahore	31°26′29.0″ N 74°19′05.2″ E	2	+ in 59a, − in 59b
60	Soil	Gujranwala	32°06′37.7″ N 74°12′35.7″ E	3	+ in 60c
61	Soil	Pioneer Cement Ltd., Lahore	31°31′29.6″ N 74°19′27.4″ E	2	− in all
62	Soil	Qazi Town, Gujranwala	32°06′41.2″ N 74°12′02.0″ E	2	+ in 62a, − in 62b
63	Soil	Glow Paints Factory, Rawalpindi	33°37′22.5″ N 73°05′46.1″ E	2	− in all
64	Soil	Corn Field, Gujranwala	32°08′02.0″ N 74°13′30.7″ E	2	− in all
65	Soil	Al-Aziz Plastic Industry, Gujranwala	32°11′37.9″ N 74°09′38.5″ E	2	+ in 65a, − in 65b

[1] Labeling scheme is carried out by numbering the sample and giving each isolate a small alphabetical notation with the number. For example, if Sample 1 has five isolates, they will be named as 1a, 1b, 1c, 1d, and 1e. [2] Coordinates of some locations are approximate. [3] PHA accumulation is determined by the presence of orange fluorescence when illuminated under UV at 315 nm: + for PHA presence, − for PHA absence.

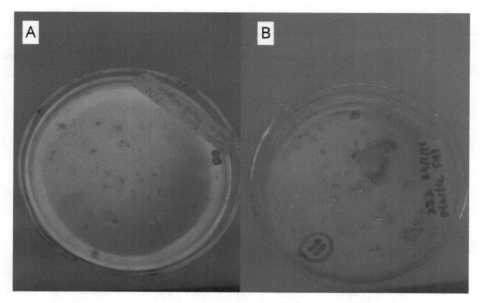

Figure 1. PHA producing samples inoculated on PHA detecting agar, exhibiting fluorescence under UV light. (**A**) Shows a petri plate of sample 8 and (**B**) shows growth on the petri plate by sample 16.

PHAs are insoluble storage granules that microorganisms accumulate in stressful environmental conditions, under excess of carbon and deficiency of other essential nutrients [23]. Thus, industrial soil samples and effluents used for the isolation of PHA producing bacteria were appropriate for the purpose. Most of the samples were collected from the industries of paint, paper, and plastic.

Any substance that becomes useless and defective after its primary use is considered as "waste" [24]. Liquid waste from industrial sites, agricultural processes or domestic sewage is "effluent waste". Effluents can be harmful to the environment if released untreated because of their polluting chemical nature [25], as they contain partially degraded organic matter with minimum nutrients. They serve as an ideal source of isolation for microorganisms adapted to survive in oligotrophic conditions [26]. Products difficult to degrade such as plastic, paint residues, cardboard, and paper residues may also be a source of isolation of novel microorganisms that degrade these difficult products with simultaneous PHA production. Whereas soil samples collected from agricultural areas, such as croplands, compost, and landfill sites are rich in carbon sources [27]. The choice of collecting the samples from these habitats has made possible the isolation of a large number of PHA positive isolates.

For a more visual detection of PHA producing bacteria, staining and fluorescence come into play. This is achieved through the use of Nile blue A stain in the agar medium. When the cells of some microbial colonies grow in a medium containing this stain, the stain is absorbed within the cytoplasm of their cells [28]. This dye subsequently enters into the PHA inclusions. It is a basic oxazine dye [29] containing Nile blue sulfate and Basic Blue 12. These compounds when excited by UV light of 312 nm, reflect orange color hence the orange fluorescence is generated [28]. Apart from being a highly sensitive method to detect PHAs, it also indicates the difference in the amount of accumulated PHAs [29]. The greater the intensity of fluorescence, the more the accumulation of PHA granules [30]. In this way, the isolate 16a showed the brightest and the most intense fluorescence as seen in Figure 1B. It was concluded that this strain held the highest potential to produce maximum PHA content as compared to other strains.

2.2. Secondary Screening of PHA Producers

Primarily screened bacterial colonies were further purified by quadrant streaking and analyzed for PHA production by submerged fermentation at 37 °C, 150 rpm for 72 h. After incubation, maximum

PHA production was shown by the isolate 16a i.e., 69.72 ± 0.17%. However, isolates 3a, 4a, 5a, 8a, 11a, 12a, 14a, 16b, 18b, 23a, 33a, 39a, 45a, 47a, 48a, 53a, and 65a showed significant PHA production ranging from 36% to 55%, approximately. While the remaining strains did not show any significant PHA production, as evident from Table 2. Bacterial isolate 16a was selected for further studies.

Table 2. Percentage PHA accumulation by isolated strains after submerged fermentation with glucose as a carbon source.

Sample Label (No. of Isolates with +PHA Production)	Isolate	Average PHA Production of the Isolate (%PHA)	Sample Label (No. of Isolates with +PHA Production)	Isolate	Average PHA Production of the Isolate (%PHA)
1 (2)	1a	19.93 ± 0.25	33 (1)	33a	38.22 ± 0.24
	1b	16.15 ± 0.19	34 (2)	34a	17.25 ± 0.13
				34c	12.02 ± 0.28
2 (2)	2a	35.66 ± 0.16	36 (2)	36a	25.22 ± 0.22
				36b	21.36 ± 0.15
	2c	29.27 ± 0.48	38 (3)	38a	31.25 ± 0.36
				38b	26.12 ± 0.23
				38c	29.55 ± 0.37
3 (1)	3a	43.76 ± 0.18	39 (1)	39a	36.25 ± 0.51
4 (1)	4a	37.59 ± 0.22	40 (1)	40a	45.23 ± 0.23
5 (1)	5a	41.2 ± 0.18	41 (2)	41a	15.96 ± 0.29
				41b	17.69 ± 0.35
7 (1)	7a	15.22 ± 0.28	43 (1)	43a	34.51 ± 0.21
8 (1)	8b	52.78 ± 0.23	45 (1)	45a	38.22 ± 0.45
9 (1)	9c	22.98 ± 0.50	46 (2)	46b	30.55 ± 0.31
				46c	28.63 ± 0.26
11 (1)	11a	39.68 ± 0.21	47 (2)	47a	39.21 ± 0.38
				47b	35.25 ± 0.48
12 (3)	12a	37.00 ± 0.15	48 (1)	48a	53.24 ± 0.15
	12b	8.87 ± 0.17			
	12c	2.09 ± 0.18			
13 (1)	13c	19.37 ± 0.39	49 (1)	49a	21.03 ± 0.14
14 (2)	14a	43.75 ± 0.35	50 (1)	50a	15.96 ± 0.23
	14b	16.79 ± 0.46			
16 (2)	16a	69.72 ± 0.17	51 (2)	51a	11.23 ± 0.53
	16b	55.98 ± 0.19		51b	13.24 ± 0.27
17 (1)	17a	24.33 ± 0.21	52 (1)	52b	8.03 ± 0.19
18 (2)	18a	8.64 ± 0.34	53 (2)	53a	36.66 ± 0.16
	18b	38.97 ± 0.27		53b	31.26 ± 0.22
19 (2)	19a	11.56 ± 0.42	54 (1)	54a	17.36 ± 0.18
	19b	27.67 ± 0.18			
20 (1)	20a	35.92 ± 0.31	57 (1)	57a	20.36 ± 0.35
21 (2)	21a	27.83 ± 0.29	59 (1)	59a	33.29 ± 0.12
	21b	28.07 ± 0.23			
23 (3)	23a	48.68 ± 0.46	60 (1)	60c	25.25 ± 0.35
	23b	26.88 ± 0.28			
	23c	15.68 ± 0.39			

Table 2. *Cont.*

Sample Label (No. of Isolates with +PHA Production)	Isolate	Average PHA Production of the Isolate (%PHA)	Sample Label (No. of Isolates with +PHA Production)	Isolate	Average PHA Production of the Isolate (%PHA)
25 (1)	25a	12.77 ± 0.38	62 (1)	62a	29.25 ± 0.11
27 (2)	27a	6.87 ± 0.40	65 (1)	65a	38.15 ± 0.41
	27b	12.68 ± 0.20			
29 (1)	29a	15.68 ± 0.39			
30 (3)	30a	30.36 ± 0.27			
	30b	25.99 ± 0.19			
	30c	29.11 ± 0.23			
31 (2)	31a	19.00 ± 0.11			
	31c	18.90 ± 0.28			

Since the isolate 16a produced maximum PHA, it could be quite possible that this strain could quickly and efficiently utilize carbon source. The source of isolate 16a was plastic industry soil. Mohammed et al. [30], Kosseva and Rusbandi [31], and Sangakharak and Prasertsan [32] reported the isolation of bacteria from similar plastic sources, such as plastic pieces, plastic chairs, and plastic waste landfill sites, which provided high PHA yields up to 0.5 g/100 mL. High PHA production rate is associated with the bacterial ability to utilize plastic as a substrate as the soils enriched with plastic pieces are their indigenous habitats. These strains already have modified metabolism rates to sustain in oligotrophic conditions so they adapt to utilize the carbon available in plastic [33]. This shows an enhanced ability of indigenous bacteria to survive in nutrient-deficient conditions than non-indigenous bacteria [34]. In this research, the outcome of considerable differences in the PHA yields of bacteria isolated from plastic habitats can be due to the differences in the surrounding habitats of the plastic sources. Puglisi et al. [35] studied and proved the hypothesis that different polyethylene (PE) plastic waste samples harbor different bacterial communities. The structure and physiological capabilities of these communities are dependent on the physico-chemical properties of the plastic waste and the environment in which they dwell.

2.3. Molecular Identification

Sequencing of isolate 16a was carried out by Macrogen sequencing company, Seoul, Korea. Primers used for PCR and 16S rRNA sequencing are given in (Table 3).

Table 3. Primers used for PCR as well as 16S rRNA sequencing.

PCR Primers	Sequencing Primers
27F 5′ (AGA GTT TGA TCM TGG CTC AG) 3′	785F 5′ (GGA TTA GAT ACC CTG GTA) 3′
1492R 5′ (TAC GGY TAC CTT GTT ACG ACT T) 3′	907R 5′ (CCG TCA ATT CMT TTR AGT TT) 3′

The sequencing results were put into Basic Local Alignment Search Tool (BLAST) for homology analysis and first ten homologues were selected (Figure 2) for phylogenetic tree construction using the Jalview application (Figure 3) [36,37]. The results of the BLAST revealed that the gene sequence of isolate 16a is having identity with *Stenotrophomonas maltophilia* strain IAM 12423. Therefore, it was determined that the strain used in this research is *S. maltophilia*. GenBank sequence of the identified strain was submitted under the name of *Stenotrophomonas maltophilia* HA-16 with the accession number MN240936. The genus *Stenotrophomonas* is phylogenetically placed in the Gammaproteobacteria, was first described with the type species *Stenotrophomonas maltophilia* [38]. The *Stenotrophomonas* genus is a gram-negative genus with at least ten species [39]. It belongs to the family Xanthomonadaceae [40].

Description	Max Score	Total Score	Query Cover	E value	Per. Ident	Accession
Stenotrophomonas maltophilia strain IAM 12423 16S ribosomal RNA, partial sequence	2841	2841	100%	0.0	100.00%	NR_041577.1
Stenotrophomonas maltophilia strain NCTC10257 genome assembly, chromosome: 1	2832	11317	99%	0.0	99.93%	LT906480.1
Stenotrophomonas maltophilia strain 13637 genome	2832	5664	99%	0.0	99.93%	CP008838.1
Stenotrophomonas maltophilia strain 2681 16S ribosomal RNA gene, partial sequence	2828	2828	99%	0.0	99.87%	HQ185399.1
Stenotrophomonas maltophilia strain FDAARGOS_507 chromosome, complete genome	2826	11291	99%	0.0	99.87%	CP033829.1
Stenotrophomonas sp. db-1 16S ribosomal RNA gene, partial sequence	2826	2826	99%	0.0	99.87%	KF059260.1
Stenotrophomonas maltophilia K279a complete genome, strain K279a	2826	11272	99%	0.0	99.87%	AM743169.1
Uncultured bacterium clone RS-B31 16S ribosomal RNA gene, partial sequence	2822	2822	99%	0.0	99.80%	KC541085.1
Stenotrophomonas maltophilia strain SVIA2 chromosome	2820	3027	99%	0.0	99.80%	CP033586.1
Stenotrophomonas maltophilia strain NCTC10498 genome assembly, chromosome: 1	2820	8451	99%	0.0	99.80%	LS483406.1

Figure 2. Nucleotide Basic Local Alignment Search Tool (BLAST) results for the isolated strain: 16a. The first ten homologues were selected for phylogenetic tree construction.

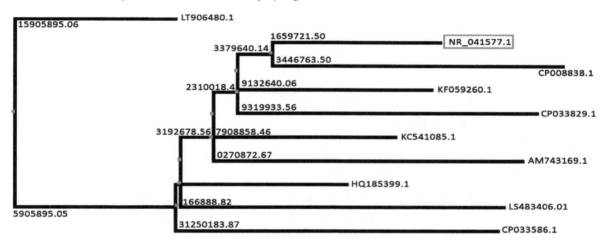

Figure 3. An illustration of the neighbor-joining phylogenetic tree of identified isolate 16a: *Stenotrophomonas maltophilia* strain IAM 12423. The numbers indicate the evolutionary distance, whereas the labels at the end of the arms represent the accession numbers of the BLAST homologues. The blue pointers indicate nodes.

Horiike [41] dictates the importance of phylogenetic trees in molecular identification. He emphasizes that phylogenetic trees are much helpful in predicting the evolutionary basis of relationships between various species. These trees can help us predict the effects of evolutionary patterns in different habitats and their effects on a strain's metabolic products. They also tell us about the evolution of metabolic products from ancestors to their descendants and how they are improved or differentiated over time [42].

2.4. Characterization of the Extracted Polymer

The PHA extracted from *S. maltophilia* HA-16 was characterized by FTIR spectroscopy. Functional group analysis was done through signal peaks recorded in the range of 4000–400 cm^{-1}. The peak was plotted using OriginLab 8.5 as shown in Figure 4 [43].

Abid et al. [44] compared his extracted polymer with FTIR peaks of standard Polyhydroxybutyrate (PHB) purchased from Sigma Aldrich© (St. Louis, MO, USA). We compared our polymer with the FTIR peaks of the standard PHB sample used by Abid et al. [44]. The Fingerprinting region is usually the region of the spectrum in the range of 670–400 cm^{-1}. The polymer under study expressed a considerable peak at 566.9 cm^{-1}, which showed that there were many stretches of carbonyl groups (C=O) present. Comparatively, the fingerprinting region of the standard PHB had multiple sharp peaks, the sharpest one being at 514.03 cm^{-1}. Absorption in the range of 1200–800 cm^{-1} indicated the presence of multiple C-C stretches with medium length bands which corresponded to an alkane group [45]. The extracted polymer expressed a sharp peak at 1017 cm^{-1}, whereas the standard showed a sharp peak at 1044.67 cm^{-1}. The sharpness of the peaks in both the polymers was almost the same which signified the same length of C-C stretches.

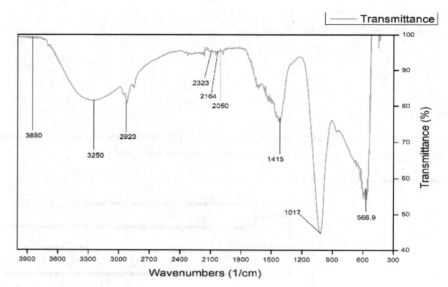

Figure 4. FTIR peaks for the extracted polymer within the transmittance range of 400–4000 cm^{-1}. The labels indicate the peaks through which functional groups are analyzed and compared with the peaks of the standard polymer, in this case, PHB.

Pérez-Arauz et al. [46] mentioned that absorption peaks in the range of 3000–2850 cm^{-1} show a strong C-H bond stretch. The extracted polymer displayed a small peak at 2923 cm^{-1} in comparison to the standard which expressed a same sized peak at 2933.36 cm^{-1}. Since the C-O peaks are not very prominent in the main region, this difference could be due to the disturbances in the polymer structure occurred during extraction [47]. The comparison of the size and the location of peaks in between the standard PHB and our extracted polymer showed that our polymer might not be as pure. The presence of similar length patterns in the peaks at some places can indicate that the biosynthesized polymer is a medium chain length (mcl) PHA copolyester [48].

2.5. Optimization of Cultural Conditions

Incubation time, incubation temperature, pH of the fermentation medium, and carbon sources were optimized for maximum PHA production through duplicate fermentation experiments. The average percentage amount of PHA produced with intracellular cell dry weight (CDW g/100 mL), per experiment per parameter is expressed in Table 5.

Table 4. Comparison of maximum PHA produced during optimization of various cultural conditions.

Parameter	Optimized at	Average PHA Production (%PHA)	Intracellular CDW (g/100 mL)
Incubation time	24 h	30.02 ± 0.27	0.02
	48 h	39.64 ± 0.16	0.07
	72 h	65.39 ± 0.42	0.52
	96 h	47.56 ± 0.37	0.38
Temperature of incubation	25 °C	41.81 ± 0.39	0.06
	30 °C	47.32 ± 0.12	0.07
	37 °C	78.85 ± 0.23	0.59
	40 °C	30.41 ± 0.47	0.28

Table 5. Comparison of maximum PHA produced during optimization of various cultural conditions.

Parameter	Optimized at	Average PHA Production (%PHA)	Intracellular CDW (g/100 mL)
pH of the fermentation medium	6.0	20.14 ± 0.26	0.05
	6.5	41.70 ± 0.29	0.11
	7.0	78.85 ± 0.11	0.19
	7.5	32.95 ± 0.33	0.07
	8.0	30.09 ± 0.18	0.07
Carbon source	Glucose (standard)	78.08 ± 0.19	0.50
	Plastic bag	68.24 ± 0.27	0.19
	Wood chips	53.15 ± 0.17	0.16
	Cardboard cutouts	51.76 ± 0.48	0.15
	Waste shredded polystyrene cups	43.75 ± 0.30	0.17
	Plastic bottle cutouts	38.19 ± 0.22	0.14

2.5.1. Effect of Time of Incubation

Effect of incubation time (24, 48, 72, and 96 h) was studied with glucose as a carbon source for PHA production. PHA production was observed to increase from 24 h and kept on increasing for 72 h where maximum PHA production (65.39 ± 0.42%) was expressed. Contrarily, at 96 h, PHA production declined to 47.56 ± 0.37%. As a result, 72 h incubation time was optimized for further experiments (Figure 5). PHA percentage seems to be directly related to the amount of intracellular CDW (g/100 mL). Nonetheless, at 72 h incubation, 0.52 g/100 mL of intracellular CDW was extracted which was recorded the highest in this pool of experiments.

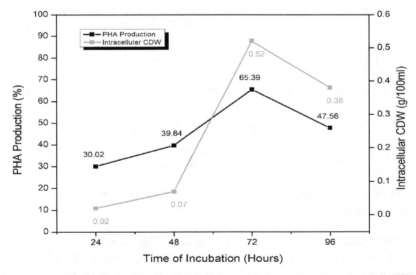

Figure 5. Optimization of time of incubation for PHA production by *S. maltophilia* HA-16 with glucose as a carbon source.

Contrarily, Munir et al. [49] expressed a different trend for PHA production by *Stenotrophomonas* genus with highest PHA yield achieved after 48 h. They used 2% glucose as the sole carbon source compared to our study where only 0.45% of glucose was used. They recorded increasing growth until 72 h, but PHA production increased until 48 h, only and after that, it started declining. Additionally, this difference can be backed by the work of Alqahtani [50], where she describes 48 h as the optima for the highest metabolic activity for PHA production. Shaaban and Mowafy [51] described that the maximum PHA production by *S. maltophilia* occurred at 96 h and stayed stable until 144 h. They utilized 1% glucose in the medium. These differences in the optimum incubation time to produce maximum PHA might be attributed to differences in the nutrients and the carbon source (which in our case was glucose).

2.5.2. Effect of Temperature of Incubation

PHA production was optimized at 25, 37, 30, and 40 °C with glucose as a carbon source to find out the temperature optima for *S. maltophilia* HA-16.37 °C was the optimum temperature with the highest PHA production i.e., 78.85 ± 0.23% as compared to other temperatures (Figure 6). At the same temperature, intracellular CDW was calculated to be the highest i.e., 0.59 g/100 mL. PHA content was low at 25 °C with a slight increase at 30 °C and a sharp decline after 37 °C with the lowest PHA production at 40 °C. PHA production at 25 and 30 °C can be considered as satisfactory with 41.81 ± 0.39 and 47.32 ± 0.12%, respectively. However, PHA yield poorly declined at temperatures above 37 °C, producing only 30.41 ± 0.47% PHA at 40 °C. A strange observation in this experiment was recorded in the intracellular CDW at this temperature. Usually, CDW is observed to be directly related to the amount of PHA produced. Yet, CDW at 40 °C was recorded to be 0.2 g/100 mL, which was still higher than the CDW recorded at 25 °C and 30 °C. Hence, it cannot be taken as a thumb rule that the higher the CDW, the higher the PHA content, because the CDW can also include the dry weight of things other than PHA.

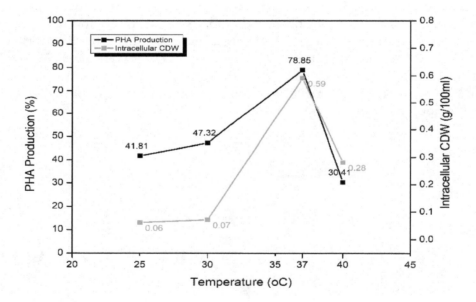

Figure 6. Optimization of the temperature of incubation for PHA production by *S. maltophilia* HA-16 with glucose as a carbon source.

Temperature optimization is important concerning the microorganism used for PHA production [52]. Scientists at the American Tissue Culture Center (ATCC) also confirmed 37 °C as the temperature optima for *S. maltophilia*. On the other hand, Singh and Parmar [13] reported PHA production with the same bacteria but different strains (*S. maltophilia* AK21 and *S. maltophilia* 13635L) at considerably low temperatures of 25 and 30 °C. This deviation might be due to the isolation of bacterial strains from different habitats. Alqahtani [50], in her work, demonstrated that extremely warm (55 °C) and extremely cold (4 °C) temperatures affect the growth of *S. maltophilia*. Her study further solidifies the results of this research by similar results, displaying the optimum growth at 37 °C. Guerrero and others [53] reported that PHA producing enzymes do not work efficiently above 37 °C and resulted in low yield.

2.5.3. Effect of pH of the Fermentation Medium

The fermentation medium was prepared at five different pH levels (6.0, 6.5, 7.0, 7.5, and 8.0) to find the pH optima for PHA production. A significant PHA yield of 78.85 ± 0.11% was recorded at 7.0 pH as compared to other pH levels. At acidic pH i.e., 6.0, PHA yield was poor i.e., only 20.14 ± 0.26% which increased to 41.70 ± 0.29% at slightly less acidic pH of 6.5. At slightly basic pH i.e., 7.5, 32.95 ± 0.33% of PHA production was observed. This yield is almost equal to the PHA yield recorded at 8.0 pH (Figure 7).

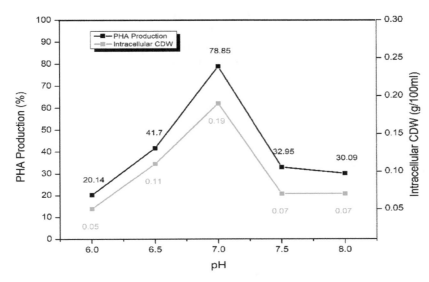

Figure 7. Optimization pH of fermentation medium for PHA production by *S. maltophilia* HA-16 with glucose as carbon source.

Shaaban with colleagues [54] found pH 7.0 as the optimum for PHB production by *S. maltophilia*. They further elaborated that at pH 6.0 and 8.0, PHB production was not significant which also related with the current study. Raj and his team [55] also reported pH 7.0 as the optimum pH for PHA production by *S. maltophilia*. Lathwal et al. [56] demonstrated the same results for PHA production at different pH of fermentation media. They were able to verify that PHA production was maximum at pH 7.0. Their work also validated the current results that on pH 6.0, PHA production was low, however, at pH 8.0, PHA production was not low and still significant.

2.5.4. Optimization of Carbon Source for PHA Production

Some unconventional carbon sources such as undegraded wood chips, cardboard cutouts, shredded plastic bottles, wasted polystyrene cups, and waste plastic bags were used for increased PHA production. These carbon sources were used without any pre-treatments. Surprisingly, plastic bags proved the most optimum among these carbon sources and produced PHA content of 68.24 ± 0.27%. Whereas the other two plastic carbon sources (shredded polystyrene cups and plastic bottle cutouts) did not prove to be much efficient in PHA production. They produced the lowest PHA content of 43.75 ± 0.30% and 38.19 ± 0.22%, respectively. PHA content extracted from the wood chips and cardboard was almost equal i.e., 53.15 ± 0.17% and 51.76 ± 0.48%, respectively (Figure 8).

The cost of carbon sources is one of the prime difficulties in PHA commercialization as these sources contribute to more than 50% of the total industrial production costs. Glucose, among the optimized carbon sources, proved to be the best carbon source for PHA production, as also confirmed by Singh and Parmar [13]. However, the use of industrially produced glucose adds to the costs considerably. Replacing glucose in the fermentation medium with waste sources in the current research was an attempt to address this concern. When the same approach will be industrially adapted, the costs can be further cut down by integrating the production lines with waste streams of paper industries, plastic industries and packaging industries. However, the collection of plastic at the end of PHA production cycle does not ensure its complete breakdown. There are still concerns that need more attention. Since the use of plastics to produce PHAs is a new approach, extensive research to study all related industrial parameters are needed to make it a reality. Jimenez's team [57] found *S. maltophilia* associated with the gut of Bark Beetle *Dendroctonus rhizophagus* (Curculionidae: Scolytinae). In its gut, it plays a role in the degradation, hydrolysis, fermentation, and oxidation of lignin and cellulose derived aromatic products. Their findings can elaborate the current results of PHA production by *S. maltophilia* HA-16 (MN240936) through wood chips and cardboard cutouts. In another study by Kirtania et al. [58], it was found that *S. maltophilia* has a significant ability to naturally degrade cellulose and hemicellulosic materials. Furthermore, Ali Wala'a et al. [59] reported numerous pretreated cellulosic and lignocellulosic sources giving maximum PHA production yield up to 90% which does not align with this research. In the current study, cellulosic and lignocellulosic materials gave lower yields as compared to one synthetic source i.e., plastic bag.

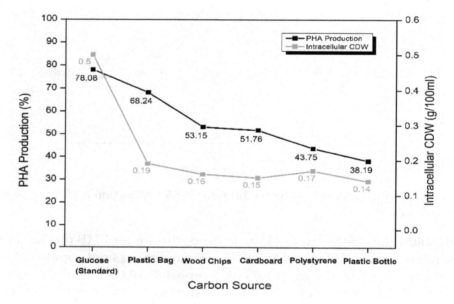

Figure 8. Optimization waste agro-industrial carbon sources for PHA production by *S. maltophilia* HA-16 with glucose as carbon source.

Plastics are of various types. One of the most abundant types of plastics is polyethylene terephthalate (PET). PET plastics are made of repeating units of polymer ethylene terephthalate [60]. Plastic bags are another very common plastic products which are made up of low density poly ethylene (LDPE) and/or high density poly ethylene (HDPE) [61]. The reason why PET and most plastics do not easily biodegrade is because the entire plastic structure has very strong C-C bonds that require too much energy to breakdown and plastics do not dissolve in water [62]. However, with the increased plastic accumulation in our environment, microbial life forms have evolved to degrade plastic products to some extent. There are many bacteria that have the ability to degrade PET, Polyethylene (PE), and Polystyrene (PS) but their enzymes only have been able to give moderate turnouts. The enzymes involved in PET degradation are known as PET hydrolases [20].

While working with synthetic plastics, Kenny and his research team [63] used pre-degraded PET as a carbon source and received a yield of only up to 21% with *Pseudomonas frederiksbergensis* GO23. Whereas, *S. maltophilia* HA-16 (MN240936) produced up to 38% of PHA with undegraded PET bottles. In another study, Dai and Reusch [64] utilized synthetic plastics by using the pyrolysis oil of polystyrene and reported PHA production up to 48% in tryptic soy broth (TSB) medium with *Cupriavidus necator* H16. Contrarily, *S. maltophilia* HA-16 gave a yield of 43% with undegraded or pretreated polystyrene fragments, which is still impressive. These yields indicate an evident ability of

S. maltophilia HA-16 to degrade PET bottles and polystyrene. Current research is the first case of PHA production reported from the strain *S. maltophilia* HA-16 (MN240936).

A surprising finding of this study is the high yield of PHA, i.e., 68.24% in plastic bags. *S. maltophilia* is a gram-negative, non-fermentative bacterium that is present ubiquitously in various anthropogenic and natural environmental habitats [36]. It is a frequent colonizer of the rhizosphere and, hence, this species is present in various types of soils worldwide [65]. *S. maltophilia* has also been found as a part of the natural microbiome of various amoebal genera that are free-living [66]. *S. maltophilia* holds bioremediation capability of sites polluted with hydrocarbons and various xenobiotics [67]. What is more surprising about *S. maltophilia* is that, not only it is a potential PHA producer, but it is also a potential PHB degrader, as demonstrated by Wani et al. [68]. This bacterium is kind of like hitting two targets with one bullet, where it degrades its creation as well. It can immensely increase recycling in PHA production processes, hence, promoting a circular economy. The use of plastic bags as a carbon source for PHA production by any bacteria is yet unreported. This study could be the first of its kind to report such impressive PHA yields by the undegraded plastic bag as a carbon source, opening up new horizons in the field of plastic bag biodegradation and bioconversion by *S. maltophilia* HA-16 (MN240936). However, their effectiveness still needs to be improved with the aid of genetic engineering [69].

2.6. PHA Film Preparation

PHA polymer film was prepared by adding chloroform into the extracted polymer and evaporating it at 60 °C. The film obtained after preparation with chloroform was somewhat in the shape of fragments as compared to the quality of standard PHB film. Its durability still needs further improvement.

The film obtained after drying with chloroform was very brittle and fragile (Figure 9). Mohammed et al. [30] reported a similar kind of brittle film made entirely of PHB polymers. He further added that these films are delicate unless made in combination with copolymers. Our lab-scale experiment was carried out with minimum resources so the copolymerization to achieve a proper film out of the extracted polymer was not possible.

Figure 9. Brittle and Fragile film of extracted polymer.

3. Materials and Methods

3.1. Reagents Preparation

3.1.1. Trace Elements Solutions

Trace elements solution (500 mL) was prepared by weighing 5 g of $FeSO_4 \cdot 7H_2O$, 1.1245 g of $ZnSO_4 \cdot 7H_2O$, 0.5 g of $CuSO_4 \cdot 5H_2O$, 0.25 g of $MnSO_4 \cdot 5H_2O$, 1 g of $CaCl_2 \cdot 2H_2O$, 0.115 g of $Na_2B_4O_7 \cdot 10H_2O$ and 0.05 g of $(NH)_4MO_7O_{24}$ using electric weight balance (Type AUW 2200 No. D 450013067, SHIMADZU

Corporation Japan) and mixing them in 100 mL of distilled water and the final volume was raised to 500 mL [42].

3.1.2. HCl (=35%) Solution Preparation

HCl (=35%) solution was prepared by adding 35 mL of concentrated HCl in 50 mL of distilled water and the volume was raised to 100 mL by distilled water.

3.2. Media Preparation

3.2.1. PHA Detecting Agar

PHA detecting agar (500 mL) was prepared by adding 4 g of nutrient broth, 10 g of nutrient agar, and 0.25 mg of Nile blue A dye in 100 mL of distilled water, and the final volume was raised to 500 mL with distilled water. The media was sterilized in the autoclave (Model: WAC-60, Wisd, WiseStri, Germany) at 121 °C, 15 psi for 20 min. After sterilization, the media was poured aseptically in petri plates, which were pre-sterilized in Digital Oven (SNB-100, hot air sterilizer, Memmert) and stored in a Varioline Intercool cold cabinet at 4 °C until further use [70,71].

3.2.2. Nutrient Broth and Nutrient Agar

The nutrient broth was prepared by dissolving 0.4 g of nutrient broth in 20 mL of distilled water and the final volume was raised to 50 mL with distilled water. The broth was sterilized by autoclaving at 121 °C, 15 psi for 20 min [71].

Nutrient agar (100 mL) was prepared by adding 0.8 g of nutrient broth and 1.5 g of agar in 50 mL of distilled water and was homogenized. The final volume of the medium was raised to 100 mL by distilled water. The medium was autoclaved at 121 °C, 15 psi for 20 min [72].

3.2.3. Fermentation Medium

The fermentation medium was prepared by adding 0.318 g of Na_2HPO_4, 0.135 g of KH_2PO_4, 0.235 g of $(NH_4)_2SO_4$, 0.0195 g of $MgSO_4$, 0.05 g of nutrient broth and 0.45 g of pre autoclaved solution of glucose in 10 mL of distilled water. The volume was raised to 50 mL by adding sterile distilled water. Separately autoclaved solution of trace elements was also added in the medium in a concentration of 1 mL/L. The medium was autoclaved at 121 °C, 15 psi for only 5 min to prevent glucose from caramelization [73].

3.2.4. Seed Culture Preparation

Twenty-four hours before fermentation, each colony was aseptically inoculated in 50 mL of autoclaved nutrient broth and incubated at 37 °C/150 rpm in a shaking incubator (Innova® 43 Incubator Shaker Series) for 24 h to get an overnight old seed culture for the main fermentation process [74].

3.3. Sample Collection

Soil, compost, solid waste landfill soil, and industrial effluent samples were collected from paint and plastic industries, sugar mills, food, paper, pulp and cardboard industries, etc. Soil samples were collected aseptically in sterilized polythene bags while effluent samples were collected in sterilized plastic containers from Lahore, Gujrat, Mandi-bahauddin, Narowal, and other areas of the province Punjab, Pakistan [75]. Table 1 shows the geographical distribution of the areas from where the samples were collected.

3.4. Waste Collection

Wastes, such as wood chips, cardboard, plastic bottle, wasted polystyrene cups, and waste plastic bags were collected aseptically in sterilized polythene bags from local landfills and dumps of Lahore,

Pakistan. The waste was used as a carbon source for the production of PHA [64]. Before inoculation, the waste was shredded into smaller pieces followed by sterilization under the UV hood [76] (Figure 10). Collected plastic bottles were made of PET (Polyethylene terephthalate). It's an aliphatic polyester made of monomers obtained by either esterification of terephthalic acid with ethylene glycol or transesterification of ethylene glycol with dimethyl terephthalate [77]. The presence of a large aromatic ring in the PET repeating units gives the polymer notable stiffness and strength, especially when the polymer chains are aligned with one another in an orderly arrangement [78]. Thick solid colored plastic bags like the ones used in this study are linear HDPE polymers. PE is a polymer formed by radical polymerization of ethylene. Due to the linear structure of HDPEs, they are flexible, durable and tough [79]. Foamed PS used in this research commercially also known as Styrofoam© is a synthetic aromatic hydrocarbon polymer formed of styrene monomers. Its alternating C centers are attached to phenyl groups through σ bonds [80]. Styrofoam is a syndiotactic polymer, which has phenyl groups positioned to alternating sides of the C chain. This structure gives it a highly crystalline and hence a brittle structure [81].

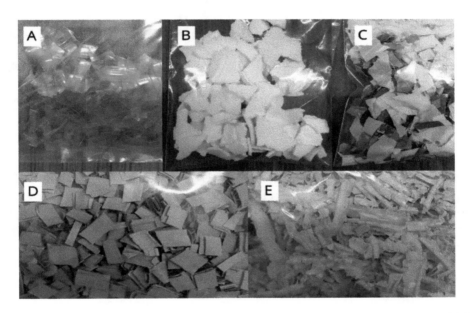

Figure 10. Photographs of carbon sources used for optimization of PHA production. (**A**) Plastic cutouts of a plastic bottle, (**B**) shredded waste polystyrene cups, (**C**) shredded plastic bag, (**D**) cardboard carton cutouts, (**E**) wood chips.

3.5. Isolation and Screening

Qualitative isolation and screening of PHA producing bacteria was initially done by culturing the samples on PHA detecting agar containing Nile blue A stain by the method of Bhuwal et al. [76]. The plates after incubation were illuminated under UV light at 312 nm in UVP Mini Benchtop Transilluminator (Model: TM-10E) for the presence of bright orange fluorescence which indicates PHA accumulation in cells [75]. The initial screening experiments for each sample were run in duplicates. Colonies with orange fluorescence were further selected for isolation by quadrant streaking onto PHA detecting agar plates followed by incubation at 37 °C for 24 h [82].

3.6. Submerged Fermentation for PHA Production

The PHA positive colonies after isolation were then subjected to submerged fermentation. Seed culture (24 h old) was aseptically inoculated in 50 mL of autoclaved fermentation medium and incubated at 37 °C/150 rpm in shaking incubator for 72 h [64]. The experiment for each sample was run in duplicates.

3.7. Extraction of PHA Produced during Fermentation

After 72 h, the PHA content was extracted by sodium hypochlorite and chloroform digestion method of Kumar [71] with a few modifications. The procedure consisted of multiple rounds of centrifugation at 6000 rpm/15 min, drying, and suspension of weighed dried pellets in solutions of 4% sodium hypochlorite and chloroform. It was later proceeded by incubation at 37 °C/150 rpm in a shaking water bath (Daigger Scientific Inc., Wisd, Model: WSB-30) for 1 h. Finally, 1:1 solution of acetone and ethanol was added and pellets were dried in the thermal oven at 60 °C until the liquid content was evaporated. The extracted PHA was collected by filtration on pre-weighed filter paper followed by drying at 60 °C until the achievement of constant weight [83].

3.8. Quantification of Produced PHA

Extracted PHA was quantified as percentage production (%PHA) in cell dry weight (CDW). PHA content was determined as a ratio between the total dry weight of extracted PHA to CDW [84]. The following formulas were applied for the quantification [49]:

$$\text{Cell dry weight (CDW)} = \text{weight of falcon tube with dried pellets} - \text{weight of empty falcon tube}$$

$$\text{Dry weight of extracted PHA} = \text{weight of filter paper with dried filtered PHA} - \text{weight of empty filter paper}$$

$$\%\text{PHA} = \frac{\text{Dry weight of extracted PHA}}{\text{CDW}} \times 100$$

3.9. Characterization of the Extracted PHA by FTIR

The PHA extracted from the most efficient PHA producing microbial isolate was sent for Fourier-Transform Infrared spectroscopy (FTIR) (model: IR-Prestige) characterization to Center for Advanced Studied in Physics (CASP), Government College University, Lahore for functional group analysis through signal peaks recorded in the form of percentage transmittance in the range of 4000–400 cm^{-1} [85].

3.10. Molecular Identification of the Most Efficient PHA Producing Strain

The most efficient PHA producing bacterial strain was identified by 16S rRNA sequencing by sending samples to Macrogen sequencing company, Seoul, Korea. The sequence of the bacterial strain was further subjected to homology analysis through Basic Local Alignment Search Tool (BLAST) [36]. The phylogenetic tree of the identified bacterial isolate was created using Jalview based upon the BLAST results [86]. After the BLAST, the sequence of the identified strain was submitted into the GenBank nucleotide sequence database via BankIt (MN240936) [87].

3.11. Optimization of Cultural Conditions

Four parameters of cultural conditions were optimized for PHA production. The first three parameters included the time of incubation (24–96 h), incubation temperature (25–40 °C), and the pH of the fermentation medium (6.0–8.0). Waste, such as wood chips, cardboard, plastic bottles, wasted polystyrene cups, and waste plastic bags were used as alternate carbon sources instead of glucose in the fermentation medium. Optimization of these sources was done to analyze their potential to be used as cheap, sustainable, and alternate carbon sources [11].

3.12. PHA Film Preparation

PHA film was prepared by adding about 5 mL of chloroform in 0.5 g of extracted dried PHA in a 15 mL falcon tube and set for evaporation in a drying oven at 60 °C until chloroform was completely evaporated [30].

3.13. Statistical Analysis

Computer software CoStat, cs6204W.exe application was applied to carry out the statistical analysis [88]. All of the experiments were run in duplicates to determine the standard error margins in the final yields. Replicates significant differences were presented as Duncan's multiple range tests in the form of probability (*p*) values. These calculations were done to validate the reproducibility of the experiments.

4. Conclusions

Paint industry soil resulted in the isolation of indigenous and unique PHA producing bacteria *Stenotrophomonas maltophilia* HA-16 (MN240936). Fermentation cultural parameters, such as incubation time, pH, temperature, and carbon sources were found to have a significant effect on PHA production as they increased the PHA yield by 1.16 folds. Moreover, utilizing an untreated plastic bag as a carbon source instead of glucose for PHA production was a standout finding with a yield difference of less than 1.1 folds. However, as the research on this unique strain is extremely limited, it still requires extensive studies to turn this bacteria beneficial for industrial use.

Author Contributions: Conceptualization, N.R.; methodology, A.N.; software, H.J.; validation, H.M.; formal analysis, I.-U.-H.; investigation, K.A.S.; data curation, H.K., writing—original draft preparation, H.J., S.M.N., K.U.; writing—review and editing, H.J., A.R., G.M., S.S.; visualization, R.M.; graphic design/infographics, H.J.; supervision, A.N.; project administration, A.N., I.K. All authors have read and agreed to the published version of the manuscript.

References

1. Global Plastics & Rubber Products Manufacturing Report. Available online: https://www.thebusinessresearchcompany.com/report/plastics-and-rubber-products-manufacturing-global-market-report (accessed on 1 September 2020).
2. Grand View Research Inc. *Plastics Market Size, Share & Trends Analysis Report By Product (PE, PP, PU, PVC, PET, Polystyrene, ABS, PBT, PPO, Epoxy Polymers, LCP, PC, Polyamide), By Application, and Segment Forecasts, 2019–2025*; Grand View Research Inc., 2019; 200p.Available online: https://www.grandviewresearch.com/industry-analysis/global-plastics-market (accessed on 11 January 2020).
3. Ecommerce Gateway Pakistan. Pakistan—The Regional Business Hub. Plastic, Packaging & Print Asia. In Proceedings of the 12th Plastic Machinery Technology Show, Karachi, Islamabad, and Lahore: Plastic, Packaging & Print Asia Secretariat, Karachi, Pakistan, 12–14 September 2020; E-commerce Gateway: Karachi, Pakistan, 2020.
4. Geyer, R.; Jambeck, J.R.; Law, K.L. Production, use, and the fate of all plastics ever made. *Sci. Adv.* **2017**, *3*, e1700782. [CrossRef] [PubMed]
5. Plastic Pollution. Available online: https://coastalcare.org/2009/11/plastic-pollution/ (accessed on 2 February 2020).
6. Ellen MacArthur Foundation. *The New Plastics Economy: Catalysing Action*; Ellen MacArthur Foundation and World Economic Forum: Isle of Wight, UK, 2017.
7. Ebere, E.C.; Wirnkor, V.A.; Ngozi, V.E.; Chizoruo, I.F.; Collins, A. Airborne Microplastics: A Review Study on Method for Analysis, Occurrence, Movement, and Risks. *PrePrints* **2019**, *191*, 668. [CrossRef]
8. Allen, S.; Allen, D.; Phoenix, V.R.; Roux, G.L.; Jiménez, P.D.; Simonneau, A.; Binet, S.; Galop, D. Atmospheric transport and deposition of microplastics in a remote mountain catchment. *Nat. Geosci.* **2019**, *12*, 339–344. [CrossRef]
9. Nielson, T.D.; Hasselbalch, J.; Holmberg, K.; Stripple, J. Politics & Plastic crisis: A review throughout the plastic lifecycle. *WIREs Energy Environ.* **2020**, *9*, 360. [CrossRef]
10. Ebnesajjad, S. *Handbook of Biopolymers and Biodegradable Plastics: Properties, Processing, and Applications*; Elsevier Inc.: Oxford, UK, 2013.
11. Kaur, L.; Khajuria, R.; Parihar, L.; Singh, G.D. Polyhydroxyalkanoates: Biosynthesis to commercial production: A review. *J. Microbiol. Biotechnol. Food Sci.* **2017**, *6*, 1098–1106. [CrossRef]
12. Koller, M. Advances in Polyhydroxyalkanoate (PHA) Production, Volume 2. *Bioengineering* **2020**, *7*, 24. [CrossRef]

13. Singh, P.; Parmar, N. Isolation and characterization of two novel polyhydroxybutyrate (PHB)-producing bacteria. *Afr. J. Biotechnol.* **2011**, *10*, 4907–4919.

14. Goh, Y.S.; Tan, I.K.P. Polyhydroxyalkanoate production by antarctic soil bacteria isolated from Casey Station and Signy Island. *Microbiol. Res.* **2012**, *167*, 211–219. [CrossRef]

15. Muhammadi; Shabina; Afzal, M.; Hameed, S. Bacterial polyhydroxyalkanoates-eco-friendly next generation plastic: Production, biocompatibility, biodegradation, physical properties and applications. *Green Chem. Lett. Rev.* **2015**, *8*, 56–77. [CrossRef]

16. Paduani, M. Microplastics as novel sedimentary particles in coastal wetlands: A review. *Mar. Pollut. Bull.* **2020**, *161*, 111739. [CrossRef]

17. Raza, Z.A.; Abid, S.; Banat, I.M. Polyhydroxyalkanoates: Characteristics, production, recent developments and applications. *Int. Biodeterior. Biodegrad.* **2018**, *126*, 45–56. [CrossRef]

18. Cabrera, F.; Torres, Á.; Campos, J.L.; Jeison, D. Effect of Operational Conditions on the Behaviour and Associated Costs of Mixed Microbial Cultures for PHA Production. *Polymers* **2019**, *11*, 191. [CrossRef] [PubMed]

19. Koller, M.; Maršálek, L.; Dias, M.D.S.; Braunegg, G. Producing microbial polyhydroxyalkanoate (PHA) biopolyesters in a sustainable manner. *New Biotechnol.* **2017**, *37*, 24–38. [CrossRef] [PubMed]

20. Danso, D.; Chow, J.; Streit, W.R. Plastics: Environmental and Biotechnological Perspectives on Microbial Degradation. *Appl. Environ. Microbiol.* **2019**, *85*, e01095-19. [CrossRef]

21. Plastic, Composition, Uses, Types and Facts, Britannica. Available online: https://www.britannica.com/science/plastic (accessed on 26 October 2020).

22. Global Production Capacities of Bioplastics 2018–2023. Available online: www.european-bioplastics.org (accessed on 2 February 2020).

23. Ciesielski, S.; Mozejko, J.; Przybylek, G. The influence of nitrogen limitation on mcl-PHA synthesis by two newly isolated strains of Pseudomonas sp. *J. Ind. Microbiol. Biotechnol.* **2010**, *37*, 511–520. [CrossRef]

24. Doron, A. *Waste of a Nation: Garbage and Growth in India*; Harvard University Press: Cambridge, MA, USA, 2018.

25. Effluent Waste, InforMEA. Available online: https://www.informea.org/en/terms/effluent-waste#:~{}:text=Definition(s),of%20their%20polluting%20chemical%20composition.%20 (accessed on 26 October 2020).

26. Koller, M.; Gasser, I.; Schmid, F.; Berg, G. Linking ecology with economy: Insights into polyhydroxyalkanoate-producing microorganisms. *Eng. Life Sci.* **2011**, *11*, 222–237. [CrossRef]

27. Barnes, D.K.A.; Galgani, F.; Thompson, R.C.; Barlaz, M. Accumulation and fragmentation of plastic debris in global environments. *Philos. Trans. R. Soc. B Biol. Sci.* **2009**, *364*, 1985–1998. [CrossRef]

28. Ostle, G.A.; Holt, J.G. Nile Blue A as a Fluorescent Stain for Poly-3-Hydroxybutyrate. *Appl. Environ. Microbiol.* **1982**, *44*, 238–241. [CrossRef]

29. Spiekermann, P.; Rehm, B.H.; Kalscheuer, R.; Baumeister, D.; Steinbüchel, A. A sensitive, viable-colony staining method using Nile red for direct screening of bacteria that accumulate polyhydroxyalkanoic acids and other lipid storage compounds. *Arch. Microbiol.* **1999**, *171*, 73–80. [CrossRef]

30. Mohammed, S.; Panda, A.N.; Ray, L. An investigation for recovery of polyhydroxyalkanoates (PHA) from Bacillus sp. BPPI-14 and Bacillus sp. BPPI-19 isolated from plastic waste landfill. *Int. J. Biol. Macromol.* **2019**, *134*, 1085–1096. [CrossRef]

31. Kosseva, M.R.; Rusbandi, E. Trends in the biomanufacture of polyhydroxyalkanoates with focus on downstream processing. *Int. J. Biol. Macromol.* **2018**, *107*, 762–778. [CrossRef] [PubMed]

32. Sangkharak, K.; Prasertsan, P. Screening and identificationof polyhydroxyalkanoates producing bacteria and bio-chemical characterization of their possible application. *J. Gen. Appl. Microbiol.* **2012**, *58*, 173–182. [CrossRef] [PubMed]

33. Shovitri, M.; Nafi'Ah, R.; Antika, T.R.; Alami, N.H.; Kuswytasari, N.D.; Zulaikha, E. Soil burial method for plastic degradation performed by Pseudomonas PL-01, Bacillus PL-01, and indigenous bacteria. *AIP Conf. Proc.* **2011**, *1854*, 020035. [CrossRef]

34. Han, L.; Zhao, D.; Li, C. Isolation and 2,4-D-degrading characteristics of Cupriavidus campinensis BJ71. *Braz. J. Microbiol.* **2015**, *46*, 433–441. [CrossRef] [PubMed]

35. Puglisi, E.; Romaniello, F.; Galletti, S.; Boccaleri, E.; Frache, A.; Cocconcelli, P.S. Selective bacterial colonization processes on polyethylene waste samples in an abandoned landfill site. *Sci. Rep.* **2019**, *9*, 14138. [CrossRef]

36. Boratyn, G.M.; Schäffer, A.A.; Agarwala, R.; Altschul, S.F.; Lipman, D.J.; Madden, T.L. Domain enhanced lookup time accelerated BLAST. *Biol. Direct* **2012**, *7*, 12. [CrossRef]

37. Waterhouse, A.M.; Procter, J.B.; Martin, D.M.A.; Clamp, M.; Barton, G.J. Jalview Version 2-a multiple sequence alignment editor and analysis workbench. *Bioinformatics* **2009**, *25*, 1089–1191. [CrossRef]

38. Ryan, R.P.; Monchy, S.; Cardinale, M.; Taghavi, S.; Crossman, L.; Avison, M.B.; Berg, G.; Lelie, D.; Dow, M. The versatility and adaptation of bacteria from the genus *Stenotrophomonas*. *Nat. Rev. Microbiol.* **2009**, *7*, 514–525. [CrossRef]

39. Gilligan, P.H.; Lum, G.; VanDamme, P.A.R.; Whittier, S. Burkholderia, Stenotrophomonas, Ralstonia, Brevundimonas, Comamonas, Delftia, Pandoraea, and Acidivorax. In *Manual of Clinical Microbiology*, 8th ed.; Murray, P.R., Baron, E.J., Jorgensen, J.H., Pfaller, M.A., Yolken, R.H., Eds.; ASM Press: Washington, DC, USA, 2003; pp. 729–748.

40. Patil, P.P.; Midha, S.; Kumar, S.; Patil, P.B. Genome Sequence of Type Strains of Genus Stenotrophomonas. *Front. Microbiol.* **2016**, *7*, 309. [CrossRef]

41. Horiike, T. An introduction to molecular phylogenetic analysis. *Rev. Agric. Sci.* **2016**, *4*, 36–45. [CrossRef]

42. Ziemert, N.; Jensen, P.R. Phylogenetic approaches to natural product structure prediction. *Methods Enzymol.* **2015**, *517*, 161–182. [CrossRef]

43. OriginLab Corporation. *Origin (Pro), Version Number (e.g., "Version 8.5")*; OriginLab Corporation: Northampton, MA, USA, 2010.

44. Abid, S.; Raza, A.Z.; Hussain, T. Production kinetics of polyhydroxyalkanoates by using Pseudomonas aeruginosa gamma ray mutant strain EBN-8 cultured on soybean oil. *3 Biotech* **2016**, *6*, 142. [CrossRef] [PubMed]

45. Reis, K.C.; Pereira, P.; Smith, A.C.; Carvalho, C.W.P.; Wellner, N.; Yakimets, I. Characterization of polyhydroxybutyrate-hydroxyvalerate (PHB-HV)/maize starch blend films. *J. Food Eng.* **2008**, *89*, 361–369. [CrossRef]

46. Pérez-Arauz, A.O.; Aguilar-Rabiela, A.E.; Vargas-Torres, A.; Rodríguez-Hernández, A.I.; Chavarría-Hernández, N.; Vergara-Porras, B.; López-Cuellar, M.R. Production and characterization of biodegradable films of a novel polyhydroxyalkanoate (PHA) synthesized from peanut oil. *Food Packag. Shelf. Life* **2019**, *20*, 100297. [CrossRef]

47. Ribitsch, D.; Heumann, S.; Karl, W.; Gerlach, J.; Leber, R.; Birner-Gruenberger, R.; Gruber, K.; Eiteljoerg, I.; Remler, P.; Siegert, P.; et al. Extracellular serine proteases from Stenotrophomonas maltophilia: Screening, isolation and heterologous expression in E. coli. *J. Biotechnol.* **2012**, *157*, 140–147. [CrossRef]

48. Khare, E.; Chopra, J.; Arora, N.K. Screening for mcl-PHAproducing fluorescent Pseudomonads and comparison of mclPHA production under iso-osmotic conditions induced by PEG and NaCl. *Curr. Microbiol.* **2014**, *68*, 457–462. [CrossRef]

49. Munir, S.; Iqbal, S.; Jamil, N. Polyhydroxyalkanoates (PHA) production using paper mill wastewater as carbon source in comparison with glucose. *J. Pure Appl. Microbiol.* **2015**, *9*, 453–460.

50. Alqahtani, N. The Effect of Varying Environmental Conditions on the Growth of Stenotrophomonas Maltophilia. Master's Thesis, University of the Incarnate Word, San Antonio, TX, USA, 2016.

51. Shaaban, M.T.; Mowafy, E.I. Studies on incubation periods, scale up and biodisintegration of poly-β-hydroxybutyrate (PHB) synthesis by *Stenotrophomonas (pseudomonas) maltophilia* and *Pseudomonas putida*. *Egypt. J. Exp. Biol. (Bot.)* **2012**, *8*, 133–140.

52. Dai, D.; Reusch, R.N. Poly-3-hydroxybutyrate Synthase from the Periplasm of Escherichia coli. *Biochem. Biophys. Res. Commun.* **2008**, *374*, 485–489. [CrossRef]

53. Guerrero, C.E.; De la Rosa, G.; Gonzalez Castañeda, J.; Sánchez, Y.; Castillo-Michel, H.; Valdez-Vazquez, I.; Balcazar, E.; Salmerón, I. Optimization of Culture Conditions for Production of Cellulase by Stenotrophomonas maltophilia. *BioResources* **2018**, *13*, 8358–8372. [CrossRef]

54. Shaaban, M.T.; Attia, M.; Mowafy, E.I. Production of some biopolymers by some selective Egyptian soil bacterial isolates. *J. Appl. Sci. Res.* **2012**, *8*, 94–105.

55. Raj, A.; Kumar, S.; Singh, S.K. A highly thermostable xylanase from Stenotrophomonas maltophilia: Purification and Characterization. *Enzyme Res.* **2013**, *8*, 429305. [CrossRef]

56. Lathwal, P.; Nehra, K.; Singh, M.; Jamdagni, P.; Rana, J.S. Optimization of Culture Parameters for Maximum Polyhydroxybutyrate Production by Selected Bacterial Strains Isolated from Rhizospheric Soils. *Pol. J. Microbiol.* **2015**, *64*, 227–239. [CrossRef] [PubMed]

57. Morales-Jimenez, J.; Zuñiga, G.; Ramırez-Saad, H.C.; Hernandez-Rodrıguez, C. Gut-Associated Bacteria Throughout the Life Cycle of the Bark Beetle Dendroctonus rhizophagus (Curculionidae: Scolytinae) and Their Cellulolytic Activities. *Microb. Ecol.* **2012**, *64*, 268–278. [CrossRef] [PubMed]

58.   Kirtania, K.; Tanner, J.; Kabir, K.; Rajendran, S.; Bhattacharya, S. In situ Snychrotron IR study relating temperature and heating rate to surface functional group changes in biomass. *Bioresour. Technol.* **2014**, *151*, 36–42. [CrossRef] [PubMed]

59.   Ali Wala'a, S.; Zaki, H.N.; Obaid, S.Y.N. Production of bioplastic by bacteria isolated from local soil and organic wastes. *Curr. Res. Microbiol. Biotechnol.* **2017**, *5*, 1012–1017.

60.   Speight, J.G.; Lange, N.A. *Lange's Handbook of Chemistry*, 16th ed.; McGraw-Hill: New York, NY, USA, 2005; pp. 2758–2807.

61.   Plastics Europe. *Plastics: The Facts*; Plastics Europe: Brussels, Belgium, 2018.

62.   Why Doesn't Plastic Biodegrade? Live Science. Available online: https://www.livescience.com/33085-petroleum-derived-plastic-non-biodegradable.html (accessed on 27 October 2020).

63.   Kenny, S.T.; Runic, J.N.; Kaminsky, W.; Woods, T.; Babu, R.P.; Keely, C.M.; Blau, W.; O'Connor, K.E. Up-cycling of PET (Polyethylene Terephthalate) to the biodegradable plastic PHA (Polyhydroxyalkanoate). *Environ. Sci. Technol.* **2008**, *42*, 7696–7701. [CrossRef]

64.   Johnston, B.; Radecka, I.; Hill, D.; Chiellini, E.; Ilieva, V.I.; Sikorska, W.; Musioł, M.; Zięba, M.; Marek, A.A.; Keddie, D.; et al. The microbial production of Polyhydroxyalkanoates from Waste polystyrene fragments attained using oxidative degradation. *Polymers (Basel)* **2018**, *10*, 957. [CrossRef]

65.   Berg, G.; Eberl, L.; Hartmann, A. The rhizosphere as a reservoir for opportunistic human pathogenic bacteria. *Environ. Microbiol.* **2005**, *7*, 1673–1685. [CrossRef]

66.   Denet, E.; Coupat-Goutaland, B.; Nazaret, S.; Pelandakis, M.; Favre-Bonte, S. Diversity of free-living amoebae in soils and their associated human opportunistic bacteria. *Parasitol. Res.* **2017**, *116*, 3151–3162. [CrossRef]

67.   Somaraja, P.K.; Gayathri, D.; Ramaiah, N. Molecular Characterization of 2-Chlorobiphenyl Degrading Stenotrophomonas maltophilia GS-103. *Bull. Environ. Contam. Toxicol.* **2013**, *91*, 148–153. [CrossRef] [PubMed]

68.   Wani, S.J.; Shaikh, S.S.; Tabassum, B.; Thakur, R.; Gulati, A.; Sayyed, R.Z. Stenotrophomonas sp. RZS 7, a novel PHB degrader isolated from plastic contaminated soil in Shahada, Maharashtra, Western India. *3 Biotech* **2016**, *6*, 179. [CrossRef] [PubMed]

69.   Mukherjee, P.; Roy, P. Genomic Potential of *Stenotrophomonas maltophilia* in Bioremediation with an Assessment of Its Multifaceted Role in Our Environment. *Front. Microbiol.* **2016**, *7*, 967. [CrossRef] [PubMed]

70.   Gamal, R.F.; Abdelhady, H.M.; Khodair, T.A.; El-Tayeb, T.S.; Hassan, E.A.; Aboutaleb, K.A. Semi-scale produFction of PHAs from waste frying oil by Pseudomonas fluorescens S48. *Braz. J. Microbiol.* **2013**, *44*, 539–549. [CrossRef]

71.   Kumar, A.M. Wood waste—Carbon source for Polyhydroxyalkanoates (PHAs) production. *Int. J. For. Wood Sci.* **2017**, *4*, 36–40.

72.   Jorgensen, J.H.; Pfaller, M.A.; Carroll, K.C.; Funke, G.; Landry, M.L.; Richter, S.S.; Warnock, D.W. *Manual of Clinical Microbiology*, 11th ed.; American Society for Microbiology Press: Washington, DC, USA, 2015.

73.   Ren, Y.; Ling, C.; Hajnal, I.; Wu, Q.; Chen, G.Q. Construction of Halomonas bluephagenesis capable of high cell density growth for efficient PHA production. *Appl. Microbiol. Biotechnol.* **2018**, *102*, 4499–4510. [CrossRef]

74.   Fauzi, A.H.M.; Chua, A.S.M.; Yoon, L.W.; Nittami, T.; Yeoh, H.K. Enrichment of PHA-accumulators for sustainable PHA production from crude glycerol. *Process Saf. Environ. Prot.* **2019**, *122*, 200–208. [CrossRef]

75.   Desouky, S.E.; Abdel-rahman, M.A.; Azab, M.S.; Esmael, M.E. Batch and fed-batch production of polyhydroxyalkanoates from sugarcane molasses by Bacillus flexus Azu-A 2. *J. Innov. Pharm. Biol. Sci.* **2017**, *4*, 55–66.

76.   Bhuwal, A.K.; Singh, G.; Aggarwal, N.K.; Goyal, V.; Yadav, A. Isolation and screening of Polyhydroxyalkanoates producing bacteria from pulp, paper, and cardboard industry wastes. *Int. J. Biomater.* **2013**, *2013*, 752821. [CrossRef]

77.   PET Plastic. Available online: https://omnexus.specialchem.com/selection-guide/polyethylene-terephthalate-pet-plastic (accessed on 7 November 2020).

78.   Polyethylene Terephthalate. Available online: https://www.britannica.com/science/polyethylene-terephthalate (accessed on 7 November 2020).

79.   Polyethylene (PE) Plastic. Available online: https://omnexus.specialchem.com/selection-guide/polyethylene-plastic (accessed on 7 November 2020).

80.   Scheirs, J.; Priddy, D. *Modern Styrenic Polymers: Polystyrenes and Styrenic Copolymers*; John Wiley & Sons: Hoboken, NJ, USA, 2003; p. 3.

81. XAREC Syndiotactic Polystyrene—Petrochemicals—Idemitsu Kosan Global. Available online: https://www.idemitsu.com/business/ipc/engineering/polystyrene.html (accessed on 7 November 2020).

82. Gopi, K.; Balaji, S.; Muthuvelan, B. Isolation, purification and screening of biodegradable polymer PHB producing cyanobacteria from Marine and Fresh Water Resources. *Iran. J. Energy Environ.* **2014**, *5*, 94–100. [CrossRef]

83. Liu, C.; Zhang, L.; An, J.; Chen, B.; Yang, H. Recent strategies for efficient production of polyhydroxyalkanoates by micro-organisms. *Lett. Appl. Microbiol.* **2016**, *62*, 9–15. [CrossRef] [PubMed]

84. Sathiyanarayanan, G.; Bhatia, S.K.; Song, H.S.; Jeon, J.M.; Kim, J.; Lee, Y.K.; Kim, Y.G.; Yang, Y.H. Production and characterization of medium-chain-length polyhydroxyalkanoate copolymer from Arctic psychrotrophic bacterium Pseudomonas sp. PAMC 28620. *Int. J. Biol. Macromol.* **2017**, *97*, 710–720. [CrossRef] [PubMed]

85. Mohapatra, S.; Sarkar, B.; Samantaray, D.P.; Daware, A.; Maity, S.; Pattnaik, S.; Bhattacharjee, S. Bioconversion of fish solid waste into PHB using Bacillus subtilis based submerged fermentation process. *Environ. Technol.* **2017**, *38*, 3201–3208. [CrossRef] [PubMed]

86. Shah, K.R. FTIR analysis of polyhydroxyalkanoates by a locally isolated novel Bacillus sp. AS 3-2 from soil of Kadi region, North Gujarat, India. *J. Biochem. Technol.* **2012**, *3*, 380–383.

87. Fetchko, M.; Kitts, A. A user's guide to BankIt. In *The GenBank Submissions Handbook*; National Centre for Biotechnology Information US: Bethesda, MD, USA, 2011.

88. Snedecor, G.W.; Cochran, W.G. *Statistical Methods*, 7th ed.; Iowa State University Press: Ames, IA, USA, 1980.

3

Biosynthesis of Polyhydroxyalkanoate from Steamed Soybean Wastewater by a Recombinant Strain of *Pseudomonas* sp. 61-3

Ayaka Hokamura, Yuko Yunoue, Saki Goto and Hiromi Matsusaki *

Department of Food and Health Sciences, Faculty of Environmental and Symbiotic Sciences,
Prefectural University of Kumamoto, Kumamoto 862-8502, Japan; a_hokamura@pu-kumamoto.ac.jp (A.H.);
aocharirider@yahoo.co.jp (Y.Y.); g1675001@pu-kumamoto.ac.jp (S.G.)
* Correspondence: matusaki@pu-kumamoto.ac.jp

Academic Editor: Martin Koller

Abstract: *Pseudomonas* sp. 61-3 accumulates a blend of poly(3-hydroxybutyrate) [P(3HB)] homopolymer and a random copolymer, poly(3-hydroxybutyrate-*co*-3-hydroxyalkanoate) [P(3HB-*co*-3HA)], consisting of 3HA units of 4–12 carbon atoms. *Pseudomonas* sp. 61-3 possesses two types of PHA synthases, PHB synthase (PhbC) and PHA synthases (PhaC1 and PhaC2), encoded by the *phb* and *pha* loci, respectively. The P(94 mol% 3HB-*co*-6 mol% 3HA) copolymer synthesized by the recombinant strain of *Pseudomonas* sp. 61-3 (*phbC::tet*) harboring additional copies of *phaC1* gene is known to have desirable physical properties and to be a flexible material with moderate toughness, similar to low-density polyethylene. In this study, we focused on the production of the P(3HB-*co*-3HA) copolymer using steamed soybean wastewater, a by-product in brewing *miso*, which is a traditional Japanese seasoning. The steamed soybean wastewater was spray-dried to produce a powder (SWP) and used as the sole nitrogen source for the synthesis of P(3HB-*co*-3HA) by the *Pseudomonas* sp. 61-3 recombinant strain. Hydrolyzed SWP (HSWP) was also used as a carbon and nitrogen source. P(3HB-*co*-3HA)s with relatively high 3HB fractions could be synthesized by a recombinant strain of *Pseudomonas* sp. 61-3 (*phbC::tet*) harboring additional copies of the *phaC1* gene in the presence of 2% glucose and 10–20 g/L SWP as the sole nitrogen source, producing a PHA concentration of 1.0–1.4 g/L. When HSWP was added to a nitrogen- and carbon-free medium, the recombinant strain could synthesize PHA without glucose as a carbon source. The recombinant strain accumulated 32 wt% P(3HB-*co*-3HA) containing 80 mol% 3HB and 20 mol% medium-chain-length 3HA with a PHA concentration of 1.0 g/L when 50 g/L of HSWP was used. The PHA production yield was estimated as 20 mg-PHA/g-HSWP, which equates to approximately 1.0 g-PHA per liter of soybean wastewater.

Keywords: polyhydroxyalkanoate; PHA; copolymer; soybean wastewater

1. Introduction

Polyhydroxyalkanoates (PHAs) are accumulated in many bacteria as intracellular carbon and energy storage materials under nutrient-limited conditions in the presence of excess carbon [1–3]. PHAs are regarded as important environmentally compatible materials because of their potential use as biodegradable plastics with properties similar to petroleum-based plastics. PHAs can be divided into three groups based on their monomer structure. Short-chain-length PHAs (scl-PHAs) consisting of monomers with 3 to 5 carbon atoms, medium-chain-length PHAs (mcl-PHAs) consisting of monomers with 6 to 14 carbon atoms and scl-mcl-PHA copolymers consisting of both scl and mcl monomer units [4]. Poly(3-hydroxybutyrate) [P(3HB)], the principal member of the scl-PHAs, is both stiff and brittle. In contrast, mcl-PHA, consisting of mcl-3-hydroxyalkanoate (mcl-3HA) units with 6 to 14 carbon atoms, is generally amorphous because of its low crystallinity. PHA copolymers consisting of scl and mcl monomers synthesized by recombinant bacteria show various properties ranging from stiff to flexible depending on the monomer composition. Therefore, the composition ratio of scl and mcl monomers is crucial in PHA properties. In particular, the P(94% 3HB-co-6% 3HA) copolymer consisting of 3HA units of 6 to 12 carbon atoms possesses properties similar to low-density polyethylene (LDPE) [5]. However, the commercial development of PHAs has been limited because of their high production costs. Therefore, it is desirable that PHAs are economically produced from inexpensive carbon sources such as waste substrates.

In this regard, several studies have reported PHA production using waste substrates as carbon sources. P(3HB) homopolymer has been reported to be synthesized by *Cupriavidus necator* H16 (formerly *Ralstonia eutropha* H16) from plant oils as the sole carbon sources [6]. Furthermore, a random copolymer, P(3HB-co-3-hydroxyhexanoate), with a high PHA content has been synthesized from plant oils using a recombinant strain of *C. necator* PHB‾4 (a PHA-negative mutant) harboring the PHA synthase gene from *Aeromonas cavie* [6]. Wong and Lee reported that P(3HB) could be synthesized from whey using the *Escherichia coli* strain GCSC 6576 harboring the PHA-biosynthetic operon from *C. necator* and the *ftsZ* gene from *E. coli* [7]. In a recent study, P(3HB) homopolymer and a P[3HB-co-3-hydroxyvalerate (3HV)] copolymer have been reported to be produced from waste such as oil extracted from spent coffee grounds [8] and waste from the olive oil industry [9]. *Cupriavidus* sp. KKU38 has been reported to synthesize P(3HB) from cassava starch hydrolysate [10]. A number of reviews on the topic have been published [11–13]. While there are numerous reports of the production of scl-PHA or mcl-PHA from waste, there are very few reports on the production of scl-mcl PHA from waste, although Wang et al. have reported scl-mcl PHA production from glycerol, a by-product of the biodiesel industry, using engineered *Escherichia coli* [13]. The production of scl-mcl PHAs consisting of 3HB and mcl-3HA from biomass sources is desirable for the dissemination of PHA as biodegradable plastics because the copolymer is expected to have various properties, ranging from stiff to flexible, depending on the monomer composition as described above.

Pseudomonas sp. 61-3 synthesizes two kinds of PHAs, a P(3HB) homopolymer and a random copolymer, P(3HB-co-3HA), consisting of 3-hydroxyalkanoate (3HA) units of 4–12 carbon atoms [14–16]. *Pseudomonas* sp. 61-3 possesses two types of PHA synthases, PHB synthase (PhbC) and PHA synthases (PhaC1 and PhaC2), encoded at the *phb* and *pha* loci, respectively [16]. PhbC shows substrate specificities for short-chain-length 3HA units, whereas PhaC1 and PhaC2 are able to incorporate a wide range of 3HA units of 4–12 carbon atoms into PHA. It has also been reported previously that PhaC1 is the major PHA providing enzyme in *Pseudomonas* sp. 61-3 [17].

Soybeans are used as raw materials in numerous Japanese foods such as *miso* (fermented soybean pastes), *shoyu* (soy sauce), *natto* (fermented soybean) and *tofu* (soybean protein curd), all of which produce wastewater during the manufacturing process. *Miso* is a traditional Japanese seasoning and many Japanese eat *miso* soup every day. However, the steamed soybean wastewater in produced in *miso* processing is a problem. The wastewater must be treated by a wastewater treatment facility as an activated sludge since the soybean wastewater still contains a large amount of organic compounds, resulting in an enormous cost. Following the production of one ton of *miso*, *shoyu*, or *tofu*, 740, 50, or 18 liters of wastewater is generated, respectively [18]. Their chemical oxygen demand is 32,000, 29,000 and 15,000 ppm, respectively, although they are more than 95% water. In Japan, over 100 million liters of wastewater is generated annually from soybean processed foods such as *miso*, *shoyu*, *natto* and *tofu*. Therefore, the utilization of this soybean wastewater is desirable. For example, there have been several reports describing the recovery of oligosaccharides from steamed soybean wastewater in *tofu* processing [19], the recovery of isoflavone aglycones from soy whey wastewater [20], and the use of the soybean-derived waste as biomass [21–24]. In this study, PHA production using steamed soybean wastewater as a nitrogen and/or carbon source was performed using a recombinant strain of *Pseudomonas* sp. 61-3. This is the first report describing scl-mcl-PHA production from steamed soybean wastewater.

2. Materials and Methods

2.1. Preparation and Hydrolysis of Steamed Soybean Wastewater and Starch

Steamed soybean wastewater was collected from barley *miso* (made from barley and soybean) brewery factory in Kumamoto prefecture, Japan and was spray dried to powder. The nutrient composition of the soybean wastewater powder (SWP) was analyzed by Japan Food Research Laboratories, one of the world's largest and most diversified testing services providers (Table 1). Since SWP contains a sufficient amount of protein, it was first used as a nitrogen source (1, 5, 10, 20 and 50 g/L) for PHA production without further treatment. According to the report by Kimura et al., the constituent sugars of the polysaccharides contained in soybean were arabinose (21.6 wt%), galactose (48.5 wt%), uronic acid (15.0 wt%), and xylose (or rhamnose) (14.9 wt%) [18]. Therefore, polysaccharides that can be used as carbon sources were also considered to be contained in the SWP. In order to investigate the use of a carbon source, an SWP hydrolysate was also prepared using the following two methods. One method involved hydrolysis using 0.6 N H_2SO_4 at 80 °C for 5 h. The other hydrolysis method used 5 N H_2SO_4 at 90 °C for 1 h. SWP or cornstarch (2.5 g, Kanto Chemical Co., Inc., Japan), as a carbon source control, was treated with 5 mL H_2SO_4 (0.6 N or 5 N) for 5 h or 1 h. After hydrolysis, the pH was adjusted to 7.0 using NaOH, and the hydrolysates were subsequently filter-sterilized or autoclaved. The hydrolyzed soybean wastewater powder (HSWP) was used for PHA production as the carbon and the nitrogen source at the concentrations of 10, 20, 30, 40, 50, 75, and 100 g/L. The cornstarch hydrolysate was used as the control of a carbon source.

Table 1. Nutrient composition of SWP (wt%).

Moisture	Crude Protein	Crude Fat	Fiber	Ash	Starch	Non-Fibrous Carbohydrate
12.37	10.91	0.17	0.13	12.83	1.75	61.84

2.2. Bacterial Strain, Plasmid, and Culture Conditions

Pseudomonas sp. 61-3 (*phbC::tet*), which is a *phbC*-negative mutant [16], and the recombinant strain were grown at 28 °C in a nutrient-broth (NB) medium consisting of 1% meat extract (Kyokuto Pharmaceutical Industrial Co., Ltd., Tokyo, Japan), 1% Bactopeptone (Difco Laboratories, Division of Becton Dickinson Company, Sparks, MD, USA) and 0.5% NaCl (pH 7.0). *Pseudomonas* sp. 61-3 (*phbC::tet*) harboring the pJKSc54-*phab*, carrying *phaC1* under the control of the *pha* promoter from *Pseudomonas* sp. 61-3 and *phbAB* under the control of the *phb* promoter from *C. necator*, was used to synthesize P(3HB-*co*-3HA) copolymer as described previously [5]. When needed, kanamycin (50 mg/L) and tetracycline (12.5 mg/L) were added to the medium for plasmid maintenance of this *Pseudomonas* sp. 61-3 recombinant strain.

2.3. Production and Analysis of PHA

Pseudomonas sp. 61-3 (*phbC::tet*) harboring pJKSc54-*phab* was grown on NB medium and transferred to 500 mL shaking flasks containing 100 mL of a nitrogen-free MS medium containing 0.9 g $Na_2HPO_4 \cdot 12H_2O$, 0.15 g of KH_2PO_4, 0.02 g of $MgSO_4 \cdot 7H_2O$ and 0.1 mL of trace element solution [15]. The culture with an initial absorbance at 600 nm of 0.05 was cultivated on a reciprocal shaker (130 strokes/min) at 28 °C for 48 h or 72 h. SWP as the sole nitrogen source was added to the medium and autoclaved. Filter-sterilized glucose (2 wt%) was aseptically added to the autoclaved medium as the sole carbon source. NH_4Cl (0.05%), as a control nitrogen source, was also added to the medium instead of SWP. HSWP and hydrolyzed cornstarch were filtered (ADVANTEC filter paper No.1, Toyo Roshi Kaisha, Ltd., Tokyo, Japan) to remove the residues and the filtrates were autoclaved, followed by they were aseptically added to the medium as both the nitrogen and the carbon sources, or the carbon source, respectively. Determination of cellular PHA composition by gas chromatography (GC) was performed as reported previously [13]. Approximately 30 mg of dry cells were subjected to methanolysis and the converted methyl esters were subjected to gas chromatography analysis on a Shimadzu GC-17A system equipped with an Inert Cap 1 capillary column (30 m × 0.25 mm, GL Sciences Inc., Tokyo, Japan) and a flame-ionization detector. Methyl caprylate was used as the internal standard. All cultivations were performed in triplicate.

3. Results

3.1. Nutrient Composition of SWP

SWP was obtained after spray drying of steamed soybean wastewater obtained from the processing of *miso*. The nutrient composition of SWP is shown in Table 1. Nutrients such as crude protein, crude fat, starch and non-fibrous carbohydrate present in the SWP could therefore be utilized for PHA production as nitrogen/carbon sources.

3.2. Utilization of SWP for PHA Production

Pseudomonas sp. 61-3 (*phbC::tet*) harboring the pJKSc54-*phab* plasmid was cultivated at 28 °C for either 48 h or 72 h in a nitrogen-free MS medium supplemented with 2% glucose and various concentrations (1, 5, 10, 20 and 50 g/L) of SWP as the nitrogen source. P(3HB-*co*-3HA) copolymers were synthesized by the recombinant strain using the SWP as the sole nitrogen source (Table 2). Compared with PHA production using 0.5 g/L of NH_4Cl as the sole nitrogen source, the PHA content (62 wt%) and the monomer composition of P(3HB-*co*-3HA) synthesized from 1 g/L of SWP were almost same but the dry cell weight (0.4–0.6 g/L) was much lower than the control (NH_4Cl). Because SWP was considered not to contain sufficient nitrogen source for cell growth, the concentration of SWP added to the medium was then increased. Increasing the concentration of SWP, caused an increase in the dry cell weight, whereas both the PHA content and the 3HB fraction in P(3HB-*co*-3HA) decreased. When the concentration of SWP was increased to 20 g/L or more, the SWP could not be completely dissolved in the medium since a precipitate was observed in the cell pellet after centrifugation of the culture broth. Overall, it was concluded that 10–20 g/L of SWP was suitable for PHA production, and that a PHA concentration of 1.0 to 1.4 g/L was produced. Thus, SWP could be used as the sole nitrogen source instead of NH_4Cl for P(3HB-*co*-3HA) production by this recombinant strain despite the fact that the PHA concentrations were lower than in the control experiment where 0.5 g/L of NH_4Cl was used as the sole nitrogen source.

Table 2. Biosynthesis of PHA *Pseudomonas* sp. 61-3 (*phbC::tet*) harboring pJKSc54-*phab* from SWP as the sole nitrogen source.

Nitrogen Source	Cultivation Time (h)	Dry Cell Weight (g/L)	PHA Content (wt%)	PHA Conc. (g/L)	PHA Composition (mol%)					
					3HB (C4)	3HHx (C6)	3HO (C8)	3HD (C10)	3HDD (C12)	3H5DD (C12')
0.5 g/L NH_4Cl	48	3.2	62	2.0	87	trace	2	6	2	3
1 g/L SWP	48	0.4	60	0.2	87	trace	1	7	2	3
	72	0.6	66	0.4	87	1	1	6	2	3
5 g/L SWP	48	0.9	52	0.5	80	trace	2	9	4	5
	72	1.2	55	0.7	81	1	2	8	4	4
10 g/L SWP	48	1.4	42	0.6	80	trace	2	10	4	4
	72	1.8	58	1.0	76	1	3	11	4	5
20 g/L SWP	48	2.4	39	1.0	76	1	2	11	5	5
	72	3.0	46	1.4	73	1	3	12	5	6
50 g/L SWP	48	4.4	23	1.0	78	trace	2	10	6	4
	72	5.5	32	1.8	68	1	4	14	7	6

Cells were cultivated at 28 °C for 48 or 72 h in nitrogen-free MS medium containing glucose (2%) and SWP (0.1, 0.5, 1, 2 or 5%) as the sole carbon and nitrogen source, respectively. SWP, soybean waste powder; 3HB, 3-hydroxybutyrate; 3HHx, 3-hydroxyhexanoate; 3HO, 3-hydroxyoctanoate; 3HD, 3-hydroxydecanoate; 3HDD, 3hydroxydodecanoate; 3H5DD, 3-hydroxy-*cis*-5-dodecenoate.

We next investigated whether SWP could be utilized as the carbon source as well as the nitrogen source for the production of P(3HB-*co*-3HA). To achieve this, hydrolysis of SWP was carried out by two methods. First, hydrolysis of SWP was attempted at 80 °C for 5 h with 0.6 N H_2SO_4 and the hydrolysate (HSWP) was subsequently used for PHA production as both the carbon/nitrogen source. As a result, the recombinant strain accumulated P(3HB-*co*-3HA) using this HSWP (Table 3). The P(3HB-*co*-3HA) produced by the recombinant strain reached levels of 1.3 g/L using 75 g/L of HSWP. Similar to what was observed for SWP, the 3HB fraction of P(3HB-*co*-3HA) decreased with increasing concentration of HSWP. The composition ratio of scl- and mcl-monomers has a considerable influence on PHA qualities, in particular P(3HB-*co*-3HA) with too low a 3HB fraction is an amorphous polymer [15]. At a concentration of 20 to 75 g/L of HSWP, the recombinant strain accumulated 32–35 wt% PHA, which is a relatively high PHA content.

Table 3. Biosynthesis of PHA in *Pseudomonas* sp. 61-3 (*phbC::tet*) harboring pJKSc54-*phab* from HSWP (by 0.6 N H_2SO_4).

HSWP (g/L)	Dry Cell Weight (g/L)	PHA Content (wt%)	PHA Conc. (g/L)	PHA Composition (mol%)						
				3HB (C4)	3HV (C5)	3HHx (C6)	3HO (C8)	3HD (C10)	3HDD (C12)	3H5DD (C12′)
10	0.6	26	0.2	88	1	trace	1	5	3	2
20	1.3	33	0.4	85	1	trace	1	6	4	3
30	2.0	35	0.7	84	1	trace	2	7	3	3
40	2.6	33	0.9	81	1	trace	2	7	5	4
50	3.1	32	1.0	80	1	trace	2	8	5	4
75	4.0	32	1.3	76	1	trace	2	10	6	5
100	4.2	21	0.9	78	1	1	3	8	6	3

Cells were cultivated at 28 °C for 48 h in carbon- and nitrogen-free MS medium containing HSWP (10, 20, 30, 40, 50, 75 or 100 g/L) as the nitrogen and the carbon sources. SWP was hydrolyzed by 0.6 N H_2SO_4 at 80 °C for 5 h and neutralized by 5 N NaOH. HSWP, Hydrolyzed soybean waste powder; 3HB, 3-hydroxybutyrate; 3HV, 3-hydroxyvalerate; 3HHx, 3-hydroxyhexanoate; 3HO, 3-hydroxyoctanoate; 3HD, 3-hydroxydecanoate; 3HDD, 3-hydroxydodecanoate; 3H5DD, 3-hydroxy-*cis*-5-dodecenoate.

Subsequently, SWP was hydrolyzed at 90 °C for 1 h with 5 N H_2SO_4 to hydrolyze it completely and the hydrolysate was used as the nitrogen and carbon source for PHA production in a similar manner as described above. The recombinant strain accumulated 26 wt% P(3HB-*co*-3HA) and produced 0.2–0.3 g/L of PHA from 10–20 g/L of HSWP (Table 4). However, PHA was barely produced at 40–50 g/L of HSWP, and the recombinant strain no longer grew at 75 g/L or more HSWP; this is likely due to the high concentration of salts formed by neutralization after hydrolysis. In conclusion, it was found that HSWP could be utilized as both the nitrogen and carbon source. In addition, the data suggested that hydrolysis of SWP using 0.6 N H_2SO_4 was better than using 5 N H_2SO_4 for the cell growth and PHA production.

Table 4. Biosynthesis of PHA in *Pseudomonas* sp. 61-3 (*phbC::tet*) harboring pJKSc54-*phab* from HSWP (by 5 N H_2SO_4).

HSWP (g/L)	Dry Cell Weight (g/L)	PHA Content (wt%)	PHA Conc. (g/L)	PHA Composition (mol%)						
				3HB (C4)	3HV (C5)	3HHx (C6)	3HO (C8)	3HD (C10)	3HDD (C12)	3H5DD (C12′)
10	0.7	26	0.2	76	3	1	2	9	5	4
20	1.1	26	0.3	66	2	1	4	14	7	6
30	1.1	11	0.1	61	4	1	5	15	9	5
40	1.2	2	0.02	39	11	trace	5	21	19	5
50	0.9	2	0.02	42	16	trace	5	21	16	0
75	0	-	-	-	-	-	-	-	-	-
100	0	-	-	-	-	-	-	-	-	-

Cells were cultivated at 28 °C for 48 h in carbon- and nitrogen-free MS medium containing HSWP (10, 20, 30, 40, 50, 75 or 100 g/L) as the nitrogen and the carbon sources. SWP was hydrolyzed by 5 N H_2SO_4 at 90 °C for 1 h and neutralized by 5 N NaOH. HSWP, Hydrolyzed soybean waste powder; 3HB, 3-hydroxybutyrate; 3HV, 3-hydroxyvalerate; 3HHx, 3-hydroxyhexanoate; 3HO, 3-hydroxyoctanoate; 3HD, 3-hydroxydecanoate; 3HDD, 3-hydroxydodecanoate; 3H5DD, 3-hydroxy-*cis*-5-dodecenoate.

We attempted to produce PHA from hydrolyzed cornstarch as a carbon source to compare with SWP, using 0.05% NH_4Cl as the nitrogen source. First, hydrolysis of cornstarch (10 g/L) was attempted at 80 °C for 5 h with 0.6 N H_2SO_4. However, no cell growth was observed, which was attributed to the lack of time for hydrolysis or the lack of an amount of cornstarch added to the medium (Table 5). Therefore, hydrolysis of cornstarch (100 g/L) was subsequently performed for 8 h. As a result, the recombinant strain accumulated 71 wt% P(3HB-co-3HA) containing 86 mol% 3HB and 12 mol% 3HA (C_6-C_{12}) units from 100 g/L of hydrolyzed cornstarch (Table 5). Additionally, the recombinant strain grew well and the PHA produced reached levels of 2.9 g/L, which was the highest production level among the P(3HB-co-3HA)s reported so far. The recombinant strain also grew well and produced 54–63 wt% P(3HB-co-3HA) with high 3HB fraction (83–84 mol%) from cornstarch (14 and 20 g/L) hydrolyzed using 5 N H_2SO_4 (Table 5), whereas the cell growth and accumulated PHA content were slightly lower than the control using 2% glucose (Table 2).

Table 5. Biosynthesis of PHA in *Pseudomonas* sp. 61-3 (*phbC::tet*) harboring pJKSc54-*phab* from hydrolyzed cornstarch.

Hydrolyzed Cornstarch (g/L)	Cultivation Time (h)	Dry Cell Weight (g/L)	PHA Content (wt%)	PHA Conc. (g/L)	PHA Composition (mol%)					
					3HB (C4)	3HHx (C6)	3HO (C8)	3HD (C10)	3HDD (C12)	3H5DD (C12')
10 [1]	48	0.2	2	0.004	34	0	0	25	17	24
	72	0.3	2	0.006	33	0	trace	25	15	27
100 [2]	48	2.9	57	1.7	87	trace	1	6	3	3
	72	4.1	71	2.9	86	trace	2	6	3	3
14 [3]	48	2.5	57	1.4	84	trace	2	7	4	3
	72	2.4	63	1.5	84	trace	2	7	4	3
20 [3]	48	2.5	54	1.4	83	trace	2	8	4	3
	72	2.5	55	1.4	83	trace	2	8	4	3

Cells were cultivated at 28 °C for 48 or 72 h in MS medium containing 0.05% NH_4Cl and hydrolyzed cornstarch (10, 14, 20 or 100 g/L) as the sole nitrogen and carbon source, respectively. [1] Cornstarch was hydrolyzed by 0.6 N H_2SO_4 at 80 °C for 5 h. [2] Cornstarch was hydrolyzed by 0.6 N H_2SO_4 at 80 °C for 8 h. [3] Cornstarch was hydrolyzed by 5 N H_2SO_4 at 90 °C for 1 h. 3HB, 3-hydroxybutyrate; 3HV, 3-hydroxyvalerate; 3HHx, 3-hydroxyhexanoate; 3HO, 3-hydroxyoctanoate; 3HD, 3-hydroxydecanoate; 3HDD, 3-hydroxydodecanoate; 3H5DD, 3-hydroxy-cis-5-dodecenoate.

Finally, PHA production using a mixture of SWP/HSWP and hydrolyzed cornstarch was attempted (Tables 6 and 7). The recombinant strain accumulated 64 wt% P(3HB-co-3HA) containing 82 mol% 3HB (0.3 g/L of PHA concentration) when 1 g/L of SWP and 13 g/L of hydrolyzed cornstarch were added to the medium as the nitrogen and the carbon sources, respectively (Table 6). From 10 g/L of SWP and 13 g/L of hydrolyzed cornstarch, the dry cell weight increased to 1.2 g/L and 0.5 g/L of PHA was obtained. However, the molar fraction of 3HB unit in the copolymer was relatively low (65 mol% 3HB). In the case where 26 g/L of hydrolyzed cornstarch was added, the dry cell weight and the PHA content decreased, resulting in a low PHA concentration (less than 0.2 g/L). When both HSWP (SWP hydrolyzed using 0.6 N H_2SO_4) and hydrolyzed cornstarch (hydrolyzed using 5 N H_2SO_4) were added to the medium, cell growth was inhibited, except when 50 g/L of HSWP only was used as the sole nitrogen and carbon source (Table 7).

Table 6. Biosynthesis of PHA in *Pseudomonas* sp. 61-3 (*phbC::tet*) harboring pJKSc54-*phab* from SWP and hydrolyzed cornstarch.

Hydrolyed Cornstarch (g/L)	SWP (g/L)	Cultivation Time (h)	Dry Cell Weight (g/L)	PHA Content (wt%)	PHA Conc. (g/L)	PHA Composition (mol%)					
						3HB (C4)	3HHx (C6)	3HO (C8)	3HD (C10)	3HDD (C12)	3H5DD (C12')
13	1	48	0.4	55	0.2	78	1	3	10	4	4
		72	0.5	64	0.3	82	1	2	8	3	4
	10	48	1.1	35	0.4	74	1	2	12	6	5
		72	1.2	44	0.5	65	1	4	16	7	7
26	1	48	0.3	46	0.1	79	1	3	9	5	3
		72	0.3	43	0.1	72	1	3	12	8	4
	10	48	0.6	15	0.1	81	trace	2	9	5	3
		72	0.7	27	0.2	73	1	3	11	7	5

Cells were cultivated at 28 °C for 48 or 72 h in nitrogen-free MS medium containing SWP and hydrolyzed cornstarch as the nitrogen and the carbon source, respectively. Cornstarch was hydrolyzed by 5 N H_2SO_4 at 90 °C for 1 h. 3HB, 3-hydroxybutyrate; 3HV, 3-hydroxyvalerate; 3HHx, 3-hydroxyhexanoate; 3HO, 3-hydroxyoctanoate; 3HD, 3-hydroxydecanoate; 3HDD, 3-hydroxydodecanoate; 3H5DD, 3-hydroxy-*cis*-5-dodecenoate.

Table 7. Biosynthesis of PHA in *Pseudomonas* sp. 61-3 (*phbC::tet*) harboring pJKSc54-*phab* from HSWP and hydrolyzed cornstarch.

HSWP (g/L)	Hydrolyzed Cornstarch (g/L)	Cultivation Time (h)	Dry Cell Weight (g/L)	PHA Content (wt%)	PHA Conc. (g/L)	PHA Composition (mol%)						
						3HB (C4)	3HV (C5)	3HHx (C6)	3HO (C8)	3HD (C10)	3HDD (C12)	3H5DD (C12')
5	20	48	0.5	47	0.2	80	1	trace	2	9	4	4
		72	0.6	56	0.3	75	1	1	3	11	5	4
	40	48	0.2	20	0.04	73	5	1	4	9	6	2
		72	0.2	22	0.04	73	3	1	4	9	7	3
10	10	48	1.0	51	0.5	79	1	1	2	9	4	4
	20	48	0.7	42	0.3	78	2	trace	2	9	5	4
50	0	48	2.2	34	0.7	85	1	trace	1	6	4	3
		72	2.3	39	0.9	84	1	trace	2	7	3	3
	15	48	0.04	0	0	0	0	0	0	trace	trace	0
		72	1.2	15	0.2	86	trace	trace	1	6	5	2
	30	48	0.02	1	trace	0	0	trace	0	100	0	0
		72	0.02	0	0	0	0	0	0	0	0	0

Cells were cultivated at 28 °C for 48 or 72 h in nitrogen-free MS medium containing the HSWP (5, 10 or 50 g/L) and cornstarch (0, 15, 20, 30 or 40 g/L) as the carbon and the nitrogen sources. Cornstarch was hydrolyzed by 5 N H_2SO_4. SWP was hydrolyzed by 0.6 N H_2SO_4. HSWP, Hydrolyzed soybean waste powder; 3HB, 3-hydroxybutyrate; 3HV, 3-hydroxyvalerate; 3HHx, 3-hydroxyhexanoate; 3HO, 3-hydroxyoctanoate; 3HD, 3-hydroxydecanoate; 3HDD, 3-hydroxydodecanoate; 3H5DD, 3-hydroxy-*cis*-5-dodecenoate.

4. Discussion

In order to effectively produce PHAs, they should be produced from inexpensive carbon sources such as waste substrates. For this reason, we elected to focus on the use of steamed soybean wastewater generated as a by-product of the soybean processing industry. In this study, P(3HB-*co*-3HA) was synthesized from SWP, HSWP and/or hydrolyzed cornstarch as the nitrogen and/or the carbon sources using *Pseudomonas* sp. 61-3 (*phbC::tet*) harboring the *phaC1* gene from *Pseudomonas* sp. 61-3 and *phbAB* genes from *C. necator*. *Pseudomonas* sp. 61-3 (*phbC::tet*) harboring pJKSc54-*phab* accumulated the P(3HB-*co*-3HA) copolymer from SWP and glucose as the sole nitrogen and carbon sources, respectively, but the 3HB fraction in the copolymer decreased at the amount of added SWP increased. This would be responsible for the expression level of PHA synthase gene (*phaC1*) of *Pseudomonas* sp. 61-3. The additional copies of the *phaC1* gene have been previously reported to result in an increase in the 3HB fraction in the copolymer synthesized when glucose was used as the sole carbon source [5]. This is due to the low substrate specificity of PhaC1 for (*R*)-3HB-CoA. PhaC1 synthase has been reported to have the highest activity toward (*R*)-3-hydroxydecanoate (3HD)-CoA among the C_4-C_{12} substrates [25]. Therefore, a decrease in the expression level of the *phaC1* gene leads to an increase in the 3HA fraction, especially with the 3HD unit being the main component, in the copolymer synthesized through *de novo* fatty acid synthesis pathway when unrelated carbon sources such as glucose were used. To obtain a higher 3HB composition in the copolymer, additional copies of *phaC1* are required, together with the *phbAB* genes, since PHA synthase activity has been reported to affect monomer composition in the copolymer as well as monomer supply by PhbA and PhbB when sugars were used as the sole carbon source [5]. In contrast, when fatty acids were used as carbon source, the 3HO (3-hydroxyoctanoate) fraction increased in the copolymer synthesized by a recombinant *Pseudomonas* sp. 61-3 (*phbC::tet*) strain carrying an additional *phaC1* gene compared with the strain containing only the vector [5]. The *phaC1* gene would be expected to be expressed under nitrogen-limited conditions. A sequence resembling the consensus sequence of the *Escherichia coli* σ^{54}-dependent promoter involved in expression under nitrogen-limited conditions has been found upstream of the *phaC1* gene [16] and P(3HB-*co*-3HA) was synthesized by *Pseudomonas* sp. 61-3 only under nitrogen-limited conditions [14]. When the initial molar ratio of nitrogen and carbon sources (C/N) was low, the copolymer content was also low [5]. The content (45 wt%) and the concentration (1.13 g/L) of PHA accumulated by this recombinant strain with a C/N molar ratio of 71 were the highest obtained, as reported in a previous study [5]. Thus, limitation of nitrogen source is necessary for the biosynthesis of P(3HB-*co*-3HA) by recombinant strains of *Pseudomonas* sp. 61-3, although nitrogen is essential for bacterial growth. In order to synthesize P(3HB-*co*-3HA) with a high 3HB fraction, similarly, a high expression level of *phaC1* is required. In fact, introduction of only the *phaC1* gene into *Pseudomonas* sp. 61-3 (*phbC::tet*) increased the 3HB fraction in P(3HB-*co*-3HA) from 27 mol% to 55 mol% [5]. As the amount of SWP added to the medium increased, that is, as the C/N molar ratio decreased, the 3HB fraction in the copolymer decreased (Table 2). This indicates that SWP can be utilized as a nitrogen source for PHA production by the recombinant *Pseudomonas* sp. 61-3 strain.

In this study, we also attempted to use HSWP for PHA production. Acid hydrolysis of SWP was attempted using two methods; either 0.6 N or 5 N H_2SO_4. As a result, 20-75 g/L of HSWP prepared using 0.6 N H_2SO_4, resulted in the synthesis of 32–35 wt% P(3HB-*co*-3HA) at the levels of 0.4–1.3 g/L of PHA (Table 3). However, the 3HB fraction in the copolymer decreased with increasing HSWP concentration, presumably due to the low C/N molar ratio as described previously [5]. With regard to PHA concentration and the monomer composition of the P(3HB-*co*-3HA) copolymer synthesized by the recombinant strain, 50 g/L of HSWP appeared to be the optical concentration for PHA production (1.0 g/L of PHA). In addition, the 3HB fraction in this copolymer was relatively high (80 mol%), which suggests that it would be expected to have good mechanical properties. The PHA production yield under the culture condition used here was estimated to be 20 mg-PHA/g-(H)SWP, which equates to approximately 1.0 g-PHA per liter of soybean wastewater. On the other hand, when HSWP, prepared using 5 N H_2SO_4, was used as both the nitrogen and carbon sources, the recombinant strain accumulated 26 wt% P(3HB-*co*-3HA) from 10–20 g/L of HSWP and the PHA concentration was 0.2–0.3 g/L (Table 4). Furthermore, the addition of more than 40 g/L of HSWP to the medium inhibited cell growth. This is probably due to the high concentration of salts formed by neutralization following hydrolysis. Thus, we found that both HSWP prepared using either 0.6 N or 5 N H_2SO_4 could be utilized as a nitrogen and carbon source, and HSWP prepared by hydrolysis with 0.6 N H_2SO_4 was suitable for cell growth and PHA production.

We also investigated PHA production using hydrolyzed cornstarch as a carbon source. As a result, 2.9 g/L of P(3HB-*co*-3HA) containing 86 mol% 3HB unit could be produced by the recombinant strain from 100 g/L of hydrolyzed cornstarch (Table 5).

Based on these data, we concluded that SWP and HSWP did not contain sufficient carbon to produce PHA, since the PHA contents (<35 wt%) obtained using HSWP were lower than the PHA content under control conditions (glucose and NH_4Cl) (Tables 2 and 3). Therefore, we attempted to add a mixture of SWP/HSWP and hydrolyzed cornstarch to the medium for PHA production (Tables 6 and 7). After 72 h of cultivation, the recombinant strain accumulated 64 wt% P(3HB-*co*-3HA) containing 82 mol% 3HB unit from 10 g/L of SWP and 13 g/L of hydrolyzed cornstarch as nitrogen and carbon sources, respectively (Table 6). However, the dry cell weights (less than 0.7 g/L) and PHA concentration (less than 0.2 g/L) decreased when 26 g/L of hydrolyzed cornstarch was used. In the case when both HSWP, prepared using 0.6 N H_2SO_4, hydrolyzed cornstarch were added to the medium, cell growth was inhibited with increasing concentrations of hydrolyzed cornstarch

(Table 7). The inhibition of cell growth is likely caused by the high concentration of salts formed by neutralization after hydrolysis. Therefore, the HSWP and hydrolyzed cornstarch should be desalted before use in PHA production. One alternative solution to this salt problem maybe through the use of enzymatic hydrolysis of SWP and cornstrach instead of acid hydrolysis treatment. Interestingly, the 3-hydroxyvalerate (3HV) unit was detected in the copolymer by GC analysis when HSWP was added to the medium (Tables 3, 4 and 7). Hydrolysis of SWP may produce a substrate (e.g., fatty acids with odd numbers of carbon atoms) leading to the supply of the 3HV monomer. The reason for this remains unclear.

In conclusion, P(3HB-co-3HA) with various monomer compositions could be synthesized from SWP, HSWP, and/or hydrolyzed cornstarch as nitrogen and/or carbon sources in this study. However, the production efficiency was found to be unsatisfactory. Although SWP could be used as a nitrogen source, the PHA concentration was less than 1.0 g/L (less than 35 wt% PHA) when only HSWP was added to the medium as both a carbon and nitrogen source (Table 7). This suggests that SWP/HSWP contains sufficient nitrogen for the recombinant *Pseudomonas* sp. 61-3 strain to produce PHA, but it is deficient as a carbon source. Further improvements will therefore be necessary to achieve effective PHA production from SWP/HSWP. Since treatment of steamed soybean wastewater by a treatment facility, for example an activated sludge, is expensive, effective utilization of this wastewater is required. The production of PHA reported here is one proposed use of this wastewater. In the future, the molecular weight and mechanical properties of the copolymer synthesized should be further investigated since the molecular weight of the polymer affects its mechanical properties in addition to the monomer composition. The utilization of waste substrates, such as steamed soybean wastewater as a nitrogen and the carbon source, could contribute significantly to reducing the costs of PHA production as well as reducing the cost of waste treatment while at the same time promoting environmental conservation.

Acknowledgments: We would like to thank Editage (www.editage.jp) for English language editing.

Author Contributions: Ayaka Hokamura analyzed data and wrote about 70% of the articles. Yuko Yunoue designed and performed the experiments, and analyzed data. Saki Goto analyzed data and wrote about 10% of the articles. Hiromi Matsusaki supervised the work and wrote about 20% of the articles.

References

1. Anderson, A.J.; Dawes, E.A. Occurrence, metabolism, metabolic role, and industrial uses of bacterial polyhydroxyalkanoates. *Microbiol. Rev.* **1990**, *54*, 450–472. [PubMed]
2. Müller, H.M.; Seebach, D. Poly(hydroxyalkanoates): A fifth class of physiologically important organic biopolymers? *Angew. Chem. Int. Ed. Engl.* **1993**, *32*, 477–502. [CrossRef]
3. Madison, L.L.; Huisman, G.W. Metabolic engineering of poly(3-hydroxyalkanoates): from DNA to plastic. *Microbiol Mol. Biol. Rev.* **1999**, *63*, 21–53. [PubMed]
4. Nomura, C.T.; Tanaka, T.; Eguen, T.E.; Appah, A.S.; Matsumoto, K.; Taguchi, S.; Ortiz, C.L.; Doi, Y. FabG mediates polyhydroxyalkanoate production from both related and nonrelated carbon sources in recombinant *Escherichia coli* LS5218. *Biotechnol. Prog.* **2008**, *24*, 342–351. [CrossRef] [PubMed]
5. Matsusaki, H.; Abe, H.; Doi, Y. Biosynthesis and properties of poly(3-hydroxybutyrate-co-3-hydroxyalkanoate) by recombinant strains of *Pseudomonas* sp. 61-3. *Biomacromolecules* **2000**, *1*, 17–22. [CrossRef] [PubMed]
6. Fukui, T.; Doi, Y. Efficient production of polyhydroxyalkanoates from plant oils by *Alcaligenes eutrophus* and its recombinant strain. *Appl. Microbiol. Biotechnol.* **1998**, *49*, 333–336. [CrossRef] [PubMed]
7. Wong, H.H.; Lee, S.Y. Poly-(3-hydroxybutyrate) production from whey by high-density cultivation of recombinant *Escherichia coli*. *Appl. Microbiol. Biotechnol.* **1998**, *50*, 30–33. [CrossRef] [PubMed]
8. Obruca, S.; Benesova, P.; Kucera, D.; Petrik, S.; Marova, I. Biotechnological conversion of spent coffee grounds into polyhydroxyalkanoates and carotenoids. *New Biotechnol.* **2015**, *32*, 569–574. [CrossRef] [PubMed]

9. Alsafadi, D.; AI-Mashaqbeh, O. A one-stage cultivation process for the production of poly-3-(hydroxybutyrate-co-hydroxyvalerate) from olive mill wastewater by *Haloferax mediterranei*. *New Biotechnol.* **2017**, *34*, 47–53. [CrossRef] [PubMed]

10. Poomipuk, N.; Reungsang, A.; Plangklang, P. Poly-β-hydroxyalkanoates production from cassava starch hydrolysate by *Cupriavidus* sp. KKU38. *Int. J. Biol. Macromol.* **2014**, *65*, 51–64. [CrossRef] [PubMed]

11. Valentino, F.; Morgan-Sagastume, F.; Campanari, S.; Villano, M.; Werker, A.; Majone, M. Carbon recovery from wastewater through bioconversion into biodegradable polymers. *New Biotechnol.* **2017**, *37*, 9–23. [CrossRef] [PubMed]

12. Koller, M.; Maršálek, L.; de Sousa Dias, M.M.; Braunegg, G. Producing microbial polyhydroxyalkanoate (PHA) biopolyesters in a sustainable manner. *New Biotechnol.* **2017**, *37*, 24–38. [CrossRef] [PubMed]

13. Zhu, C.; Chiu, S.; Nakas, J.P.; Nomura, C.T. Bioplastics from waste glycerol derived from biodiesel industry. *J. Appl. Polym. Sci.* **2013**, *130*, 1–13. [CrossRef]

14. Kato, M.; Bao, H.J.; Kang, C.K.; Fukui, T.; Doi, Y. Production of a novel copolyester of 3-hydroxybutyric acid and medium-chain-length 3-hydroxyalkanoic acids by *Pseudomonas* sp. 61-3. *Appl. Microbiol. Biotechnol.* **1996**, *45*, 363–370. [CrossRef]

15. Kato, M.; Fukui, T.; Doi, Y. Biosynthesis of polyester blends by *Pseudomonas* sp. 61-3 from alkanoic acids. *Bull. Chem. Soc. Jpn.* **1996**, *69*, 515–520. [CrossRef]

16. Matsusaki, H.; Manji, S.; Taguchi, K.; Kato, M.; Fukui, T.; Doi, Y. Cloning and molecular analysis of the poly(3-hydroxybutyrate) and poly(3-hydroxybutyrate-*co*-3-hydroxyalkanoate) biosynthesis genes in *Pseudomonas* sp. strain 61-3. *J. Bacteriol.* **1998**, *180*, 6459–6467. [PubMed]

17. Matsumoto, K.; Matsusaki, H.; Taguchi, K.; Seki, M.; Doi, Y. Isolation and characterization of polyhydroxyalkanoates inclusions and their associated proteins in *Pseudomonas* sp. strain 61-3. *Biomacromolecules* **2002**, *3*, 787–792. [CrossRef] [PubMed]

18. Kimura, I.; Matsubara, Y.; Shimasaki, H. Utilization of waste water from soy-bean cooking (in Japanese). *J. Brew. Jpn.* **1997**, *7*, 478–485. [CrossRef]

19. Matsubara, Y.; Iwasaki, K.; Nakajima, M.; Nabetani, H.; Nakao, S. Recovery of oligosaccharides from steamed soybean waste water in *tofu* processing by reverse osmosis and nanofiltration membranes. *Biosci. Biotech. Biochem.* **1996**, *60*, 421–428. [CrossRef] [PubMed]

20. Liu, W.; Zhang, H.X.; Wu, Z.L.; Wang, Y.J.; Wang, L.J. Recovery of isoflavone aglycones from soy whey wastewater using foam fractionation and acidic hydrolysis. *J. Agric. Food Chem.* **2013**, *61*, 7366–7372. [CrossRef] [PubMed]

21. Hongyang, S.; Yalei, Z.; Chunmin, Z.; Xuefei, Z.; Jinpeng, L. Cultivation of *Chlorella pyrenoidosa* in soybean processing wastewater. *Bioresour. Technol.* **2011**, *102*, 9884–9890. [CrossRef] [PubMed]

22. Zhu, G.-F.; Li, J.-Z.; Liu, C.-X. Fermentative hydrogen production from soybean protein processing wastewater in an anaerobic baffled reactor (ABR) using anaerobic mixed consortia. *Appl. Biochem. Biotechnol.* **2012**, *168*, 91–105. [CrossRef] [PubMed]

23. Zhu, J.; Zheng, Y.; Xu, F.; Li, Y. Solid-state anaerobic co-digestion of hay and soybean processing waste for biogas production. *Bioresour. Technol.* **2014**, *154*, 240–247. [CrossRef] [PubMed]

24. Liu, S.; Zhang, G.; Zhang, J.; Li, X.; Li, J. Performance, 5-aminolevulic acid (ALA) yield and microbial population dynamics in a photobioreactor system treating soybean wastewater: Effect of hydraulic retention time (HRT) and organic loading rate (OLR). *Bioresour. Technol.* **2016**, *210*, 146–152. [CrossRef] [PubMed]

25. Takase, K.; Matsumoto, K.; Taguchi, S.; Doi, Y. Alteration of substrate chain-length specificity of type II synthase for polyhydroxyalkanoate biosynthesis by in vitro evolution: in vivo and in vitro enzyme assays. *Biomacromolecules* **2004**, *5*, 480–485. [CrossRef] [PubMed]

4

Bioprocess Engineering Aspects of Sustainable Polyhydroxyalkanoate Production in Cyanobacteria

Donya Kamravamanesh [1,2], Maximilian Lackner [2,3] and Christoph Herwig [1,*]

[1]   Institute of Chemical, Environmental and Bioscience Engineering, Research Area Biochemical Engineering, Technische Universität Wien, 1060 Vienna, Austria; donya.kamravamanesh@tuwien.ac.at
[2]   Lackner Ventures and Consulting GmbH, Hofherr Schrantz Gasse 2, 1210 Vienna, Austria; kontakt@drlackner.com
[3]   Institute of Industrial Engineering, University of Applied Sciences FH Technikum Wien, Höchstädtplatz 6, 1200 Vienna, Austria
*   Correspondence: christoph.herwig@tuwien.ac.at

Abstract: Polyhydroxyalkanoates (PHAs) are a group of biopolymers produced in various microorganisms as carbon and energy reserve when the main nutrient, necessary for growth, is limited. PHAs are attractive substitutes for conventional petrochemical plastics, as they possess similar material properties, along with biocompatibility and complete biodegradability. The use of PHAs is restricted, mainly due to the high production costs associated with the carbon source used for bacterial fermentation. Cyanobacteria can accumulate PHAs under photoautotrophic growth conditions using CO_2 and sunlight. However, the productivity of photoautotrophic PHA production from cyanobacteria is much lower than in the case of heterotrophic bacteria. Great effort has been focused to reduce the cost of PHA production, mainly by the development of optimized strains and more efficient cultivation and recovery processes. Minimization of the PHA production cost can only be achieved by considering the design and a complete analysis of the whole process. With the aim on commercializing PHA, this review will discuss the advances and the challenges associated with the upstream processing of cyanobacterial PHA production, in order to help the design of the most efficient method on the industrial scale.

Keywords: polyhydroxyalkanoate (PHA), bioprocess design; carbon dioxide; cyanobacteria; upstream processing

---

1. Introduction

Petroleum-based polymers are relatively inert, versatile, and durable; therefore, they have been used in industry for more than 70 years [1]. However, they bear negative properties such as CO_2 emissions from incineration, toxicity from additives, and accumulated toxic substances in the environment, particularly in marine as microplastics, recalcitrance to biodegradation, and massive waste accumulation into the marine environment and the landfills [2,3]. With the limited fossil fuel resources and the environmental impact associated with the products, the research for an alternative seems essential in order to reduce our dependencies on non-renewable resources [4,5].

Biodegradable polymers, due to their eco-friendly nature, offer one of the best solutions to environmental problems caused by synthetic polymers [5]. Polyhydroxyalkanoates (PHAs) are a class of naturally occurring polymers produced by microorganisms [1,6,7], among which poly (3-hydroxybutyrate) (PHB) is the most studied biodegradable polymer that accumulates in bacteria in the form of inclusion bodies as carbon reserve material when cells grow under stress conditions [5,8]. PHB with a high crystallinity represents properties similar to synthetic polyesters and also to polyolefins such as polypropylene [6,9,10]. In addition, due to biocompatibility and biodegradability, PHB possesses extensive interesting functions and can replace fossil-based plastics in many applications [7]. However, the low elongation and break and the brittleness of PHB are limitations that can be overcome using other PHA, like blends of copolymers such as polyhydroxyvalerate (PHV) and poly (3-hydroxybutric acid-co-3-hydroxyvaleric acid) (PHBV). The copolymer can either be directly biosynthesized under varying cultivation conditions or be chemically produced in vitro. Apart from short-chain length PHA, there are medium- and long-chain-length polymers which can help to tailor the material properties [11]. The strategies to overcome these limitations are studied in various wild-type and recombinant cyanobacteria, reviewed by Lackner et al. and Balaji et al. [11,12].

Today, PHB is commercially produced by heterotrophic bacteria, such as *Cupriavidus necator* (*C. necator*), and recombinant *Escherichia coli* (*E. coli*) [6,13,14]. High production cost, when compared with petroleum-based polymers, is one major challenge for extensive production and commercialization of PHB [5]. Major contributors to the overall cost being the expensive substrates, continuous oxygen supply, equipment depreciation, high energy demand, and chemicals used for downstream processing [15–17].

Attention has been focused to reduce the production cost, mostly by selecting more economically feasible and efficient carbon substrates for PHB production such as whey, hemicellulose, sugar cane, agricultural wastes, and molasses [5,18–20]. In this context, PHB production using cyanobacteria from more sustainable resources, such as CO_2, has gained importance. Cyanobacteria are an ideal platform for the production of biofuels and bulk chemicals through efficient and natural CO_2 fixation [21].

Other reviews have mainly discussed the potential of cyanobacteria for PHA production, cultivation conditions, and cyanobacterial metabolism, as well as the applications and industrial prospects of the synthesis of this biopolymer [22–26]. Minimization of the PHA production cost can only be achieved by considering the design and a complete analysis of the whole process [27]. In this work, the authors will discuss the bioprocess engineering aspects that focus on upstream processing and advances of sustainable PHA production from cyanobacteria, concentrating primarily on the unit operations of the upstream processing. The authors believe that a proper-time resolved quantification of the process will aid in a better understanding for process manipulation and optimization of industrial production.

2. Cyanobacteria: The Future Host in Biotechnology

Cyanobacteria are gram-negative bacteria with a long evolutionary history and are the only prokaryotes capable of plant-like oxygenic photosynthesis [28]. Unlike heterotrophic organisms, cyanobacteria require only greenhouse gas CO_2 and sunlight, along with minimal nutrients for growth, eliminating the cost of carbon source and complex media components [28]. Cyanobacteria are equipped with superior photosynthetic machinery, showing higher biomass production rates compared to plants and can convert up to 3–9% of the solar energy into biomass [28,29]. Moreover, in contrast to plants, cultivation of cyanobacteria requires less land area, therefore cyanobacteria do not compete for arable land used for agriculture [30]. Some cyanobacteria can produce PHB, when essential nutrients for growth, such as nitrogen and phosphorus, is limiting. From an economic point of view, however, photosynthetic PHB production in cyanobacteria has two major disadvantages: little productivity and slow growth [25]. Therefore, in order to promote photosynthetic PHB production on an industrially relevant scale, the productivity needs to improve significantly. Productivity is defined as the amount of PHA produced by unit volume in unit time [27]. In spite of all the efforts done, so far, very few reports have shown an actual improvement in the cyanobacterial PHB production process, while there are various challenges associated with cultivation, engineering and large-scale production of autotrophic cyanobacterial biomass.

3. Challenges in Cyanobacterial Bioprocess Technology

Figure 1 represents a bioprocess development chain for cyanobacterial PHA production consisting of strain and bioprocess developments and the downstream processing. The workflow shows the strain selection, strain improvement, process understanding, and strategies for scale-up and then down-stream processing, representing the separation and purification of the final product. In order to obtain an optimal scalable bioprocess and in-depth understanding of the bioprocess kinetics, the following points have to be known. First, the holistic knowledge of enzymatic and metabolic pathways of the PHB biosynthesis in cyanobacteria. Second, the selection or development of the optimal strain with maximum productivity. Third, the selection of inexpensive substrate and optimization of media components for the particular production strain. Fourth, the design of the bioreactor system and the optimization of process parameters for scale-up, using the statistical design of experiments (DoE). Fifth, the usage of past data and advanced mathematical models for monitoring and control. Lastly, the development of novel strategies for PHB recovery with a minimum cost of energy and chemical requirement. This review will mainly focus on the bioprocess engineering aspects of photosynthetic PHB production providing an overview of the PHB production process chain starting from a single cell.

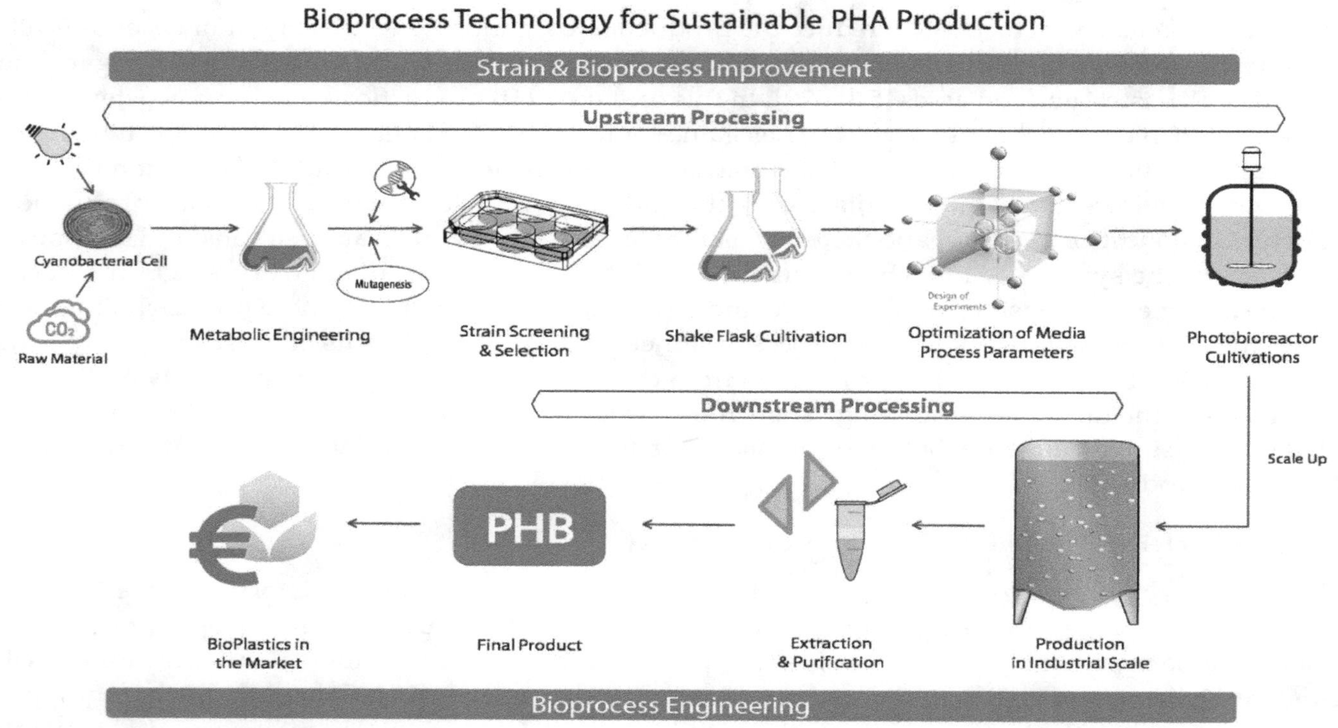

Figure 1.   Represents the work-flow in bioprocess technology for the production of polyhydroxyalkanoates (PHAs) using cyanobacteria as the host system.

3.1. Process Design and Optimization

3.1.1. Existing Wild-Type Strains and Their Reported PHB Content

Cyanobacteria are indigenously the only organisms that produce PHA biopolymers using oxygenic photosynthesis [24]. Cyanobacteria grow mainly under autotrophic conditions, nevertheless, supplementation of sugars or organic acids in some species increases growth and PHB accumulation [24], which contributes to the production cost. To date, a few cyanobacterial strains have been identified for photosynthetic PHB accumulation. Table 1 presents the wild-type cyanobacterial strains, their PHB content in dry cell weight (DCW), and the carbon source used for the production.

As is indicated in Table 1, the PHB production process using CO_2 as the only carbon source shows a lower product content than using organic acids or sugars as substrate.

Table 1. Examples of wild-type cyanobacterial strains, with reported poly (3-hydroxybutyrate) (PHB) content, the carbon source, and growth conditions used for the production of the polymer.

Cyanobacteria	PHB Content (% DCW)	Substrate	Production Condition	Polymer Composition	Reference
Synechocystis sp. PCC 6803	38	Acetate	P limitation and gas exchange limitation [1]	PHB	[31]
Synechocystis sp. PCC 6714	16	CO_2	N [2] and P [3] limitation	PHB	[32]
Spirulina platensis	6.0	CO_2	Not given	PHB	[33]
Spirulina platensis UMACC 161	10	acetate and CO_2	N starvation	PHB	[34]
Spirulina maxima	7–9	CO_2	N and P limitation	PHB	[35]
Gloeothece sp. PCC 6909	9.0	acetate	Not given	Not specified	[36]
Nostoc moscorum Agardh	60	acetate and valerate	N deficiency	PHB-co-PHV	[37]
Nostoc moscorum	22	CO_2	P starvation	PHB	[38]
Alusira fertilisima CCC444	77	fructose and valerate	N deficiency	PHB-co-PHV	[39]
Alusira fertilisima CCC444	85	citrate and acetate	P deficiency	PHB	[40]
Synechocystis PCC 7942	3	CO_2	N limitation	PHB	[41]
Synechocystis PCC 7942	25.6	acetate	N limitation	PHB	[41]
Synechocystis sp. CCALA192	12.5	CO_2	N limitation	PHB	[42]
Anabaena Cylindrica	<0.005	CO_2	Balanced growth	PHB	[43]
Anabaena cylindrica	2.0	propionate	N limitation	PHB + PHV	[43]
Synechococcus elongatus	17.2	CO_2 and sucrose	N deficiency	Not specified	[44]
Caltorix scytonemicola TISTR 8095	25	CO_2	N deficiency	PHB	[45]

[1] gas exchange limitation = limitation of gas transfer to the culture vessel. [2] N = nitrogen. [3] P = phosphorus.

3.1.2. More Competent Cyanobacterial Cell Lines

Cyanobacteria are considered a sustainable and alternative host for PHB production due to their photoautotrophic nature [46]. Despite the fact that cyanobacterial PHB has been the subject of research for many years, it has not found its way to the market. One of the main challenges for cyanobacterial products to enter the market is that cyanobacterial strains are not yet optimized as cell factories for industrial processes. Intensive research has been done over the past 20 years for cyanobacterial strain improvement, research that has aimed to increase PHB productivity, mainly by overexpression of PHB biosynthetic genes. However, these attempts have rarely shown success regarding increased volumetric or specific polymer content for commercial production of cyanobacterial PHB. Recently, Katayama et al. reviewed the production of bioplastic compounds using genetically modified and metabolically engineered cyanobacteria [47]. In this study, we provide a list of genetically modified cyanobacteria with their PHB content and the tools used for the metabolic engineering of the strain.

3.1.3. Genetic Engineering of Cyanobacteria for PHB Production

Being prokaryotic, cyanobacteria possess a relatively simple genetic background which eases their manipulation [48]. Most cyanobacterial studies on metabolic engineering and PHB biosynthesis have been conducted with a limited number of model strains, of which *Synechocystis* sp. PCC 6803 is the most widely studied species for cyanobacterial research. The research, which has been done for decades on photosynthesis and the genome annotations, has resulted in a wide range of metabolic engineering tools and extensive biological insight for this species [48,49]. PHB production in cyanobacteria occurs mainly via three biosynthetic steps, where two molecules of acetyl-CoA form one molecule of acetoacetyl-CoA using the enzyme 3-ketothiolase encoded with the *phaA* gene [50]. Later, acetoacetyl-CoA is reduced by PhaB to hydroxybutyryl-CoA, utilizing NADPH as an electron donor [51]. In the end, the PHA synthase comprises of PhaC and PhaE polymerizes (R)-3-hydroxybutyryl-CoA to PHB [50,52]. Table 2 summarizes the efforts to overcome the bottlenecks in PHB biosynthetic pathway in cyanobacteria.

Table 2. Strategies to increase PHB biosynthesis yield.

Cyanobacterial Strain (Recombinant)	PHB Content (% DCW)	Genetic Tool Used	Production Conditions	References
Synechococcus sp. PCC 7942	1.0	Defective in glycogen synthesis	CO_2	[53]
Synechococcus sp. PCC 7942	26	Introducing PHA biosynthetic genes from *C. necator*	Acetate and nitrogen limitation	[41]
Synechocystis sp. PCC 6803	26	Overexpression of native *pha* genes	CO_2 and nitrogen deprivation	[46]
Synechocystis sp. PCC 6803	11	Introducing PHA biosynthetic genes from *C. necator*	Acetate and nitrogen limitation	[54]
Synechocystis sp. PCC 6803	14	Overexpression of PHA synthase	Direct photosynthesis	[55]
Synechocystis sp. PCC 6803	12	Increasing acetyl-CoA levels	CO_2	[56]
Synechococcus sp. PCC 7002	4.5	Introduction of GABA Shunt	CO_2	[57]
Synechocystis sp.	35	Optimization of acetoacetyl-CoA reductase binding site	CO_2	[58]
Synechocystis sp. PCC 6803	7.0	Transconjugant cells harboring expression vectors carrying *pha* genes	CO_2	[59]

Metabolic engineering of cyanobacteria with the aim to increase PHB content was also done by introducing the PHA synthase gene from *C. necator* into *Synechhocytis* sp. PCC 6803 [54]. The resulting recombinant *Synechhocytis* sp. PCC 6803 showed increased PHA synthase activity; the total PHB content, however, did not increase [47,54]. For cyanobacterial strain *Synechocystis* sp. PCC 6803, up to 35% (DCW) PHB was obtained using phaAB overexpression and 4 mM acetate [46]. However, the specific production rates in this case also did not show a significant improvement either. Recently, the overexpression of the acetoacetyl-CoA reductase gene in *Synechocystis* was found to increase the productivity of R-3-hydroxybutyrate from CO_2 to up to 1.84 g L^{-1} [58]. The highest volumetric productivity reported in this case was 263 mg L^{-1} d^{-1}.

3.1.4. Randomly Mutated Strains with Improved PHB Content

As an alternative approach to genetic engineering, random mutagenesis can be done for the generation of a mutant library with improved phenotypes. Mutagenesis can be done by exposing the cells of interest to a mutagenic source in order to induce random mutations into the genome. This can, for instance, completely knock-out a gene function [60] or increase enzymatic activity. UV irradiation is the most frequently used mutagen, which leads to transversion in the genome. Furthermore, ethidium bromide and ethyl-methanesulfonate are used as chemical mutagens [60,61]. A major disadvantage of using random mutagenesis is the need for intensive screening to select the mutant with desired phenotypes. The cyanobacterial strain *Synechocystis* sp. PCC 6714 has a great potential as photosynthetic PHB production organism. It has shown up to 17% (DCW) PHB content under nitrogen and phosphorus limiting conditions [32]. In addition to PHB, the strain also accumulates glycogen during the early phase of nitrogen limitation [62]. The random mutagenesis approach used for *Synechocystis* sp. PCC 6714 showed an increase in productivity of up to 2.5-folds resulting in $37 \pm 4\%$ (DCW) PHB for the best mutant [63]. The UV-mutation lead to an amino acid change in the phosphate system transport protein (PstA), resulting in higher efficiency of photosynthesis and CO_2 uptake rate for the mutant MT_a24 [63].

3.1.5. CRISPR/Cas Based Genome Editing in Cyanobacteria

Cyanobacteria are promising platforms for the production of biofuels and bio-based chemicals, however, the metabolic engineering of cyanobacteria poses various challenges [64]. CRISPR/Cas technology has enabled genome modification of cyanobacteria with gene substitution, marker-less point mutations, and gene knockouts and knock-ins with improved efficiency [65]. So far, the CRISPRi system has been used to downregulate the production of PHA and glycogen production in order to increase fluxes towards other carbon storage compounds of interest [66], such as succinate [64]. However, the CRISPRi based gene editing to overexpress PHB biosynthetic genes has not been reported. While the CRISPR-based editing allows the creation of marker-less knockouts and knock-ins. Thus, in the future, the cyanobacterial strains produced might be considered commercially sustainable and safe for outdoor cultivations and CO_2 sequestration.

3.2. Process Design and Bioprocess Improvement Strategies

Nutrient deficiency or stress, mainly in terms of nitrogen or phosphorus limitation, stimulates the accumulation of PHB in cyanobacteria. Besides the strain engineering and improvement approach, various reports have discussed other factors, which can facilitate superior growth and productivity in cyanobacteria. Herein, the most important routes for improvement of PHB production in cyanobacteria are listed.

3.2.1. Media and Cultivation Conditions

Like all other bioprocesses, PHB production from cyanobacteria is mainly influenced by the cultivation parameters and nutrient supply. The importance of defined cultivation conditions used to obtain highly productive process for cyanobacteria and microalgae has been previously discussed [32,67,68]. Cyanobacterial growth requires a high concentration of essential nutrients, such as nitrogen, phosphorus, sulfur, potassium, magnesium, iron, and some traces of micromolecules. The supply of nutrients like nitrogen and phosphorus in limiting concentration is important for the production of PHB. Therefore, media optimization plays an important role to maximize the PHB productivity and lower production costs. Regarding optimized nitrogen concentration in the media, it was shown by Coelho et al. that 0.05 g L^{-1} nitrogen in the media results in the production of up to 30.7% (DCW) of PHB in *Spirulina* sp. LEB 18. Further optimization of nitrogen content to 0.22 g L^{-1} in the media increased the PHB content in spirulina sp. LEB 18 to 44.2% (DCW) [69]. However, the impact of nitrogen optimization on the volumetric or specific productivities were not reported in both cases. The optimization of media components, nitrogen, and phosphorus in the case of *Synechocystis* sp. PCC 6714 increased volumetric as well as specific production rates, both in the case of biomass growth and PHB content [62].

Besides media components, other key parameters influencing growth and PHB production in cyanobacteria are cultivation conditions, such as temperature, pH, light intensity, or light/dark cycles. Furthermore, production of the copolymers can be tailor-made by using co-substrates and varying the cultivation conditions, such as temperature and pH [70]. Various studies have used the statistical design of experiments (DoEs) in order to optimize the media as well as the cultivation conditions [31,32,71]. The DoEs are used to minimize the error in determining the influential parameters, allowing systematic and efficient variation of all factors [72]. Table 3 summarizes the cultivation parameters and the nutrient limitation used for cyanobacterial PHB synthesis.

Table 3. Reported cultivation parameters and media limitation used for photoautotrophic PHB production in wild-type cyanobacterial strains.

Cyanobacterial Strain	Limiting Component	Temperature °C	pH	Light Condition	PHB Content % (DCW)	Cultivation Time (Days)	Volume (L)	References
Synechocystis sp. PCC 6803	N and P starvation	28–32	7.5–8.5	dark/light cycle	11	10	0.05	[31]
Synechocystis sp. PCC6803	N starvation	30	n.p	light	4.1	7	n.p	[73]
Synechocystis sp. PCC6803	N limitation	28	n.p	18:6	8	30	0.8	[74]
Synechocystis sp. PCC 6714	N and P limitation	28	8.5	light	16.4	16	1	[32]
Synechocystis salina CCALA192	Optimized BG-11 media [4]	n.p [5]	8.5	light	6.6	21	200	[23]
Phormidium sp. TISTR 8462	N limitation	28	7.5	light	14.8	12	n.p	[75]
Calothrix scytonemicola TISTR 8095	N deprivation	28	7.5	light	25.4	12	n.p	[75]
Nostoc muscorum	Growth associated [6]	25	8.5	14:10	8.6	21	0.05	[76]
Nostoc muscorum	P depletion	22	n.p	light	10.2	19	n.p	[38]
Spirulina sp. LEB 18	Defined media [7]	30	n.p	12:12	30.7	15	1.8	[77]
Aulosira fertilissima	P limitation	28	8.5	14:10	10	4	0.05	[40]
Anabaena sp.	n.p	25	8	14:10	2.3	n.p	0.1	[78]

[4] Optimized BG-11 media = the optimized BG-11 media contains 0.45 g L^{-1} NaNO$_3$ and leads to a self-limitation of the culture. [5] n.p = not provided. [6] Growth associated = the production of PHA was associated with growth and no media limitation was given. [7] Defined media = concentration of nitrate, phosphate and sodium bicarbonate was optimized.

Two primary challenges of entering cyanobacterial PHB into the market are the concern of the sustainability of the production process and the high production costs of fresh water and nutrients. One solution could be to use waste streams like agricultural effluents with high nitrogen and phosphorus contents. Therefore, the production of the polymer is accompanied by the removal of nutrients from the water. On the other hand, the undefined substrate may raise new challenges that then need to be resolved [79]. Various reports have shown production of cyanobacterial PHB using waste streams. Troschl et al. have summarized the list of cyanobacterial strains cultivated on agro-industrial waste streams and anaerobic digestants to produce PHB [23]. One example is the cultivation of the diazotrophic cyanobacterial strain *Aulosira fertilisima* under a circulatory aquaculture system that resulted in increased dissolved oxygen levels during the cultivation period and the complete removal of nutrients, such as ammonia, nitrite, and phosphate, within 15 days of cultivation, yielding an average PHB content of 80–92 g m^{-3} [80]. This report, along with other previously shown studies [79,81–84], clearly shows the potential of cyanobacterial PHB production for wastewater treatment facilities.

3.2.2. PHB Production Using Mixed Photosynthetic Consortia

Another approach used for PHB production is the feast-famine strategy, which uses a mixed consortium of algae and cyanobacteria [85–88]. During this regime, the feast operation consists of a mixed culture of cyanobacterial consortium cultivated in a sequencing batch reactor (SBR) without aeration using acetate as a carbon source and light as an energy source [86]. During the famine phase, the NADH or the NADPH reserves of the cell is consumed using the oxygen produced by the algae cells present in the consortia leading to accumulation of around 20% (DCW) PHA [89]. Furthermore, maximum polymer content of 60% (DCW) of PHA was produced by a photosynthetic mixed culture in a permanent feast regime using high light intensity [86]. The anaerobic dark energy generation's capability of cyanobacteria is already been known [90]. Some cyanobacteria have also been known for their fermentation capability at the expense of their carbohydrate reserves [91]. The axenic dark feast conditions facilitated the acetate uptake, increasing the productivity significantly (up to 60%) (DCW), as the famine phase was eliminated [85,86]. The anaerobic fermentation of cyanobacteria to produce PHB has a potential, while the need for sterilization and aeration is eliminated, reducing also the energy costs. However, the source and cost of the substrate used remains a cost driver issue.

3.2.3. PHB Production Using Mixed Feed Systems

Production of PHAs in cyanobacteria can occur during phototrophic growth, using CO_2 as a sole carbon source and light energy, and also during heterotrophy, when using sugar supplementation. It has been estimated that the carbon substrate in a large-scale manufacturing context would constitute approximately 37% of the total production costs [27,92]. However, in order to cope with the low phototrophic PHB productivity in cyanobacteria, various studies have used supplementation of other carbon sources. The mini-review by Singh and Mallick has summarized the wild-type and recombinant cyanobacterial strains, their PHA content, and the substrate used for the biosynthesis of the biopolymer [24]. However, in most reported cases [34,46,93–97] of heterotrophic PHB production, the biomass concentrations produced are less than 1 g L^{-1} and increases in volumetric productivities are not described. Even though the PHB productivity increases in terms of biopolymer content (%DCW) using mixed feed systems, the use of external carbon substrates increases the production costs and also raises the question of the economic feasibility of PHB production from cyanobacteria. As long as heterotrophic organisms produce PHB at much higher rates than cyanobacteria, the only sense for commercialization of cyanobacterial PHB would be the sustainability. Therefore, research must focus on improving the phototrophic PHB production with the aim of increasing CO_2 uptake rates of cyanobacteria, with the support of viable bioprocess technology tools.

3.2.4. CO_2 Sequestration

CO_2 is a major greenhouse gas; its emission into the atmosphere has gradually increased in the past decades, causing global warming and its associated problems [98]. Carbon contributes to all organic compounds and is the main constituent of cyanobacterial and all biomass, amounting to up to 65% of DCW [79]. The industrial production of PHB that uses CO_2 feedstocks helps reduce the environmental impacts of CO_2 emission. Various studies have shown that an increase in CO_2 concentration during a cyanobacterial cultivation may increase the production of carbon reserve compounds, such as PHB. Markou et al. showed that an increase in carbon content leads to the production of carbon reserve compounds, such as lipids and PHAs [79]. The increase in the concentration of the carbon source also increased biopolymer accumulation in cyanobacterial strain *Spirulina* sp. LEB 18 [77]. However, what has not been discussed in the literature thus far are the effects of the day-night cycle on the CO_2 uptake rate and the productivity of carbon reserve compounds in cyanobacteria. Since CO_2 fixation occurs during the light phase, the total productivity and CO_2 sequestration rate will be lower in outdoor cultivations. During the dark phase, CO_2 utilization is minimized and the productivities are lowered and some carbon reserve molecules, such as glycogen, degrade. Other methods would need to be used to temporarily sequester CO_2 as a carbonate species during nighttime, which could then be utilized by cyanobacteria when the light is available again [99].

4. Production Strategies

The economic efficiency of any production process is indicated by the productivity, which comprises of growth rates, specific production rates, and the biomass concentration of the culture. Therefore, the economic efficiency of the production process will increase only when the mentioned parameters are improved. Once the strain and cultivation parameters are selected and optimized for the production process, the process performance can be considered and analyzed.

4.1. Cultivation Modes

Cyanobacteria producing PHA have been classified into two groups based on the culture conditions required for efficient polymer synthesis: group one requires a limitation of an essential media component for PHA synthesis; the second has no requirement for nutrient limitation for the accumulation of the polymer [100]. For industrial production of PHB, the second group is favorable for growth as it is accompanied by polymer synthesis.

In general, cultivation of cyanobacteria for the production of PHB can be done using various cultivation modes. The most common approach is using batch cultivation, in which the production of PHB is induced by a limiting nutrient or, in an ideal case, the production of the polymer becomes growth dependent. For the batch cultivation with the group one cyanobacteria, the concentration of nitrogen and phosphorus in the media play the key-role facilitating biomass growth, and their limitation trigger PHB synthesis. Thus, in such a process the cell growth is maintained without nutrient limitation, until the desired concentration is reached. Then, an essential limitation allows for efficient polymer accumulation. So far, a few studies have focused on optimizing the nutrients for the batch production of PHB in large-scale; others mainly have been done in flasks. Batch cultivation of *Synechocystis* sp. PCC 6803 using a nitrogen concentration of half of the optimal BG-11 media, showed 180 mg L^{-1} PHB from CO_2 [74]. The maximum PHB content of 125 mg L^{-1} was obtained for the non-sterile batch cultivation of *Synechocystis* sp. CCALA192 in a 200-L tubular photobioreactor [42]. Moreover, PHB was produced using optimized media in a one-liter tank reactor for the wild-type cyanobacterial strain *Synechocystis* sp. PCC 6714 obtained 640 mg L^{-1} of polymer [62].

The other common strategy for the cyanobacterial PHB production is using the SBR mode of operation, where growth and production occur in different reactors. In the growth vessel, media components are provided in abundance to facilitate maximum biomass production. In the induction photobioreactor, one or more media components are limited facilitating PHB biosynthesis. During the induction, residual biomass concentration remains more or less constant, while cell concentration increases only by intracellular polymeric accumulation [101]. In order to facilitate higher productivities, both reactors can operate as chemostats. Thus far, no reports of cyanobacterial PHB production in SBR or chemostat mode have been described.

4.2. Cultivation Systems

Another challenge in commercial cyanobacterial production is associated with biomass production. There are three main production systems used for large-scale cultivation of microalgae and cyanobacteria. The most basic approach for the cultivation of photosynthetic organisms is the use of large natural locations, which is mostly done for microalgae such as *Dunaliella*. Releasing into natural locations is regarded as a deliberate release into the environment, since there are no effective protective measures to prevent the microalgae from entering the surroundings [60]. The other approach is the use of open raceway pond systems, which has been commonly applied worldwide. When these raceway ponds are used outdoors, the cultivation are regarded as a deliberate release, so the spread of genetically modified organisms cannot be excluded in this case [60]. Even though these systems are economically feasible, the maintenance of monocultures and improving productivity are the main bottlenecks associated with such cultivations. The use of an open pond system has so far been reported in a wastewater treatment facility, containing fish pond discharge that uses the cyanobacterial strain *Aulosira fertilissima*, which shows a PHB productivity of up to 92 g m^{-3} [80]. The third system is the sophisticated, closed production system: photobioreactors (PBR). These systems can be both placed in greenhouses to obtain more defined cultivation conditions or be installed outdoors. PBR systems are more flexible for the needs of the cultivation process and the desired species. The industrial-scale production of photoautotrophic cyanobacterial PHB has not been widely reported in photobioreactors. The various photobioreactor systems used to cultivate cyanobacteria is given by Koller et al. [102]. Yet Troschl et al. has described the cultivation of *Synechocystis salina* CCALA192 in a 200-L tubular photobioreactor for the production of PHB from CO_2 [23]. The maximum PHB productivity obtained under nitrogen limitation was 6.6% (DCW), while the volumetric and specific productivities were not reported in this case. Moreover, a mixed consortium of wastewater born cyanobacteria was cultivated in a 30-L PBR, showing a maximum productivity of 104 mg L^{-1} under phosphorus limitation [103].

5. Process Monitoring and Control

Today the most commonly used method for accurate determination of PHAs in bacterial cultivations is gas chromatography (GC) [104] or high-performance liquid chromatography (HPLC) [105,106]. These methods involve hydrolysis, subsequent methanolysis, or propanolysis of the PHAs in whole cells, in the presence of sulfuric acid and chloroform [107]. These extraction methods are laborious, time-consuming, and the optimum time of harvest might be lost due to the time needed for the analysis. Other methods for PHA analysis include gravimetric, infrared spectroscopy of chemically extracted PHB, fluorimetry, and cell carbon analysis [107–109]. It is necessary to develop viable analytics to help the development of an efficient commercial production process that enables monitoring and control of production, along with a rapid feedback on the state of the process.

Fourier transform infrared (FTIR) spectroscopy has been applied to determine the chemical composition of cyanobacteria with major cellular analytes, such as proteins, lipids, polysaccharide, nucleic acids, and PHAs [107,110]. It has been shown that FTIR spectroscopy can monitor water-soluble extracellular analytes in fermentation systems, as well as being an indirect method to determine the stage of fermentation by monitoring the physiological state of the cells [111]. Various studies have shown the potential of FTIR spectroscopy for determination of intracellular PHA contents in various microorganisms [107,112]. In the same direction, Jarute et al. have introduced an automated approach for on-line monitoring of the intracellular PHB in a process with recombinant *E. coli*, which uses stopped-flow attenuated total reflection FTIR spectroscopy [113]. In the case of cyanobacteria, there exists no such studies reporting on-line or at-line determination of intracellular carbon compounds, such as PHAs and glycogen. The measurements used and the parameters controlled in microalgae processes are specific in-line probes, such as pO_2, pCO_2, pH, and temperature. In cyanobacterial industrial processes, spectroscopic measurement techniques, such as FTIR, can be used for monitoring, controlling the production, and determining the time of harvest. The on-line determination can also identify the limitation time and the limiting components based on the cell physiology, thus helping to make the cyanobacterial PHA production robust and manageable.

6. Production Scenarios

In order to compete with synthetic and other starch-based polymers in the market, the cost of cyanobacterial PHB needs to be reduced significantly. Yet, no economic analysis has been done to estimate the production costs of phototrophic PHB production. It has been reported that the cost of PHB production from heterotrophic organisms is in the range of 2–5 € kg^{-1} [114]. This value is still much higher than the estimated cost of petrochemical-based mass polymers like PE, PP, or PET, which is around 1.2 € kg^{-1} [114] and less. Taking into account the much lower time space yield and the biomass productivity in cyanobacteria and complications associated with the downstream processing, the cost associated with the production of PHB in cyanobacteria could be higher than that of heterotrophic microorganisms (>5 € kg^{-1}). Typically, more than 4.3 kg of sugar is needed to produce 1 kg of PHB [85]. Nevertheless, higher yields of product per substrate consumed have also been reported, showing values of 3.1 kg sucrose/kg PHB and of 3.33 kg glucose/kg of the polymer [15,115]. In this context, the substrate costs can be avoided by photoautotrophically produced PHB by cyanobacteria. However, the lower productivities of cyanobacteria will still increase the costs significantly. Among the main factors contributing to the cost of PHA production are equipment-related costs, such as direct-fixed-capital-dependent items, overheads, and some labor-dependent factors, which considerably increase with a decrease in productivity [27,101]. Therefore, for the production of the same amount of PHA per year, the process with lower productivity requires larger equipment [27,116]. To that end, one approach could be to reduce the costs associated with the building of photobioreactors. This can be accomplished by simplifying the design and the material used for the production of photobioreactors and their energy consumption [22]. Another alternative to increase the size of the facility or reactor while also reducing production costs is to use open pond raceways and wastewater born cyanobacteria instead of fresh water strains. However, it should be taken into account that the increase in volume will directly increase the effort associated with the downstream processing [25].

Moreover, it has been shown that using industrial flue gases may reduce the production cost of cyanobacterial biomass to around 2.5 € kg^{-1}, while using wastewater can decrease the costs further, to less than 2 € kg^{-1} [22,117]. Therefore, as already discussed, wastewater streams with high carbon, nitrogen, and phosphorus that are mix-fed with CO_2 from industrial flue gases, can be used to make the PHB production from cyanobacteria more efficient. Furthermore, producing several chemicals from the same microalgae feedstock could potentially make the production of multiple commodity chemicals from a biological resource economically viable [92].

7. The Remaining Challenges in Photosynthetic PHB Production

Current industrial PHA production processes rely mostly on the availability of agricultural resources, which are unsustainable (compare the food versus fuel discussion with first-generation biofuels) and leave a large ecological footprint [65].

In the case of cyanobacterial, PHA production research has mainly focused on genetic engineering to increase productivity, which mainly reports as higher % DCW polymer content. The studies have rarely reported an increase in photosynthetic efficiency or an increase in the specific growth rates and production rates. So far, very few studies have shown the use of wastewater-open pond systems for the production of PHAs. For a recent review of PHA production, see Koller et al. [118].

8. Outlook

Currently, the global research efforts directed towards individual aspects of cyanobacterial PHA production mainly focus on improved strains and recovery processes. Although various challenges are associated with the efficiency of the cyanobacterial PHA productivity and the extraction and purification of PHAs, optimization of each step separately will waste considerable effort and result in overall sub-optimality [27]. With respect to commercialization and scale-up of the cyanobacterial PHA production, the view of the whole processes needs to be considered. More attention towards sustainable and viable upstream processing may help to reach an economic PHA production point. Cyanobacterial PHA production, from an economic point of view, will only make sense if a continuous process can be achieved, especially using waste streams as a carbon source and for the media. The process can then be coupled with the bioremediation of agricultural and industrial effluents. Thus far, some wild-type and improved cyanobacterial strains are reported with PHB content which is mostly cultivated under controlled, defined, and sterile lab conditions. For production in industrial scale that is done under unsterile conditions, only *Nostoc moscorum* as an example is reported. Other strains are not tested or can hardly tolerate the harsh outdoor conditions. Although we have emphasized the importance of optimized media and cultivation conditions on PHA productivity, sustainable and viable commercial processes conducted under unsterile conditions using waste streams and open systems are required. Research needs to focus on screening for more robust strains, such as wastewater born mixed-cultures that can tolerate fluctuations in cultivation conditions like pH, temperature, salinity, and media composition. Furthermore, the durable strains for which production of PHB is associated with biomass growth and therefore the time-spaced yield will be improved.

PHA shows both the advantages of biobased carbon content and full biodegradability. In addition, cyanobacterial PHA can be more sustainable and more cost effective in the marine environment and when compared to PHA from carbohydrate fermentation. It can be a carbon-negative material, making the process not only attractive for PHA converters and users, but also for CO_2 emitters, like power stations. There are plenty of medium-size CO_2 point sources, e.g., biogas production facilities, where the CO_2 could be used in an adjacent cyanobacterial PHA factory erected on the non-arable land. Preferably, a biorefinery approach would be executed, where valuable compounds such as phytohormones and pigments are extracted from the cyanobacteria; then. PHA and biomass is anaerobically digested in a biogas plant, yielding a cost-effective and fully integrated process.

It is expected that over the next two decades there will be a shift toward more recycling of fossil-based and conventional plastics, with an accompanying reduction in material variety to facilitate collection, processing, and reuse. Moreover, we will see a maturing of the bioplastics industry, with more applications being developed with bioplastics, other than "gimmick" giveaways and small household and kitchen tools. Due to the unique and interesting property set of PHA, it can be anticipated that these materials, particularly PHB and its copolymers such as PHBV, will gain significance.

Author Contributions: D.K. and C.H. designed this study; D.K. carried out the research and wrote this manuscript, with the support of M.L. and C.H; the study was initiated by M.L. and C.H.; C.H. supervised this project; D.K., M.L., and C.H. have several years of practical research experience with PHB production from cyanobacteria focused on strain improvement, bioprocess optimization, and scale-up.

Acknowledgments: The authors would like to thank the Vienna Business Agency for supporting the funding of this work.

References

1.   De Koning, G.J.M. *Prospects of Bacterial Poly[(R)-3-(Hydroxyalkanoates)]*; Technische Universiteit Eindhoven: Eindhoven, The Netherlands, 1993.

2.   Keshavarz, T.; Roy, I. Polyhydroxyalkanoates: Bioplastics with a green agenda. *Curr. Opin. Microbiol.* **2010**, *13*, 321–326. [CrossRef] [PubMed]

3.   Lackner, M. Bioplastics. In *Kirk-Othmer Encyclopedia of Chemical Technology*; John Wiley & Sons, Inc.: Hoboken, NJ, USA, 2015.

4.   Tan, G.-Y.; Chen, C.-L.; Li, L.; Ge, L.; Wang, L.; Razaad, I.; Li, Y.; Zhao, L.; Mo, Y.; Wang, J.-Y. Start a Research on Biopolymer Polyhydroxyalkanoate (PHA): A Review. *Polymers* **2014**, *6*, 706–754. [CrossRef]

5.   Getachew, A.; Woldesenbet, F. Production of biodegradable plastic by polyhydroxybutyrate (PHB) accumulating bacteria using low cost agricultural waste material. *BMC Res. Notes* **2016**, *9*, 509. [CrossRef] [PubMed]

6.   Madison, L.L.; Huisman, G.W. Metabolic Engineering of Poly(3-Hydroxyalkanoates): From DNA to Plastic. *Microbiol. Mol. Biol. Rev.* **1999**, *63*, 21–53. [PubMed]

7.   Ten, E.; Jiang, L.; Zhang, J.; Wolcott, M.P. 3—Mechanical Performance of polyhydroxyalkanoate (PHA)-based biocomposites. In *Biocomposites*; Misra, M., Pandey, J.K., Mohanty, A.K., Eds.; Woodhead Publishing: Sawston, UK, 2015; pp. 39–52.

8.   Galia, M.B. Isolation and analysis of storage compounds. In *Handbook of Hydrocarbon and Lipid Microbiology*; Timmis, K.N., Ed.; Springer: Berlin, Germany, 2010.

9.   Barham, P.J.; Organ, S.J. Mechanical properties of polyhydroxybutyrate-hydroxybutyrate-hydroxyvalerate copolymer blends. *J. Mater. Sci.* **1994**, *29*, 1676–1679. [CrossRef]

10. Harding, K.G.; Dennis, J.S.; Blottnitz, H.; Harrison, S.T. Environmental analysis of plastic production processes: Comparing petroleum-based polypropylene and polyethylene with biologically-based poly-beta-hydroxybutyric acid using life cycle analysis. *J. Biotechnol.* **2007**, *130*, 57–66. [CrossRef] [PubMed]

11. Lackner, M.; Markl, E.; Grünbichler, H. Cyanobacteria for PHB Bioplastics Production: A Review. *Nov. Tech. Nutr. Food Sci.* **2018**, *2*, 4.

12. Balaji, S.; Gopi, K.; Muthuvelan, B. A review on production of poly β hydroxybutyrates from cyanobacteria for the production of bio plastics. *Algal Res.* **2013**, *2*, 278–285. [CrossRef]

13. Grothe, E.; Chisti, Y. Poly (ß-hydroxybutyric acid) thermoplastic production by Alcaligenes latus: Behavior of fed-batch cultures. *Bioprocess Biosyst. Eng.* **2000**, *22*, 441–449. [CrossRef]

14. Schubert, P.; Steinbüchel, A.; Schlegel, H.G. Cloning of the Alcaligenes eutrophus genes for synthesis of poly-beta-hydroxybutyric acid (PHB) and synthesis of PHB in *Escherichia coli*. *J. Bacteriol.* **1988**, *170*, 5837–5847. [CrossRef] [PubMed]

15. Nonato, R.; Mantelatto, P.; Rossell, C. Integrated production of biodegradable plastic, sugar and ethanol. *Appl. Microbiol. Biotechnol.* **2001**, *57*, 1–5. [PubMed]

16. Koller, M.; Salerno, A.; Reiterer, A.; Malli, H.; Malli, K.; Kettl, K.H.; Narodoslawsky, M.; Schnitzer, H.; Chiellini, E.; Braunegg, G. Sugar cane as feedstock for biomediated polymer production. In *Sugarcane: Production, Cultivation and Uses*; Nova Publishers: Hauppauge, NY, USA, 2012; pp. 105–136.

17. Steinbüchel, A. PHB and Other Polhydroxyalkanoic Acids. In *Biotechnology Set*, 2nd ed.; Wiley: Hoboken, NJ, USA, 2008; pp. 403–464.

18. Gurieff, N.; Lant, P. Comparative life cycle assessment and financial analysis of mixed culture polyhydroxyalkanoates production. *Bioresour. Technol.* **2007**, *98*, 3393–3403. [CrossRef] [PubMed]

19. Reis, M.A.M.; Serafim, L.S.; Lemos, P.C.; Ramos, A.M.; Aguiar, F.R.; Loosdrecht, M.C.M. Production of polyhydroxyalkanoates by mixed microbial cultures. *Bioprocess Biosyst. Eng.* **2003**, *25*, 377–385. [CrossRef] [PubMed]

20. Alias, Z.; Tan, I.K. Isolation of palm oil-utilising, polyhydroxyalkanoate (PHA)-producing bacteria by an enrichment technique. *Bioresour. Technol.* **2005**, *96*, 1229–1234. [CrossRef] [PubMed]

21. Oliver, N.J.; Rabinovitch-Deere, C.A.; Carroll, A.L.; Nozzi, N.E.; Case, A.E.; Atsumi, S. Cyanobacterial metabolic engineering for biofuel and chemical production. *Curr. Opin. Chem. Biol.* **2016**, *35*, 43–50. [CrossRef] [PubMed]

22. Costa, J.A.V.; Moreira, J.B.; Lucas, B.F.; Braga, V.D.S.; Cassuriaga, A.P.A.; Morais, M.G.D. Recent Advances and Future Perspectives of PHB Production by Cyanobacteria. *Ind. Biotechnol.* **2018**, *14*, 249–256. [CrossRef]

23. Troschl, C.; Meixner, K.; Drosg, B. Cyanobacterial PHA Production—Review of Recent Advances and a Summary of Three Years' Working Experience Running a Pilot Plant. *Bioengineering* **2017**, *4*, 26. [CrossRef] [PubMed]

24. Singh, A.K.; Mallick, N. Advances in cyanobacterial polyhydroxyalkanoates production. *FEMS Microbiol. Lett.* **2017**, *364*, fnx189. [CrossRef] [PubMed]

25. Drosg, B.; Fritz, I.; Gattermayer, F.; Silvestrini, L. Photo-autotrophic Production of Poly(hydroxyalkanoates) in Cyanobacteria. *Chem. Biochem. Eng. Q.* **2015**, *29*, 145–156. [CrossRef]

26. Koller, M.; Marsalek, L. Cyanobacterial Polyhydroxyalkanoate Production: Status Quo and Quo Vadis? *Curr. Biotechnol.* **2015**, *4*, 464–480. [CrossRef]

27. Choi, J.; Lee, S.Y. Factors affecting the economics of polyhydroxyalkanoate production by bacterial fermentation. *Appl. Microbiol. Biotechnol.* **1999**, *51*, 13–21. [CrossRef]

28. Lau, N.-S.; Matsui, M.; Al Abdullah, A.-A. Cyanobacteria: Photoautotrophic Microbial Factories for the Sustainable Synthesis of Industrial Products. *BioMed Res. Int.* **2015**, *2015*, 754934. [CrossRef] [PubMed]

29. Dismukes, G.C.; Carrieri, D.; Bennette, N.; Ananyev, G.M.; Posewitz, M.C. Aquatic phototrophs: Efficient alternatives to land-based crops for biofuels. *Curr. Opin. Biotechnol.* **2008**, *19*, 235–240. [CrossRef] [PubMed]

30. Case, A.E.; Atsumi, S. Cyanobacterial chemical production. *J. Biotechnol.* **2016**, *231*, 106–114. [CrossRef] [PubMed]

31. Panda, B.; Jain, P.; Sharma, L.; Mallick, N. Optimization of cultural and nutritional conditions for accumulation of poly-β-hydroxybutyrate in *Synechocystis* sp. PCC 6803. *Bioresour. Technol.* **2006**, *97*, 1296–1301. [CrossRef] [PubMed]

32. Kamravamanesh, D.; Pflügl, S.; Nischkauer, W.; Limbeck, A.; Lackner, M.; Herwig, C. Photosynthetic poly-β-hydroxybutyrate accumulation in unicellular cyanobacterium *Synechocystis* sp. PCC 6714. *AMB Express* **2017**, *7*, 143. [CrossRef] [PubMed]

33. Campbell, J.; Stevens, S.E.; Balkwill, D.L. Accumulation of poly-beta-hydroxybutyrate in *Spirulina platensis*. *J. Bacteriol.* **1982**, *149*, 361–363. [PubMed]

34. Toh, P.S.Y.; Jau, M.H.; Yew, S.P.; Abed, R.M.M.; Sidesh, K. Comparison of polyhydroxyalkanoates biosynthesis, mobilization and the effects on cellular morphology in Spirulina Platensis and *Synechocystis* sp. UNIWG. *J. Biosci.* **2008**, *19*, 21–38.

35. De Philippis, R.; Sili, C.; Vincenzini, M. Glycogen and poly-β-hydroxybutyrate synthesis in *Spirulina maxima*. *Microbiology* **1992**, *138*, 1623–1628. [CrossRef]

36. Stal, L.J.; Heyer, H.; Jacobs, G. Occurrence and Role of Poly-Hydroxy-Alkanoate in the Cyanobacterium Oscillatoria Limosa. In *Novel Biodegradable Microbial Polymers*; Dawes, E.A., Ed.; Springer: Dordrecht, The Netherlands, 1990; pp. 435–438.

37. Bhati, R.; Mallick, N. Production and characterization of poly(3-hydroxybutyrate-co-3-hydroxyvalerate) co-polymer by a N2-fixing cyanobacterium, *Nostoc muscorum* Agardh. *J. Chem. Technol. Biotechnol.* **2012**, *87*, 505–512. [CrossRef]

38. Haase, S.M.; Huchzermeyer, B.; Rath, T. PHB accumulation in *Nostoc muscorum* under different carbon stress situations. *J. Appl. Phycol.* **2012**, *24*, 157–162. [CrossRef]

39. Samantaray, S.; Mallick, N. Production of poly (3-hydroxybutyrate-co-3-hydroxyvalerate) co-polymer by the diazotrophic cyanobacterium *Aulosira fertilissima* CCC 444. *J. Appl. Phycol.* **2014**, *26*, 237–245. [CrossRef]

40. Samantaray, S.; Mallick, N. Production and characterization of poly-β-hydroxybutyrate (PHB) polymer from *Aulosira fertilissima*. *J. Appl. Phycol.* **2012**, *24*, 803–814. [CrossRef]

41. Takahashi, H.; Miyake, M.; Tokiwa, Y.; Asada, Y. Improved accumulation of poly-3-hydroxybutyrate by a recombinant cyanobacterium. *Biotechnol. Lett.* **1998**, *20*, 183–186. [CrossRef]

42. Troschl, C.; Meixner, K.; Fritz, I.; Leitner, K.; Romero, A.P.; Kovalcik, A.; Sedlacek, P.; Drosg, B. Pilot-scale production of poly-β-hydroxybutyrate with the cyanobacterium *Synechocytis* sp. CCALA192 in a non-sterile tubular photobioreactor. *Algal Res.* **2018**, *34*, 116–125. [CrossRef]

43. Lama, L.; Nicolaus, B.; Calandrelli, V.; Manca, M.C.; Romano, I.; Gambacorta, A. Effect of growth conditions on endo- and exopolymer biosynthesis in *Anabaena cylindrica* 10 C. *Phytochemistry* **1996**, *42*, 655–659. [CrossRef]

44. Mendhulkar, V.D.; Laukik, A.S. Synthesis of Biodegradable Polymer Polyhydroxyalkanoate (PHA) in Cyanobacteria Synechococcus elongates Under Mixotrophic Nitrogen- and Phosphate-Mediated Stress Conditions. *Ind. Biotechnol.* **2017**, *13*, 85–93. [CrossRef]

45. Monshupanee, T.; Nimdach, P.; Incharoensakdi, A. Two-stage (photoautotrophy and heterotrophy) cultivation enables efficient production of bioplastic poly-3-hydroxybutyrate in auto-sedimenting cyanobacterium. *Sci. Rep.* **2016**, *6*, 37121. [CrossRef] [PubMed]

46. Khetkorn, W.; Incharoensakdi, A.; Lindblad, P.; Jantaro, S. Enhancement of poly-3-hydroxybutyrate production in *Synechocystis* sp. PCC 6803 by overexpression of its native biosynthetic genes. *Bioresour. Technol.* **2016**, *214*, 761–768. [CrossRef] [PubMed]

47. Katayama, N.; Iijima, H.; Osanai, T. Production of Bioplastic Compounds by Genetically Manipulated and Metabolic Engineered Cyanobacteria. In *Synthetic Biology of Cyanobacteria*; Zhang, W., Song, X., Eds.; Springer: Singapore, 2018; pp. 155–169.

48. Koksharova, O.; Wolk, C. Genetic tools for cyanobacteria. *Appl. Microbiol. Biotechnol.* **2002**, *58*, 123–137. [PubMed]

49. Wilde, A.; Dienst, D. Tools for Genetic Manipulation of Cyanobacteria. In *Bioenergetic Processes of Cyanobacteria: From Evolutionary Singularity to Ecological Diversity*; Peschek, G.A., Obinger, C., Renger, G., Eds.; Springer: Dordrecht, The Netherlands, 2011; pp. 685–703.

50. Hauf, W.; Watzer, B.; Roos, N.; Klotz, A.; Forchhammer, K. Photoautotrophic Polyhydroxybutyrate Granule Formation Is Regulated by Cyanobacterial Phasin PhaP in *Synechocystis* sp. Strain PCC 6803. *Appl. Environ. Microbiol.* **2015**, *81*, 4411–4422. [CrossRef] [PubMed]

51. Taroncher-Oldenburg, G.; Nishina, K.; Stephanopoulos, G. Identification and analysis of the polyhydroxyalkanoate-specific β-ketothiolase and acetoacetyl coenzyme A reductase genes in the cyanobacterium *Synechocystis* sp. strain PCC6803. *Appl. Environ. Microbiol.* **2000**, *66*, 4440–4448. [CrossRef] [PubMed]

52. Hein, S.; Tran, H.; Steinbuchel, A. Synechocystis sp. PCC6803 possesses a two-component polyhydroxyalkanoic acid synthase similar to that of anoxygenic purple sulfur bacteria. *Arch. Microbiol.* **1998**, *170*, 162–170. [CrossRef] [PubMed]

53. Suzuki, E.; Ohkawa, H.; Moriya, K.; Matsubara, T.; Nagaike, Y.; Iwasaki, I.; Fujiwara, S.; Tsuzuki, M.; Nakamura, Y. Carbohydrate metabolism in mutants of the cyanobacterium *Synechococcus elongatus* PCC 7942 defective in glycogen synthesis. *Appl. Environ. Microbiol.* **2010**, *76*, 3153–3159. [CrossRef] [PubMed]

54. Sudesh, K.; Taguchi, K.; Doi, Y. Effect of increased PHA synthase activity on polyhydroxyalkanoates biosynthesis in *Synechocystis* sp. PCC6803. *Int. J. Biol. Macromol.* **2002**, *30*, 97–104. [CrossRef]

55. Lau, N.-S.; Foong, C.P.; Kurihara, Y.; Sudesh, K.; Matsui, M. RNA-Seq Analysis Provides Insights for Understanding Photoautotrophic Polyhydroxyalkanoate Production in Recombinant *Synechocystis* sp. *PLoS ONE* **2014**, *9*, e86368. [CrossRef] [PubMed]

56. Carpine, R.; Du, W.; Olivieri, G.; Pollio, A.; Hellingwerf, K.J.; Marzocchella, A.; Branco dos Santos, F. Genetic engineering of *Synechocystis* sp. PCC6803 for poly-β-hydroxybutyrate overproduction. *Algal Res.* **2017**, *25*, 117–127. [CrossRef]

57. Zhang, S.; Qian, X.; Chang, S.; Dismukes, G.C.; Bryant, D.A. Natural and Synthetic Variants of the Tricarboxylic Acid Cycle in Cyanobacteria: Introduction of the GABA Shunt into *Synechococcus* sp. PCC 7002. *Front. Microbiol.* **2016**, *7*, 1972. [CrossRef] [PubMed]

58. Wang, B.; Xiong, W.; Yu, J.; Maness, P.-C.; Meldrum, D.R. Unlocking the photobiological conversion of CO_2 to (R)-3-hydroxybutyrate in cyanobacteria. *Green Chem.* **2018**, *20*, 3772–3782. [CrossRef]

59. Hondo, S.; Takahashi, M.; Osanai, T.; Matsuda, M.; Hasunuma, T.; Tazuke, A.; Nakahira, Y.; Chohnan, S.; Hasegawa, M.; Asayama, M. Genetic engineering and metabolite profiling for overproduction of polyhydroxybutyrate in cyanobacteria. *J. Biosci. Bioeng.* **2015**, *120*, 510–517. [CrossRef] [PubMed]

60. Jaeger, L.D. *Strain Improvement of Oleaginous Microalgae*; Wageningen University: Wageningen, The Netherlands, 2015; p. 200.

61. Lee, B.; Choi, G.-G.; Choi, Y.-E.; Sung, M.; Park, M.S.; Yang, J.-W. Enhancement of lipid productivity by ethyl methane sulfonate-mediated random mutagenesis and proteomic analysis in *Chlamydomonas reinhardtii*. *Korean J. Chem. Eng.* **2014**, *31*, 1036–1042. [CrossRef]

62. Kamravamanesh, D.; Slouka, C.; Limbeck, A.; Lackner, M.; Herwig, C. Increased carbohydrate production from carbon dioxide in randomly mutated cells of cyanobacterial strain *Synechocystis* sp. PCC 6714: Bioprocess understanding and evaluation of productivities. *Bioresour. Technol.* **2019**, *273*, 277–287. [CrossRef] [PubMed]

63. Kamravamanesh, D.; Kovacs, T.; Pflügl, S.; Druzhinina, I.; Kroll, P.; Lackner, M.; Herwig, C. Increased poly-β-hydroxybutyrate production from carbon dioxide in randomly mutated cells of cyanobacterial strain *Synechocystis* sp. PCC 6714: Mutant generation and characterization. *Bioresour. Technol.* **2018**, *266*, 34–44. [CrossRef] [PubMed]

64. Li, H.; Shen, C.R.; Huang, C.-H.; Sung, L.-Y.; Wu, M.-Y.; Hu, Y.-C. CRISPR-Cas9 for the genome engineering of cyanobacteria and succinate production. *Metab. Eng.* **2016**, *38*, 293–302. [CrossRef] [PubMed]

65. Behler, J.; Vijay, D.; Hess, W.R.; Akhtar, M.K. CRISPR-Based Technologies for Metabolic Engineering in Cyanobacteria. *Trends Biotechnol.* **2018**, *36*, 996–1010. [CrossRef] [PubMed]

66. Yao, L.; Cengic, I.; Anfelt, J.; Hudson, E.P. Multiple Gene Repression in Cyanobacteria Using CRISPRi. *ACS Synth. Biol.* **2016**, *5*, 207–212. [CrossRef] [PubMed]

67. Pruvost, J.; Van Vooren, G.; Le Gouic, B.; Couzinet-Mossion, A.; Legrand, J. Systematic investigation of biomass and lipid productivity by microalgae in photobioreactors for biodiesel application. *Bioresour. Technol.* **2011**, *102*, 150–158. [CrossRef] [PubMed]

68. García-Malea, M.C.; Acién, F.G.; Del Río, E.; Fernández, J.M.; Cerón, M.C.; Guerrero, M.G.; Molina-Grima, E. Production of astaxanthin by *Haematococcus pluvialis*: Taking the one-step system outdoors. *Biotechnol. Bioeng.* **2009**, *102*, 651–657. [CrossRef] [PubMed]

69. Martins, R.G.; Goncalves, I.S.; de Morais, M.G.; Costa, J.A.V. Bioprocess Engineering Aspects of Biopolymer Production by the Cyanobacterium Spirulina Strain LEB 18. *Int. J. Polymer Sci.* **2014**, *2014*, 895237. [CrossRef]

70. Samantaray, S.; Mallick, N. Role of cultural variables in tailoring poly (3-hydroxybutyrate-co-3-hydroxyvalerate) copolymer synthesis in the diazotrophic cyanobacterium *Aulosira fertilissima* CCC 444. *J. Appl. Phycol.* **2015**, *27*, 197–203. [CrossRef]

71. Silva, C.S.P.; Silva-Stenico, M.E.; Fiore, M.F.; de Castro, H.F.; Da Rós, P.C.M. Optimization of the cultivation conditions for *Synechococcus* sp. PCC7942 (cyanobacterium) to be used as feedstock for biodiesel production. *Algal Res.* **2014**, *3*, 1–7. [CrossRef]

72. Ooijkaas, L.P.; Wilkinson, E.C.; Tramper, J.; Buitelaar, R.M. Medium optimization for spore production of coniothyrium minitans using statistically-based experimental designs. *Biotechnol. Bioeng.* **1999**, *64*, 92–100. [CrossRef]

73. Wu, G.; Wu, Q.; Shen, Z. Accumulation of poly-β-hydroxybutyrate in cyanobacterium *Synechocystis* sp. PCC6803. *Bioresour. Technol.* **2001**, *76*, 85–90. [CrossRef]

74. Carpine, R.; Olivieri, G.; Hellingwerf, K.; Pollio, A.; Pinto, G.; Marzocchella, A. Poly-β-hydroxybutyrate (PHB) Production by Cyanobacteria. *New Biotechnol.* **2016**, *33*, S19–S20. [CrossRef]

75. Kaewbai-Ngam, A.; Incharoensakdi, A.; Monshupanee, T. Increased accumulation of polyhydroxybutyrate in divergent cyanobacteria under nutrient-deprived photoautotrophy: An efficient conversion of solar energy and carbon dioxide to polyhydroxybutyrate by *Calothrix scytonemicola* TISTR 8095. *Bioresour. Technol.* **2016**, *212*, 342–347. [CrossRef] [PubMed]

76. Sharma, L.; Mallick, N. Accumulation of poly-beta-hydroxybutyrate in Nostoc muscorum: Regulation by pH, light-dark cycles, N and P status and carbon sources. *Bioresour. Technol.* **2005**, *96*, 1304–1310. [CrossRef] [PubMed]

77. Coelho, V.C.; da Silva, C.K.; Terra, A.L.; Costa, J.A.V.; de Morais, M.G. Polyhydroxybutyrate production by *Spirulina* sp. LEB 18 grown under different nutrient concentrations. *Afr. J. Microbiol. Res.* **2015**, *9*, 1586–1594.

78. Gopi, K.; Balaji, S.; Muthuvelan, B. Isolation Purification and Screening of Biodegradable Polymer PHB Producing Cyanobacteria from Marine and Fresh Water Resources. *Iran. J. Energy Environ.* **2014**, *5*, 94–100. [CrossRef]

79. Markou, G.; Vandamme, D.; Muylaert, K. Microalgal and cyanobacterial cultivation: The supply of nutrients. *Water Res.* **2014**, *65*, 186–202. [CrossRef] [PubMed]

80. Samantaray, S.; Nayak, K.J.; Mallick, N. Wastewater utilization for poly-β-hydroxybutyrate production by the cyanobacterium *Aulosira fertilissima* in a recirculatory aquaculture system. *Appl. Environ. Microbiol.* **2011**, *77*, 8735–8743. [CrossRef] [PubMed]

81. Bhati, R.; Mallick, N. Carbon dioxide and poultry waste utilization for production of polyhydroxyalkanoate biopolymers by *Nostoc muscorum* Agardh: A sustainable approach. *J. Appl. Phycol.* **2016**, *28*, 161–168. [CrossRef]

82. Chaiklahan, R.; Chirasuwan, N.; Siangdung, W.; Paithoonrangsarid, K.; Bunnag, B. Cultivation of Spirulina platensis using pig wastewater in a semi-continuous process. *J. Microbiol. Biotechnol.* **2010**, *20*, 609–614. [CrossRef] [PubMed]

83. Phang, S.M.; Miah, M.S.; Yeoh, B.G.; Hashim, M.A. Spirulina cultivation in digested sago starch factory wastewater. *J. Appl. Phycol.* **2000**, *12*, 395–400. [CrossRef]

84. Olguín, E.J.; Galicia, S.; Mercado, G.; Pérez, T. Annual productivity of Spirulina (Arthrospira) and nutrient removal in a pig wastewater recycling process under tropical conditions. *J. Appl. Phycol.* **2003**, *15*, 249–257. [CrossRef]

85. Kourmentza, C.; Plácido, J.; Venetsaneas, N.; Burniol-Figols, A.; Varrone, C.; Gavala, H.N.; Reis, M.A.M. Recent Advances and Challenges towards Sustainable Polyhydroxyalkanoate (PHA) Production. *Bioengineering* **2017**, *4*, 55. [CrossRef] [PubMed]

86. Fradinho, J.C.; Reis, M.A.M.; Oehmen, A. Beyond feast and famine: Selecting a PHA accumulating photosynthetic mixed culture in a permanent feast regime. *Water Res.* **2016**, *105*, 421–428. [CrossRef] [PubMed]

87. Fradinho, J.C.; Oehmen, A.; Reis, M.A. Photosynthetic mixed culture polyhydroxyalkanoate (PHA) production from individual and mixed volatile fatty acids (VFAs): Substrate preferences and co-substrate uptake. *J. Biotechnol.* **2014**, *185*, 19–27. [CrossRef] [PubMed]

88. Arias, D.M.; Fradinho, J.C.; Uggetti, E.; García, J.; Oehmen, A.; Reis, M.A.M. Polymer accumulation in mixed cyanobacterial cultures selected under the feast and famine strategy. *Algal Res.* **2018**, *33*, 99–108. [CrossRef]

89. Fradinho, J.C.; Domingos, J.M.; Carvalho, G.; Oehmen, A.; Reis, M.A. Polyhydroxyalkanoates production by a mixed photosynthetic consortium of bacteria and algae. *Bioresour. Technol.* **2013**, *132*, 146–153. [CrossRef] [PubMed]

90. Stal, L.J.; Moezelaar, R. Fermentation in cyanobacteria1. *FEMS Microbiol. Rev.* **1997**, *21*, 179–211. [CrossRef]

91. Stal, L.; Krumbein, W. Metabolism of cyanobacteria in anaerobic marine sediments. In Proceedings of the 2. Colloque International de Bacteriologie Marine, Brest, France, 1–4 October 1984; pp. 1–5.

92. Rahman, A.; Putman, R.J.; Inan, K.; Sal, F.A.; Sathish, A.; Smith, T.; Nielsen, C.; Sims, R.C.; Miller, C.D. Polyhydroxybutyrate production using a wastewater microalgae based media. *Algal Res.* **2015**, *8*, 95–98. [CrossRef]

93. Sharma, L.; Kumar Singh, A.; Panda, B.; Mallick, N. Process optimization for poly-beta-hydroxybutyrate production in a nitrogen fixing cyanobacterium, *Nostoc muscorum* using response surface methodology. *Bioresour. Technol.* **2007**, *98*, 987–993. [CrossRef] [PubMed]

94. Mallick, N.; Sharma, L.; Kumar Singh, A. Poly-beta-hydroxybutyrate accumulation in *Nostoc muscorum*: Effects of metabolic inhibitors. *J. Plant Physiol.* **2007**, *164*, 312–317. [CrossRef] [PubMed]

95. Mallick, N.; Gupta, S.; Panda, B.; Sen, R. Process optimization for poly(3-hydroxybutyrate-co-3-hydroxyvalerate) co-polymer production by *Nostoc muscorum*. *Biochem. Eng. J.* **2007**, *37*, 125–130. [CrossRef]

96. Bhati, R.; Samantaray, S.; Sharma, L.; Mallick, N. Poly-β-hydroxybutyrate accumulation in cyanobacteria under photoautotrophy. *Biotechnol. J.* **2010**, *5*, 1181–1185. [CrossRef] [PubMed]

97. Panda, B.; Sharma, L.; Mallick, N. Poly-β-hydroxybutyrate accumulation in *Nostoc muscorum* and *Spirulina platensis* under phosphate limitation. *J. Plant Physiol.* **2005**, *162*, 1376–1379. [CrossRef] [PubMed]

98. Google Scholar. Earth System Research Laboratory. 2017. Available online: https://esrl.noaa.gov/ (accessed on 15 April 2017).

99. Eberly, J.O.; Ely, R.L. Photosynthetic accumulation of carbon storage compounds under CO_2 enrichment by the thermophilic cyanobacterium *Thermosynechococcus elongatus*. *J. Ind. Microbiol. Biotechnol.* **2012**, *39*, 843–850. [CrossRef] [PubMed]

100. Lee, S.Y. Bacterial polyhydroxyalkanoates. *Biotechnol. Bioeng.* **1996**, *49*, 1–14. [CrossRef]

101. Lee, S.Y. Plastic bacteria? Progress and prospects for polyhydroxyalkanoate production in bacteria. *Trends Biotechnol.* **1996**, *14*, 431–438. [CrossRef]

102. Koller, M.; Khosravi-Darani, K.; Braunegg, G. *Advanced Photobioreactor Systems for the Efficient Cultivation of Cyanobacteria, in Photobioreactors Advancements, Applications and Research*; Tsang, Y.F., Ed.; Nova Science Publishers: New York, NY, USA, 2017; pp. 35–90.

103. Arias, D.M.; Uggetti, E.; García-Galán, M.J.; García, J. Production of polyhydroxybutyrates and carbohydrates in a mixed cyanobacterial culture: Effect of nutrients limitation and photoperiods. *New Biotechnol.* **2018**, *42*, 1–11. [CrossRef] [PubMed]

104. Riis, V.; Mai, W. Gas chromatographic determination of poly-?-hydroxybutyric acid in microbial biomass after hydrochloric acid propanolysis. *J. Chromatogr.* **1988**, *445*, 285–289. [CrossRef]

105. Karr, D.B.; Waters, J.K.; Emerich, D.W. Analysis of Poly-beta-Hydroxybutyrate in *Rhizobium japonicum* Bacteroids by Ion-Exclusion High-Pressure Liquid Chromatography and UV Detection. *Appl. Environ. Microbiol.* **1983**, *46*, 1339–1344. [PubMed]

106. Koller, M.; Rodríguez-Contreras, A. Techniques for tracing PHA-producing organisms and for qualitative and quantitative analysis of intra- and extracellular PHA. *Eng. Life Sci.* **2015**, *15*, 558–581. [CrossRef]

107. Kansiz, M.; Billman-Jacobe, H.; McNaughton, D. Quantitative determination of the biodegradable polymer Poly(beta-hydroxybutyrate) in a recombinant *Escherichia coli* strain by use of mid-infrared spectroscopy and multivariative statistics. *Appl. Environ. Microbiol.* **2000**, *66*, 3415–3420. [CrossRef] [PubMed]

108. Degelau, A.; Scheper, T.; Bailey, J.E.; Guske, C. Fluorometric measurement of poly-β hydroxybutyrate in Alcaligenes eutrophus by flow cytometry and spectrofluorometry. *Appl. Microbiol. Biotechnol.* **1995**, *42*, 653–657. [CrossRef]

109. Stenholm, H.; Song, S.; Eriksen, N.T.; Iversen, J.J.L. Indirect Estimation of Poly-β-Hydroxybutyric Acid by Cell Carbon Analysis. *Biotechnol. Tech.* **1998**, *12*, 451–454. [CrossRef]

110. Naumann, D.; Helm, D.; Labischinski, H. Microbiological characterizations by FT-IR spectroscopy. *Nature* **1991**, *351*, 81–82. [CrossRef] [PubMed]

111. Schuster, K.C.; Mertens, F.; Gapes, J.R. FTIR spectroscopy applied to bacterial cells as a novel method for monitoring complex biotechnological processes. *Vib. Spectrosc.* **1999**, *19*, 467–477. [CrossRef]

112. Randriamahefa, S.; Renard, E.; Guérin, P.; Langlois, V. Fourier transform infrared spectroscopy for screening and quantifying production of PHAs by Pseudomonas grown on sodium octanoate. *Biomacromolecules* **2003**, *4*, 1092–1097. [CrossRef] [PubMed]

113. Jarute, G.; Kainz, A.; Schroll, G.; Baena, J.R.; Lendl, B. On-line determination of the intracellular poly(beta-hydroxybutyric acid) content in transformed *Escherichia coli* and glucose during PHB production using stopped-flow attenuated total reflection FT-IR spectrometry. *Anal. Chem.* **2004**, *76*, 6353–6358. [CrossRef] [PubMed]

114. Song, J.H.; Murphy, R.J.; Narayan, R.; Davies, G.B.H. Biodegradable and compostable alternatives to conventional plastics. *Philos. Trans. R. Soc. B Biol. Sci.* **2009**, *364*, 2127–2139. [CrossRef] [PubMed]

115. Gerngross, T.U. Can biotechnology move us toward a sustainable society? *Nat. Biotechnol.* **1999**, *17*, 541. [CrossRef] [PubMed]

116. Choi, J.-I.; Lee, Y.S. Process analysis and economic evaluation for poly (3-hydroxybutyrate) production by fermentation. *Bioprocess Eng.* **1997**, *17*, 335–342. [CrossRef]

117. Acién, F.G.; Fernández, J.M.; Magán, J.J.; Molina, E. Production cost of a real microalgae production plant and strategies to reduce it. *Biotechnol. Adv.* **2012**, *30*, 1344–1353. [CrossRef] [PubMed]

118. Koller, M. Advances in Polyhydroxyalkanoate (PHA) Production. *Bioengineering* **2017**, *4*, 88. [CrossRef] [PubMed]

5

Utilization of Sugarcane Bagasse by *Halogeometricum borinquense* Strain E3 for Biosynthesis of Poly(3-hydroxybutyrate-*co*-3-hydroxyvalerate)

Bhakti B. Salgaonkar * and Judith M. Bragança

Department of Biological Sciences, Birla Institute of Technology and Science Pilani, K K Birla, Goa Campus, NH-17B, Zuarinagar, Goa 403 726, India; judith@goa.bits-pilani.ac.in
* Correspondence: salgaonkarbhakti@gmail.com

Academic Editor: Martin Koller

Abstract: Sugarcane bagasse (SCB), one of the major lignocellulosic agro-industrial waste products, was used as a substrate for biosynthesis of polyhydroxyalkanoates (PHA) by halophilic archaea. Among the various wild-type halophilic archaeal strains screened, *Halogeometricum borinquense* strain E3 showed better growth and PHA accumulation as compared to *Haloferaxvolcanii* strain BBK2, *Haloarcula japonica* strain BS2, and *Halococcus salifodinae* strain BK6. Growth kinetics and bioprocess parameters revealed the maximum PHA accumulated by strain E3 to be 50.4 ± 0.1 and 45.7 ± 0.19 (%) with specific productivity (qp) of 3.0 and 2.7 (mg/g/h) using NaCl synthetic medium supplemented with 25% and 50% SCB hydrolysate, respectively. PHAs synthesized by strain E3 were recovered in chloroform using a Soxhlet apparatus. Characterization of the polymer using crotonic acid assay, X-ray diffraction (XRD), differential scanning calorimeter (DSC), Fourier transform infrared (FT-IR), and proton nuclear magnetic resonance (^{1}H-NMR) spectroscopy analysis revealed the polymer obtained from SCB hydrolysate to be a co-polymer of poly(3-hydroxybutyrate-*co*-3-hydroxyvalerate) [P(3HB-*co*-3HV)] comprising of 13.29 mol % 3HV units.

Keywords: archaea; halophiles; sugarcane bagasse; polyhydroxyalkanoates; *Halogeometricum borinquense* strain E3; Soxhlet extractor

1. Introduction

Conventional plastics obtained from non-renewable petrochemical resources are creating environmental havoc due to their non-degradable nature. To solve this problem, various bio-based materials derived from renewable resources have been explored as a replacement for conventional plastics. These materials could be (i) directly extracted from biomass as polysaccharides, lignocelluloses, proteins, and lipids; (ii) chemically synthesized, e.g., by in vitro polymerization of bio-derived monomers such as lactate to produce poly(lactic acid) (PLA); or (iii) biologically synthesized by microorganisms, in vivo polymerization of hydroxyalkanoic acid (HA) units to polyhydroxyalkanoates (PHAs) [1]. PHAs are synthesized and accumulated as inclusions by microorganisms when the available nitrogen/phosphorus gets depleted while carbon is in excess. PHAs are synthesized either in the inner membrane, on a central scaffold, or in the cytoplasm of cells and aggregated in the form of globular, water-insoluble granules [2,3].

High production costs, downstream processing, and low yields are the major hurdles for the commercial production and application of PHA, making microbially synthesized PHA 5–10 times more expensive than petroleum-derived polymers [4]. Carbon sources/substrates represent half of the PHA fermentation cost [5–7]. Various strategies such as replacing commercial substrates with inexpensive renewable agro-industrial waste, finding novel high PHA-accumulating microorganisms or microbial strain improvements, and reducing the cost of PHA recovery/downstream process can make the overall fermentation process more cost-effective [4].

Sugarcane (*Saccharum officinarum*) is the world's largest cash crop, with Brazil being the leading producer, followed by India, and is exploited for the production of sugar, jaggery, ethanol, molasses, alcoholic beverages (rum), soda, etc. [8]. Sugarcane bagasse (SCB) is the leftover, fibrous residue of sugarcane stalk after the extraction of juice and is a major lignocellulosic, inexpensive byproduct of the sugarcane industry [9]. SCB needs special attention for its management, and is primarily used as a source of energy for electricity/biogas production. It also serves as a raw material for fermentation processes in the production of various products such as enzymes (cellulose, lipase, xylanase, inulinase, and amylase), animal feed (single cell protein), amino acids, organic acids, bioethanol, bioplastics, etc. [10].

SCB is comprised of cellulose (46%), hemicelluloses (27%), lignin (23%), and ash (4%) [9]. Cellulose is a homopolysaccharide comprised of a linear chain of $\beta(1\rightarrow4)$ linked D-glucose units, whereas hemicellulose is a heteropolysaccharide consisting of many different sugar monomers such as xylan (consisting units of pentose sugar, xylose), glucuronoxylan (consisting units of glucuronic acid and xylose), arabinoxylan (consisting of copolymers of two pentose sugars, arabinose and xylose), glucomannan (consisting of D-mannose and D-glucose), and xyloglucan (consisting of units of glucose and xylose). Degradation of these polymers would yield various sugars that could serve as a substrate for PHA production. The fibrous nature of SCB makes its microbial degradation slower and more difficult and hence limits its utilization. To overcome this, pre-treatment of SCB is usually carried out for improved substrate availability and to speed up the fermentation process. Acid hydrolysis is a fast and simple method, mostly performed to release sugars from the SCB that can be readily used by microorganisms, rather than feeding solid bagasse, which is time-consuming and interferes with the downstream processing [11].

Much insight has been gained into PHA synthesis by using non-halophilic microorganisms as compared to their halophilic counterparts. PHA production by halophiles has been studied among members of the families *Halobacteriaceae* (Domain *Archaea*) and *Halomonadaceae* (Domain *Bacteria*). Kirk and Ginzburg (in 1972) first documented the occurrence of PHA granules in halophilic archaea [12]. Subsequent research on members of the family *Halobacteriaceae* revealed significant PHA synthesis by the following genera: *Haloferax* (*Hfx. mediterranei*), *Haloarcula* (*Har. marismortui*), *Halobacterium* (*Hbt. noricense*), *Haloterrigena* (*Htg. hispanica*), *Halogeometricum* (*Hgm. borinquense*), *Halococcus* (*Hcc. dombrowskii*), *Natrinema* (*Natrinema pallidum*); *Halobiforma* (*Hbf. haloterrestris*), and *Halopiger* (*Hpg. aswanensis*) [13–20]. Among the halophilic bacteria, members of the genus *Halomonas* (*H. boliviensis*) are known to synthesize a large amount of PHA from a variety of substrates such as glucose, maltose, starch hydrolysate, etc. [21,22].

Overall, there are few reports on PHA production using agro-industrial waste/cheap substrates by employing halophilic microorganisms. *Haloferax mediterranei* is the most widely studied and is reported to produce a copolymer of poly(3-hydroxybutyrate-*co*-3-hydroxyvalerate) [P(3HB-*co*-3HV)] using various renewable agro-industrial waste products like extruded corn starch, rice bran, wheat bran, hydrolyzed whey, waste stillage from the rice-based ethanol industry, vinasse, etc. [23–26]. Pramanik et al. and Taran reported the synthesis of a homopolymer of hydroxybutyrate (PHB) by *Haloarcula marismortui* MTCC 1596 and *Haloarcula* sp. IRU1 by utilization of vinasse waste from the ethanol industry and petrochemical wastewater, respectively [27,28].

To date, to the best of our knowledge, there has been no report on the utilization of sugarcane bagasse for PHA production by members of the domain Archaea, especially the genus *Halogeometricum*. Therefore, in the present study, the ability of halophilic archaeal isolates to accumulate PHA from SCB hydrolysate was examined. The growth kinetics and bioprocess parameters during growth and PHA production by *Hgm. borinquense* strain E3 were examined and the polymer synthesized was characterized using physico-chemical analysis techniques.

2. Materials and Methods

2.1. HalophilicArchaeal Strains and Media Used

Four halophilic archaeal isolates (GenBank/DDBJ database accession number), *Halococcus salifodinae* strain BK6 (AB588757), *Haloferax volcanii* strain BBK2 (AB588756), and *Haloarcula japonica* strain BS2 (HQ455798), isolated from solar salterns of Ribandar in Goa, India, and *Hgm. borinquense* strain E3 (AB904833), obtained from Marakkanam in Tamil Nadu, India, were used in the present study [29,30]. The cultures were maintained on complex media (Table 1), i.e., NTYE (NaCl Tryptone Yeast Extract), NT (NaCl Tri-sodium citrate), EHM (Extremely Halophilic Medium). The NSM (NaCl Synthetic Medium) with a varying concentration of SCB hydrolysate as per the requirements was used as a production medium (Table 1).

Table 1. Composition of maintenance and production media used in the study.

Ingredients (g/L)	Maintenance Media			Production Medium
	NTYE	NT	EHM	NSM
NaCl	250.0	250.0	250.0	200.0
$MgSO_4 \cdot 7H_2O$	20.0	20.0	20.0	-
$MgCl_2 \cdot 6H_2O$	-	-	-	13.0
KCl	5.0	2.0	2.0	4.0
Tryptone	5.0	-	-	-
Yeast Extract	3.0	10.0	10.0	1.0
Tri-Sodium citrate	-	3.0	-	-
$CaCl_2 \cdot 2H_2O$	-	-	0.36	1.0
NaBr			0.23	-
$NaHCO_3$	-	-	0.06	0.2
NH_4Cl	-	-	-	0.2
KH_2PO_4	-	-	-	0.5
Peptone	-	-	5.0	-
$FeCl_3 \cdot 6H_2O$	-	-	Trace	0.005

NTYE: NaCl Tryptone Yeast Extract; NT: NaCl Tri-sodium citrate; EHM: Extremely Halophilic Medium; NSM: NaCl Synthetic Medium. NSM with various concentration of SCB was used as production media. Agar 1.8% (*w/v*) was use as solidifying agent. The pH of the medium was adjusted to 7.0–7.2 using 1M NaOH.

2.2. Procurement, Processing, and Hydrolysis of Sugarcane Bagasse

Sugarcane bagasse (SCB) was collected from a local sugarcane juice extractor, from Vasco-da-Gama, Goa, India. It was dried under sunlight for 3–5 days, cut into small ~5–10 cm pieces, followed by pulverization to a fine powder using a blender. The powdered form of the waste was subjected to dilute acid hydrolysis. Briefly, 5 gm of the SCB powder was added to 100 mL of 0.75% (*v/v*) sulfuric acid in water. The mixture was heated at 100 °C for 1 h under reflux using an allihn condenser. The hydrolysate was filtered using non-absorbent cotton to separate the solid residue from the liquid hydrolysate. The liquid hydrolysate was neutralized to pH 7.0–7.4 using NaOH, followed by sterilization at 121 °C for 10 min and storage at 4 °C [11].

2.3. Characterization of the SCB

The SCB powder was characterized for the following physical and chemical parameters. Total solids (TS) and volatile solids (VS) were estimated according to the American Public Health Association (APHA) [31]. The chemical oxygen demand (COD) was determined as described by Raposo et al. [32]. The carbon (C), hydrogen (H), nitrogen (N), and sulfur (S) contents of the SCB were determined using a CHNS Analyzer (Elementar, Rhine Main area near Frankfurt, Germany). The total carbohydrate content of the SCB hydrolysate was estimated by the phenol sulphuric acid method, as described by Dubois et al. [33]. Total Kjeldahl nitrogen (TKN) was determined as described by

Labconco [34]. All the physical and chemical characterization was performed in triplicate to determine means and standard deviations.

2.4. Screening of the Halophilic Archaeal Isolates for PHA Accumulation using SCB Hydrolysate

Halophilic archaeal isolates were screened for the production of PHA using NSM (Table 1) and Nile Red stain. Briefly, the NSM plates were prepared by adding 1.8% agar (w/v) to the medium followed by autoclaving; while still molten, the medium was supplemented with varying concentrations (0.5%–30% v/v) of SCB hydrolysate along with 50 µL of Nile Red stain [stock of 0.01% (w/v) Nile red in DMSO] such that the final concentration was 0.5 µg/mL medium. Twenty microliters of log phase (three-day-old) halophilic archaeal cultures were spot inoculated on the agar plates and incubated at 37 °C for 6–7 days. The plates were exposed to ultraviolet (UV) light using gel documentation system (BIO-RAD Laboratories, Hercules, CA, USA) and the emitted fluorescence from the culture was quantified using TotalLab Quant software [16,35].

2.5. Selection and Further Study of the Best PHA Producer Strain

Based on it having the best growth and fluorescence on NSM supplemented with SCB hydrolysate, the halophilic archaeon *Hgm. borinquense* strain E3 was selected for further study. Preliminary screening indicted the strain E3 to grow up to 30% (v/v) of SCB hydrolysate. Therefore, the concentration of SCB hydrolysate that inhibited the growth of strain E3 was determined by growing the culture on NSM agar plates containing higher concentrations of the SCB hydrolysate, i.e., 50%, 75%, and 100%. NSM with 100% SCB hydrolysate was prepared by dissolving the medium ingredients in the directly SCB hydrolysate. Based on the growth observed on NSM agar plates, growth of the strain E3 was further recorded in NSM broth with that particular concentration of SCB hydrolysate.

2.6. The Growth Kinetics and PHA Quantification

Growth and intracellular PHA content for *Hgm. borinquense* strain E3 were determined as follows. An actively growing mid-log phase culture (3–4 days) of *Hgm. borinquense* strain E3 grown in NGSM (<u>N</u>aCl <u>G</u>lucose <u>S</u>ynthetic <u>M</u>edium) containing 0.2% glucose was used as a starter culture. One percent of the starter culture was inoculated in NSM (<u>N</u>aCl <u>S</u>ynthetic <u>M</u>edium) containing 25% and 50% (v/v) SCB hydrolysate. The flasks were maintained at 37 °C, 110 rpm on a rotary shaker (Skylab Instruments, Mumbai, India). At regular intervals of 24 h, aliquots of the culture broth were aseptically withdrawn and the following parameters were monitored: (i) culture growth was monitored by recording the absorbance at 600 nm using a UV-visible spectrophotometer (UV-2450, Shimadzu, Tokyo, Japan) with the respective medium as blank, (ii) Cell Dry Mass (CDM) was determined by centrifuging 2 mL of the culture broth at 12,000 g for 15 min, washing the pellet with distilled water, and recentrifuging it at 12,000 g for 15 min, followed by drying at 60 °C until a constant weight was obtained. Since the SCB hydrolysate had some particle participation, the dry mass of the plain medium (without culture) was taken and subtracted from the culture CDM so as to avoid error. (iii) Total carbohydrates were determined colorimetrically according to Dubios et al. and compared with the standard curve [33]; (iv) the pH of the medium was monitored using a pH meter; and (v) polymer quantification was done by converting PHA to crotonic acid using concentrated sulfuric acid. The absorbance was recorded at 235 nm using a UV-visible spectrophotometer (UV-2450, Shimadzu, Tokyo, Japan) and compared with the standard curve for PHB [36]. All the experiments were performed in triplicate to determine means and standard deviations.

2.7. Extraction of the PHA

The PHA extraction from the biomass was done as described by Sánchez et al. with slight modifications [37]. Briefly, *Hgm. borinquense* strain E3 was grown in NSM containing 25% SCB hydrolysate for six days. The cells were harvested by centrifuging at 12,000 g for 10 min using Eppendorf centrifuge 5810R (Hamburg, Germany). The cell pellet was dried for 12 h at 60 °C in an oven (Bio Technics, Mumbai, India). The dried cells were ground using mortar and pestle and extracted for 8–10 h at 60 °C in a soxhlet extractor using chloroform. Up to 95% of the chloroform was recollected using rotary evaporator (Rotavapor R-210, Büchi, Switzerland) and the remaining 5% of the chloroform-containing polymer was poured into a clean glass Petri dish and kept undisturbed during the total evaporation of the chloroform to give a uniform polymer film.

2.8. Characterization of the PHA

Characterization of the polymer was done using UV-visible spectrophotometry (crotonic acid assay), X-ray diffraction (XRD) analysis, differential scanning calorimeter (DSC) analysis, Fourier transform infrared (FT-IR) spectroscopy, and proton nuclear magnetic resonance (^{1}H-NMR) spectroscopy, as described in detail by Salgaonkar and Bragança [30].

3. Results and Discussion

3.1. Sugarcane Bagasse (SCB)

The SCB used in the present study appeared greenish-brown with a sweet odor and upon pulverization had total solids (TS) of 94.3 ± 0.14%, volatile solids (VS) of 92.7 ± 0.14%, and COD of 1.18 ± 0.05 g/Kg. The carbon (C), hydrogen (H), nitrogen (N), and sulfur (S) content of the SCB was found to be C = 43.45 ± 0.12%, H = 6.0 ± 0.27%, N = 0.26 ± 0.02%, and S = 0.32 ± 0.12%. Hydrolysis of SCB using dilute sulfuric acid yielded total carbohydrates of 12.64 ± 0.7 g/L and total Kjeldahl nitrogen (TKN) of 0.7 g/L (Table S1).

3.2. Screening for PHA using SCB Hydrolysate

All four halophilic archaeal isolates were able to grow on NSM plates with Nile Red dye supplemented with SCB hydrolysate as substrate. Upon exposure of the plates to UV light, only three cultures, *Hfx. volcanii* strain BBK2, *Har. japonica* strain BS2, and *Hgm. borinquense* strain E3, showed bright orange fluorescence, indicating the accumulation of PHA (Figure S1). *Hcc. salifodinae* strain BK6 showed weak growth but failed to show any fluorescence. The intensity of fluorescence exhibited by the cultures varied, in the order *Hgm. borinquense* strain E3 > *Har. japonica* strain BS2 > *Hfx. volcanii* strain BBK2. *Hgm. borinquense* strain E3 grew faster and showed better fluorescence, which directly correlates withthe amount of polymer accumulated over a range of SCB concentrations (Figure S1). Preliminary work on *Hgm. borinquense* strain E3 proved it to be the best accumulator of PHA in an NGM medium supplemented with 2% glucose [30].

3.3. Optimization of SCB Hydrolysate Concentration

Figure 1 represents the growth of *Hgm. borinquense* strain E3 on NSM agar plates and broth containing various concentrations of SCB hydrolysate. The SCB hydrolysate optimization studies revealed that the culture could tolerate and grew up to 75% (v/v) SCB hydrolysate on NSM agar plates (Figure 1A). Interestingly, when grown in NSM broth (Figure 1B), the culture grew only up to 50% (v/v) SCB hydrolysate and failed to grow at higher concentrations.

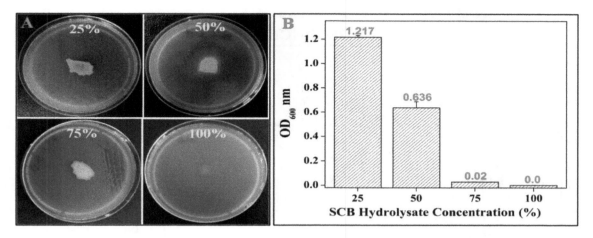

Figure 1. Growth of *Hgm. borinquense* strain E3 on (**A**) NSM agar plates and (**B**) NSM broth, containing various concentrations of SCB hydrolysate.

3.4. Growth Profile of Hgm. borinquense Strain E3 and Polymer Quantification Study

The time course of growth of *Hgm. borinquense* strain E3 in NSM containing 25% and 50% SCB hydrolysate is presented in Figure 2. Table 2 gives a comparison of the various kinetics and bioprocess parameters used to determine growth and PHA production by *Hgm. borinquense* strain E3 using SCB hydrolysate. In the presence of 25% SCB hydrolysate, isolate E3 showed a 48-h lag and reached 3.17 ± 0.19 g/L of maximum cell dry mass (CDM), containing 1.6 ± 0.09 g/L of PHA. In the presence of 50% SCB hydrolysate, isolate E3 exhibited a longer lag phase of 96 h and reached 4.15 ± 0.7 g/L of maximum CDM, containing 1.9 ± 0.3 g/L of PHA. The lag phase of the culture depends on the definite environmental conditions. This prolonged lag phase could be reduced by increasing the inoculum size and decreasing the effect of culture conditions on the growth, which can be achieved by acclimatizing the starter culture to the ingredients of production medium by pre-culturing the isolate in the presence of the respective substrates concentration. The substantial quantity of carbohydrates, i.e., 12.64 ± 0.7 g/L in the SCB hydrolysate, served as the basic essential carbon source required for the growth and synthesis of PHA by *Hgm. borinquense* strain E3. The rapid consumption of the total carbohydrates by the isolate was observed as the growth progressed and a steady drop in the pH of the production medium from 7.2 to 5.0 was also noted.

To the best of our knowledge, haloarchaea have not been explored for their potential to utilize SCB hydrolysate for the synthesis of PHA. Bioprocess parameters (Table 2) revealed the maximum PHA accumulation by *Hgm. borinquense* strain E3 to be 50.4 ± 0.1 and 45.7 ± 0.19 (%) with specific productivity (qp) of 3.0 and 2.7 (mg/g/h) using NaCl synthetic medium supplemented with 25% and 50% SCB hydrolysate, respectively. A recent investigation by Pramanik et al. reported the potential of haloarchaeon *Har. marismortui* MTCC 1596 to produce 23 ± 1.0 and 30 ± 0.3 (%) P(3HB) with specific productivity (qp) of 1.21 and 1.39 (mg/g/h) using a nutrient-deficient medium (NDM) supplemented with 10% and 100% raw and treated vinasse, respectively [27]. The specific productivity (qp) attained by *Hgm. borinquense* strain E3 using SCB hydrolysate was higher compared to *Har. marismortui* MTCC 1596 grown in the presence of vinasse. Silva et al. (2004) investigated the synthesis of P(3HB) by *Burkholderia sacchari* IPT 101 and *Burkholderia cepacia* IPT 048 by feeding the cultures with SCB hydrolysate as a carbon source. It was noted that both the cultures reached 4.4 g/L of dry biomass, containing 62% and 53% of P(3HB) in the case of *B. sacchari* IPT 101 and *B. cepacia* IPT 048, respectively. The specific production rate and yield coefficient of the PHA were 0.11 (g/L/h) and 0.39 (g/g) for *B. sacchari* IPT 101, whereas it was 0.09 (g/L/h) and 0.29 (g/g) for *B. cepacia* IPT 048 [7].

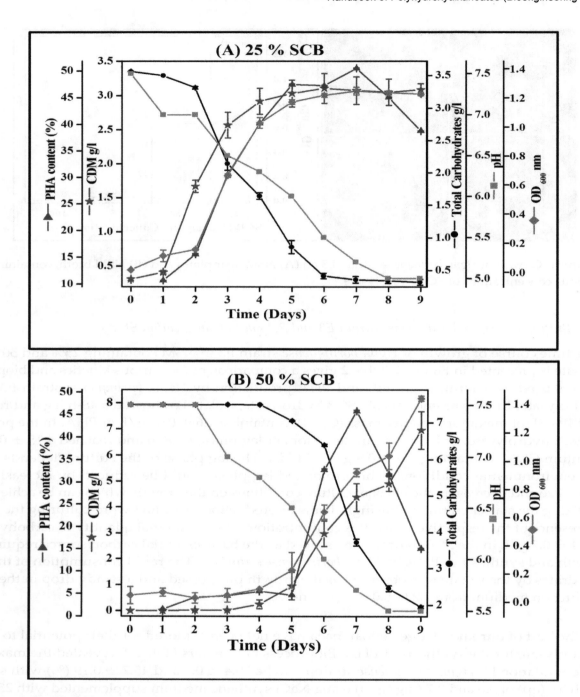

Figure 2. Growth profile and PHA production by *Hgm. borinquense* strain E3 in NSM containing (**A**) 25% and (**B**) 50% SCB hydrolysate.

Attempts have been made to reduce the fermentation cost of PHA by employing various haloarchaeal strains and examining their ability to utilize inexpensive substrates. Danis et al. showed the ability of *Natrinema pallidum*1KYS1 to produce 0.075, 0.055, 0.091, 0.039, 0.077, and 0.464 g/L of polymer by utilizing various waste products such as corn starch, sucrose, whey, melon, apple, and tomato as carbon substrates [18]. Pramanik et al. studied the ability of *Har. marismortui* to utilize 10% raw vinasse and 100% pre-treated vinasse to produce 2.8 g/L and 4.5 g/L of PHB [27]. Similarly, Bhattacharyya et al. employed *Hfx. mediterranei* to produce 19.7 g/L and 17.4 g/L from 25% and 50% pre-treated vinasse, respectively [26]. Also, 24.2 g/L PHBV biosynthesis was observed in *Hfx. mediterranei* with extruded cornstarch [23].

Utilization of Sugarcane Bagasse by Halogeometricum borinquense Strain E3 for Biosynthesis...

Table 2. Kinetic and bioprocess parameters during growth and PHA production by *Hgm. borinquense* strain E3 (present study) in comparison with *Har. marismortui* MTCC 1596 (literature).

Halophilic Archaeal Strain	Production Medium	Lag (h)	CDM (g/L)	PHA (g/L)	PHA Content (%)	μmax (1/h)	qp[a] (mg/g/h)	$Y_{P/S}$[b]	Vol. Productivity[c] (g/L/h)	Reference
Hgm. borinquense strain E3	NSM25% SCB	48	3.17 ± 0.19	1.6 ± 0.09	50.4 ± 0.1	0.017	3.0	0.448	0.0095	Present study
	NSM50% SCB	96	4.15 ± 0.7	1.9 ± 0.3	45.7 ± 0.19	0.023	2.7	0.253	0.0113	
	NGSM2% Glucose	-	5.78 ± 0.4	4.25 ± 0.045	73.5 ± 0.045	ND	4.3	0.212	0.0252	[30]*
Har. marismortui MTCC 1596	NDM10% Raw vinasse	96 ± 12	12.0 ± 0.20	2.8 ± 0.2	23 ± 1.0	0.086	1.21	2.17	0.015	[27]
	NDM100% treated vinasse	144 ± 12	15.0 ± 0.35	4.5 ± 0.2	30 ± 0.3	0.128	1.39	0.77	0.020	

NSM: NaCl Synthetic Medium; NGSM: NaCl Glucose Synthetic Medium; NDM: Nutrient Deficient Medium; ND: not determined; μmax: maximum specific growth rate; CDM: Cell Dry Mass; PHA: Polyhydroxyalkanoate; *Hgm*: *Halogeometricum*; *Har*: *Haloarcula*; MTCC: Microbial Type Culture Collection; [a] Specific production rate of PHA (qp) = PHA (g/L) / time (h) × CDM (g/L) [38]; [b] Yield coefficient of PHA ($Y_{P/S}$) = PHA (g/L) / total organic carbon (g/L). The $Y_{P/S}$ was calculated based on total carbohydrate in NSM with 25% (3.5642 g/L) and 50% SCB hydrolysate (7.494 g/L) [26]; [c] Volume productivity of PHA = PHA (g/L) / time (h) [39]. Time of growth for *Hgm. borinquense* strain E3 using NSM supplemented with glucose and SCB hydrolysate was 168 h (7 days), whereas that for *Har. marismortui* MTCC 1596 using NDM and 10% raw and 100% treated vinasse was 192 h (8 days) and 216 h (9 days), respectively [27]; * Some of the bioprocess parameters are not mentioned in the reference.

3.5. Bench Scale Polymer Production and Extraction by Hgm. borinquense Strain E3

The polymer was extracted from the cell biomass as shown in Figure S2. The dried cells of *Hgm. borinquense* strain E3 (before Soxhlet extraction), when subjected to concentrated H_2SO_4 hydrolysis, showed a clear peak at 235 nm, which is indicative of crotonic acid, indicating the presence of PHA (Figure S3A) [36]. After Soxhlet extraction, no peak at 235 nm was observed, thus confirming the complete extraction of the polymer from the cell mass (Figure S3B).

3.6. Polymer Characterization

The polymer obtained using SCB hydrolysate (Figure S2J) appeared orange due to the co-extraction of carotenoid pigment along with some cellular lipids from the cells of *Hgm. borinquense* strain E3. These impurities were taken care of by treating the polymer with acetone for 10 min [30]. This polymer was characterized using a UV-visible spectrophotometer, XRD, DSC, FT-IR, and NMR analysis.

3.6.1. UV-Visible Spectrophotometric Analysis

Concentrated H_2SO_4 hydrolysis of the polymer obtained from SCB gave a characteristic peak at 235 nm of crotonic acid, which corresponded with the standard PHB (Sigma-Aldrich, St. Louis, MO, USA) (inset of Figure S3) and also with the copolymer P(3HB-*co*-3HV) synthesized by *Hgm. borinquense* strain E3 and *Hfx. mediterranei* "DSM 1411"when fed with substrates such as glucose and raw vinasse, respectively [26,30].

3.6.2. XRD Analysis

Figure 3 represents the X-ray diffraction (XRD) patterns of the polymer obtained from SCB hydrolysate in comparison with standard PHB (Sigma-Aldrich, St. Louis, MO, USA). The profile of polymer from SCB exhibited prominent peaks at 2θ = 13.8°, 17.4°, 21.8°, 25.8°, and 30.7°, corresponding to (020), (110), (101), (121), and (002) reflections of the orthorhombic crystalline lattice. Overall, the diffraction pattern was similar to that of the PHB. However, peak shifts as well as a decrease in peak intensity were observed in the case of the SCB polymer when compared with standard PHB. It was clearly observed that the diffraction peaks between 2θ = 0–25° were broadened and drastically decreased in intensity for the SCB polymer. A broadening of the peaks indicates a decrease in crystallinity, i.e., the amorphous nature of the polymer [40]. The crystallite size L (nm) was determined for the highest peaks using the Scherrer equation, which is defined as: L(nm) = 0.94 λ / BCosθ, where λ is the wavelength of the X-ray radiation, which is 1.542 Å(wavelength of the Cu); B is the full width at half maximum (FWHM) in radians; and θ is the Bragg angle [41,42]. The crystallite size for the highest peak (020) in the case of standard PHB was found to be 22.3 nm, which decreased drastically to 10.4 nm for peak (110) of the SCB polymer. For the diffraction peak of the (110) plane of SCB polymer, an increase in the FWHM was observed as compared to the standard PHB. This clearly indicates a decrease in the crystallite size, given that the peak width is inversely proportional to the crystallite size. The crystallite size matched more or less to the one reported for the copolymer P(3HB-*co*-3HV) synthesized by *Hgm. borinquense* strain E3 by utilizing glucose, which was found to be 12.17 nm for the (110) peak [30].

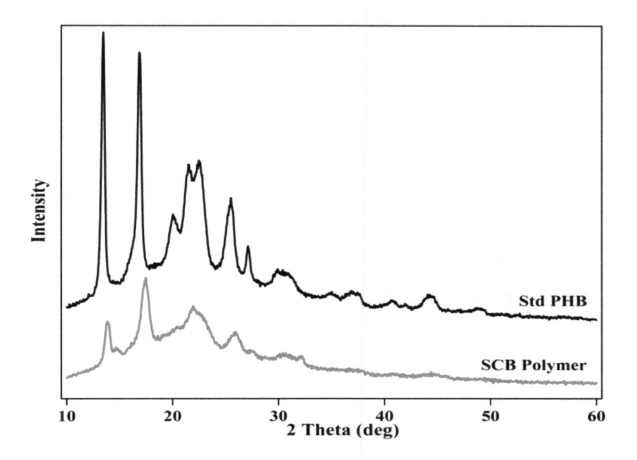

Figure 3. Comparison of X-ray diffraction patterns of standard PHB and polymer obtained from *Hgm. borinquense* strain E3 grown in NSM containing SCB hydrolysate.

3.6.3. DSC Analysis

The thermograms derived from differential scanning calorimetry (DSC) analysis for the polymer obtained using SCB hydrolysate and standard PHB (Sigma Aldrich, St. Louis, MO, USA) are represented in Figure 4. The polymer obtained from SCB hydrolysate exhibited two melting endotherms at Tm_1 = 136.5 °C and Tm_2 = 149.4 °C, whereas standard PHB displayed a single melting endotherm at Tm = 169.2 °C. The degradation temperature (Td) peaks for the SCB polymer and PHB were at 275.4 °C and 273.2 °C, respectively (Table 3). A recent study by Buzarovska et al. (2009) reported two melting endotherms for pure copolymer PHBV containing 13 mol% 3-hydroxyvalerate (3HV) [43]. The lower melting peak (Tm_1) at ~138 °C could be due to the melting of the primary formed crystallites, whereas the upper one (Tm_2) at 152 °C is mostly due to the recrystallization of species during the scan [43]. The existence of multiple melting peaks in a polymer indicates that the polymers have varying monomer units such as 3HB and 3HV units [44]. *Haloferax mediterranei* is known to produce PHA with multiple melting endotherms by utilizing various carbon substrates [45]. Chen et al. (2006) and Koller et al. (2007) showed the ability of *Haloferax mediterranei* ATCC 33500/DSM 1411 to utilize extruded cornstarch/whey sugars as carbon substrates for the production of copolymer P(3HB-*co*-3HV) containing 10.4 mol% and 6 mol% of 3-hydroxyvalerate (3HV), respectively. The P(3HB-*co*-3HV) produced by strain DSM 1411 showed two melting peaks at 150.8 °C (Tm_1) and 158.9 °C (Tm_2), whereas the melting endotherms for strain ATCC 33500 were at 129.1 °C (Tm_1) and 144.0 °C (Tm_2) [13,46].

Table 3. Comparison of the DSC data of a polymer synthesized by *Hgm. borinquense* strain E3 using SCB waste with data from the literature.

PHA from Various Substrates	Haloarchaeal Isolate	DSC Characterization (°C)			Reference
		T*m*1	T*m*2	T*d*	
SCB	*Hgm. borinquense* strain E3	136.59	149.4	275.4	Present study
Glucose		138.15	154.74	231.08	[30]
Cornstarch	*Hfx. mediterraneic* ATCC 33500	129.1	144.0	NR	[13]
Whey	*Hfx. mediterranei* DSM 1411	150.8	158.9	241	[46]

T*m*—melting temperature; T*d*—degradation temperature NR-not reported

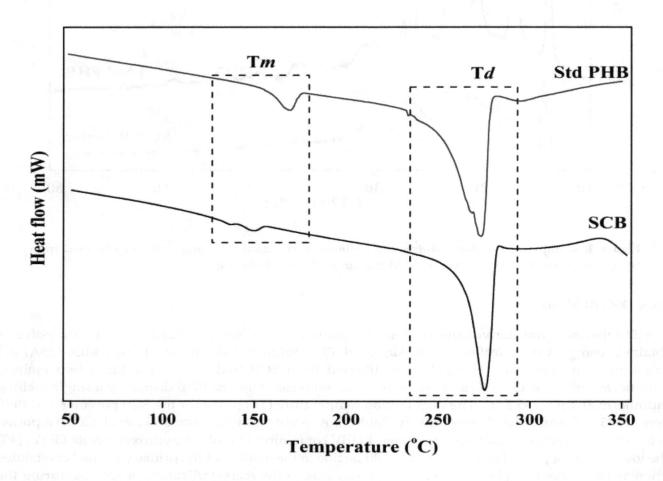

Figure 4. Comparison of DSC curves of standard PHB and polymer obtained from *Hgm. borinquense* strain E3 grown in NSM containing SCB hydrolysate.

3.6.4. FT-IR Analysis

The FT-IR spectra of polymer obtained using SCB hydrolysate were compared with those of the standard PHB (Sigma Aldrich, St. Louis, MO, USA) (Figure 5). The IR spectra of polymer obtained from SCB and standard PHB exhibited one intense absorption band at 1724 cm^{-1} and 1731 cm^{-1}, respectively, characteristic of ester carbonyl group (C=O) stretching. A band at 1281 cm^{-1} represents C–O–C stretching, whereas one in the region 3100–2800 cm^{-1}, i.e., 2983 cm^{-1} and 2981 cm^{-1}, represents

C–H stretching (Figure 5). Apart from these, other prominent bands were also observed, which may be due to interactions between the OH and C=O groups resulting in a shift of the stretching [40]. The peaks obtained for a polymer obtained from SCB hydrolysate matched well with those of the standard polymer.

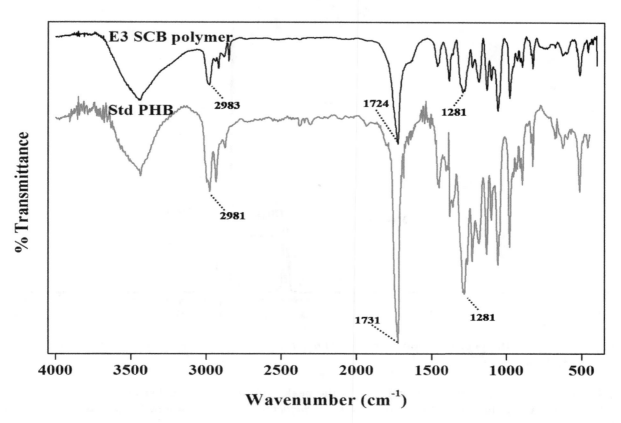

Figure 5. FT-IR spectra of standard PHB and polymer obtained from *Hgm. borinquense* strain E3 grown in NSM containing SCB hydrolysate.

3.6.5. ^{1}H-NMR Analysis

^{1}H-NMR scans of the polymer obtained from *Hgm. borinquense* strain E3 using SCB hydrolysate are represented in Figure 6. The chemical shift of the peaks and their chemical structure are represented in Table 4. Characteristic peaks at 0.889 ppm and 1.26 ppm are of methyl (CH$_3$) from 3-hydroxyvalerate (3HV) and 3-hydroxybutyrate (3HB) unit, respectively. Therefore, it can be confirmed that the polymer obtained using SCB hydrolysate is a co-polymer of Poly(3-hydroxybutyrate-*co*-3-hydroxyvalerate) [P(3HB-*co*-3HV)]. The signals obtained from ^{1}H-NMR correlated with those reported by Bhattacharyya et al. (2012) and Chen et al. (2006) for a co-polymer P(3HB-*co*-3HV)] obtained from *Hfx. mediterranei* strain DSM 1411 and strain ATCC 33500 by utilization of molasses spent wash (vinasse) and cornstarch, respectively (Figure 6, Table 4) [13,26]. Moreover, the ^{1}H NMR spectrum of homopolymer of 3HB (P3HB) showed only one prominent peak at 1.25 ppm of methyl (CH$_3$) from HB unit [35]. The co-polymer of P(3HB-*co*-3HV) comprised of 13.29% 3HV units, which was calculated as described by Salgaonkar and Bragança [30]. Interestingly, copolymer P(3HB-*co*-3HV)] containing a higher amount of 3HV (21.47% 3HV) was synthesized by the same E3 strain in NSM media with glucose as the substrate. The drastic reduction in 3HV units from 21.47% (glucose) to 13.29% (SCB) could be due to the inhibition of propionyl-coenzyme A synthesis, an important precursor of 3HV monomer by various byproducts of the SCB hydrolysate.

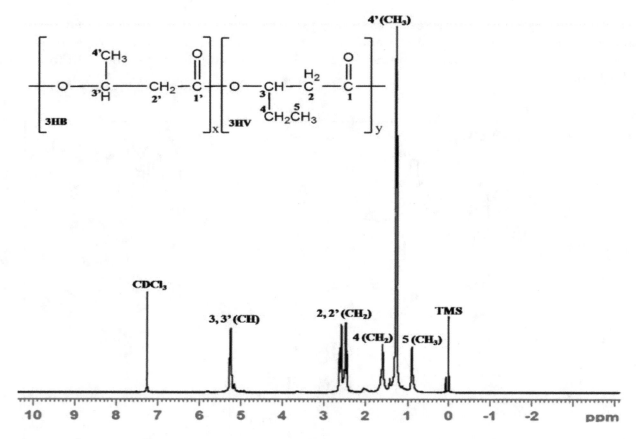

Figure 6. ^{1}H-NMR spectra of polymer from *Hgm. borinquense*strain E3 grown in NSM containing SCB hydrolysate.

Table 4. Comparison of the chemical shift of the peaks and their chemical structure, obtained from ^{1}H-NMR data of polymer synthesized by *Hgm. borinquense* strain E3 using SCB waste with data from the literature.

PHA from Various Substrates	Haloarchaeal Isolate	Relative Chemical Structure					Reference
		CH₃ (3HB)	CH₂ (3HV/3HB)	CH (3HV/3HB)	CH₃ (3HV)	CH₂ (3HV)	
		Chemical Shifts of Each Peak (ppm)					
SCB	*Hgm. borinquense* strain E3	1.26–1.27	2.44–2.63	5.22–5.27	0.889	1.618–1.635	Present study
Glucose		1.26–1.28	2.44–2.63	5.26	0.85–0.91	1.6	[30]
Cornstarch	*Hfx. mediterranei* ATCC 33500	1.2	2.5	5.2	0.9	1.6	[13]
Vinasse	*Hfx. mediterranei* DSM 1411	1.26–1.28	2.43-2.645	5.22-5.28	0.86–0.95	1.586	[26]

HB: hydroxybutyrate; HV: hydroxyvalerate; CH₃: methyl; CH₂: methylene; CH: methane.

PHA accumulation by extremely halophilic archaea and moderately halophilic and/or halotolerant bacteria, inhabiting hypersaline and marine regions of countries such as China, Turkey, Bolivia, Vietnam, India, etc., has been documented [6,16,18,21,22]. Moderately halophilic bacteria belonging to the genus *Halomonas* such as *H. boliviensis* LC1, *H. nitroreducens*, and *H. salina* have been reported to accumulate 56.0%, 33.0%, and 55.0% (*w/w*) CDM of homopolymer of 3-hydroxybutyrate (3HB), i.e., P(3HB) by utilizing versatile substrates such as starch hydrolysate, glucose, and glycerol, respectively [22,47]. Similarly, Van-Thuoc et al. (2012) reported the ability of halophilic and halotolerant bacteria *Bacillus* sp. ND153 and *Yangia pacifica* QN271 to accumulate P(3HB) (65.0 and 48.0% *w/w*

CDM)/PHBV (71.0 and 31.0% w/w of CDM) when glucose with or without propionate was provided as the carbon source [48]. However, there are very few reports on halophilic bacteria such as *H. campisalis* MCM B-1027 and *Yangia pacifica* ND199/ND218 synthesizing copolymer PHBV, irrespective of precursors like propionic/valeric acid in the culture medium [49,50]. Shrivastav et al.(2010) reported the utilization of *Jatropha* biodiesel byproduct as a substrate by *Bacillus sonorensis* strain SM-P-1S and *Halomonas hydrothermalis* strain SM-P-3M for the production of 71.8 and 75.0% (w/w) CDM of P(3HB), respectively [50].

Various members of halophilic archaea, belonging to the family *Halobacteriaceae*, such as *Halopiger aswanensis* strain 56 and *Hgm. borinquense* strain TN9 have been reported to accumulate 34.0% and 14.0% (w/w) CDM of homopolymer of P(3HB) by utilizing versatile substrates such as glucose, yeast extract, butyric acid, and sodium acetate [16,20]. Interestingly, *Hfx.* mediterranei is known to accumulate 23.0% (w/w) CDM of copolymer PHBV from glucose naturally, without any addition of precursor [23]. There are limited reports on the utilization of SCB hydrolysates as substrates for PHA production by microorganisms. Silva et al. reported the ability of two Gram-negative soil bacteria, *Burkholderia sacchari* IPT 101 and *Burkholderia cepacia* IPT 048, to accumulate poly-3-hydroxybutyrate (P3HB) when cultivated in SCB hydrolysate by submerged fermentation (SMF) [7].

Yu and Stahl reported the ability of the Gram-negative bacterium *Ralstonia eutropha* to synthesize both P(3HB) and P(3HB-co-3HV) when grown on SCB hydrolysate along with glucose as a carbon substrate. However, the bacterium failed to synthesize the polymer when grown in a hydrolysate solution devoid of glucose and was also unable to utilize pentose sugars like xylose and arabinose as a sole source of carbon [51]. Interestingly, in the present study, isolate *Hgm. borinquense* strain E3 was able to grow and synthesize PHA [P(3HB-co-3HV)] from crude SCB hydrolysate without any supplementation of carbon substrate (glucose) or prior treatment of the SCB hydrolysate for the removal of inhibitors. The strain E3 was also able to utilize arabinose and xylose when supplied as the sole source of carbon. Further studies should be done to investigate the cell and polymer yield after pre-treatment of the SCB hydrolysate to remove toxic substances. Also, the effect of glucose or other carbon substrates as supplements to SCB hydrolysate could be investigated with respect to increasing the PHA yield.

4. Conclusions

In the present study, NSM with crude SCB hydrolysate was used as a carbon substrate for the production of PHA by the extremely halophilic archaeon, *Hgm. borinquense* strain E3. The maximum PHA accumulation was observed on the seventh day, reaching a total dry biomass of 3.17 and 4.15 g/L, containing 50.4% and 45.7% PHA. The polymer exhibited two melting endotherms and was identified to be a co-polymer of P(3HB-co-3HV) comprising of 13.29% 3HV. Strain E3 accumulated a substantial quantity of PHA using crude SCB hydrolysate without any prior treatment or additional carbon substrate. Investigation into enhancing the quality and yield of P(3HB-co-3HV) from SCB hydrolysate could be achieved by: (i) standardization of the production medium, by additional supplement of carbon substrate such as glucose/xylose, and (ii) detoxification/pre-treatment of SCB hydrolysate for removal of inhibitors. Reducing the long lag phase of the culture by increasing the inoculum concentrations and optimization of other cultivation parameters such as pH, temperature, aeration, or salt can further increase the biomass and polymer yield.

The potential application of *Hgm. borinquense* strain E3 for the utilization of agro-industrial waste such as SCB has been clearly demonstrated in the present study. Since India's economy is dominated by agriculture and agro-based industries, large amounts of agro-industrial waste are being generated. Various agro-industrial waste products that can be degraded by such halophilic microbes should be explored for the production of biopolymers as this may help with both managing waste and cutting down the costs of commercial substrates. PHAs from halophilic archaea can be looked upon as a promising prospect for exploring novel bioplastics. This polymer can be further studied for various medical applications like tissue engineering and as a scaffold in organ culture.

Acknowledgments: Bhakti B. Salgaonkar thank the Council of Scientific and Industrial Research (CSIR) India for a Senior Research Fellowship (SRF) (09/919(0016)/2012-EMR-I. The authors thank the Sophisticated Instrumentation Facility (SIF), Chemistry Division, VIT University, Vellore for FT-IR and ^{1}H-NMR analysis. The authors are grateful to Narendra Nath Ghosh, Department of Chemistry, BITS Pilani Goa Campus, for the DSC analysis.

Author Contributions: Bhakti B. Salgaonkar and Judith M. Bragança conceived the idea, designed the experiments, and analyzed the data. Bhakti B. Salgaonkar performed the experiments and drafted the manuscript, which was reviewed and edited by Judith M. Bragança.

Abbreviations

APHA	American Public Health Association
CDM	cell dry mass
COD	chemical oxygen demand
DSC	Differential scanning calorimetry
EHM	Extremely Halophilic Medium
FT-IR	Fourier transform infrared
NMR	nuclear magnetic resonance
Har.	*Haloarcula*
Hbf.	*Halobiforma*
Hbt.	*Halobacterium*
Hcc.	*Halococcus*
Hfx.	*Haloferax*
Hgm.	*Halogeometricum*
Hpg.	*Halopiger*
Htg.	*Haloterrigena*
Nnm.	*Natrinema*
NSM	NaCl synthetic medium
NTYE	NaCl Tryptone Yeast extract
P(3HB-*co*-3HV)	poly(3-hydroxybutyrate-*co*-3-hydroxyvalerate)
P3HB	Poly-3-hydroxybutyrate
PHA	Polyhydroxyalkanoates
SCB	sugarcane bagasse
TKN	Total Kjeldahl nitrogen
TS	total solids
VS	volatile solids
XRD	X-ray diffraction

References

1. Chen, G.Q.; Patel, M.K. Plastics derived from biological sources: Present and future: A technical and environmental review. *Chem. Rev.* **2011**, *112*, 2082–2099. [CrossRef] [PubMed]
2. Jendrossek, D.; Pfeiffer, D. New insights in the formation of polyhydroxyalkanoate granules (carbonosomes) and novel functions of poly(3-hydroxybutyrate). *Environ. Microbiol.* **2014**, *16*, 2357–2373. [CrossRef] [PubMed]
3. Vadlja, D.; Koller, M.; Novak, M.; Braunegg, G.; Horvat, P. Footprint area analysis of binary imaged *Cupriavidus necator* cells to study PHB production at balanced, transient, and limited growth conditions in a cascade process. *Appl. Microb. Biotechnol.* **2016**, *100*, 10065–10080. [CrossRef] [PubMed]
4. Koller, M.; Maršálek, L.; de Sousa Dias, M.M.; Braunegg, G. Producing microbial polyhydroxyalkanoate (PHA) biopolyesters in a sustainable manner. *New Biotechnol.* **2017**, *37*, 24–38. [CrossRef] [PubMed]
5. Valappil, S.P.; Boccaccini, A.R.; Bucke, C.; Roy, I. Polyhydroxyalkanoates in Gram-positive bacteria: Insights from the genera *Bacillus* and *Streptomyces*. *Antonie Leeuwenhoek.* **2007**, *91*, 1–17. [CrossRef] [PubMed]

6. Han, J.; Hou, J.; Liu, H.; Cai, S.; Feng, B.; Zhou, J.; Xiang, H. Wide distribution among halophilic archaea of a novel polyhydroxyalkanoate synthase subtype with homology to bacterial type III synthases. *Appl. Environ. Microbiol.* **2010**, *76*, 7811–7819. [CrossRef] [PubMed]

7. Silva, L.F.; Taciro, M.K.; Ramos, M.M.; Carter, J.M.; Pradella, J.G.C.; Gomez, J.G.C. Poly-3-hydroxybutyrate (P3HB) production by bacteria from xylose, glucose and sugarcane bagasse hydrolysate. *J. Ind. Microbiol. Biotechnol.* **2004**, *31*, 245–254. [CrossRef] [PubMed]

8. Parameswaran, B. Sugarcane bagasse. In *Biotechnology for Agro-Industrial Residues Utilisation*; Springer: Dordrecht, The Netherlands, 2009; pp. 239–252.

9. Pippo, W.A.; Luengo, C.A. Sugarcane energy use: Accounting of feedstock energy considering current agro-industrial trends and their feasibility. *Int. J. Energy Environ. Eng.* **2013**, *4*. [CrossRef]

10. Obruca, S.; Benesova, P.; Marsalek, L.; Marova, I. Use of lignocellulosic materials for PHA production. *Chem. Biochem. Eng. Q.* **2015**, *29*, 135–144. [CrossRef] [PubMed]

11. Lavarack, B.P.; Griffin, G.J.; Rodman, D. The acid hydrolysis of sugarcane bagasse hemicellulose to produce xylose, arabinose, glucose and other products. *Biomass Bioenergy* **2002**, *23*, 367–380. [CrossRef]

12. Kirk, R.G.; Ginzburg, M. Ultrastructure of two species of halobacterium. *J. Ultrastruct. Res.* **1972**, *41*, 80–94. [CrossRef]

13. Chen, C.W.; Don, T.M.; Yen, H.F. Enzymatic extruded starch as a carbon source for the production of poly(3-hydroxybutyrate-*co*-3-hydroxyvalerate) by *Haloferax mediterranei*. *Process Biochem.* **2006**, *41*, 2289–2296. [CrossRef]

14. Han, J.; Lu, Q.; Zhou, L.; Zhou, J.; Xiang, H. Molecular characterization of the phaECHm genes, required for biosynthesis of poly(3-hydroxybutyrate) in the extremely halophilicarchaeon *Haloarculamarismortui*. *Appl. Environ. Microbiol.* **2007**, *73*, 6058–6065. [CrossRef] [PubMed]

15. Romano, I.; Poli, A.; Finore, I.; Huertas, F.J.; Gambacorta, A.; Pelliccione, S.; Nicolaus, G.; Lama, L.; Nicolaus, B. *Haloterrigena hispanica* sp. nov., an extremely halophilicarchaeon from Fuente de Piedra, Southern Spain. *Int. J. Syst. Evol. Microbiol.* **2007**, *57*, 1499–1503. [CrossRef] [PubMed]

16. Salgaonkar, B.B.; Mani, K.; Bragança, J.M. Accumulation of polyhydroxyalkanoates by halophilic archaea isolated from traditional solar salterns of India. *Extremophiles* **2013**, *17*, 787–795. [CrossRef] [PubMed]

17. Legat, A.; Gruber, C.; Zangger, K.; Wanner, G.; Stan-Lotter, H. Identification of polyhydroxyalkanoates in *Halococcus* and other haloarchaeal species. *Appl. Microbiol. Biotechnol.* **2010**, *87*, 1119–1127. [CrossRef] [PubMed]

18. Danis, O.; Ogan, A.; Tatlican, P.; Attar, A.; Cakmakci, E.; Mertoglu, B.; Birbir, M. Preparation of poly(3-hydroxybutyrate-*co*-hydroxyvalerate) films from halophilic archaea and their potential use in drug delivery. *Extremophiles* **2015**, *19*, 515–524. [CrossRef] [PubMed]

19. Hezayen, F.F.; Rehm, B.H.A.; Eberhardt, R.; Steinbuchel, A. Polymer production by two newly isolated extremely halophilic archaea: Application of a novel corrosion-resistant bioreactor. *Appl. Microbiol. Biotechnol.* **2000**, *54*, 319–325. [CrossRef] [PubMed]

20. Hezayen, F.F.; Gutierrez, M.C.; Steinbuchel, A.; Tindall, B.J.; Rehm, B.H.A. *Halopiger aswanensis* sp. nov., a polymerproducing and extremely halophilicarchaeon isolated from hypersaline soil. *Int. J. Syst. Evol. Microbiol.* **2010**, *60*, 633–637. [CrossRef] [PubMed]

21. Guzmán, H.; Van-Thuoc, D.; Martín, J.; Hatti-Kaul, R.; Quillaguamán, J. A process for the production of ectoine and poly(3-hydroxybutyrate) by *Halomonas boliviensis*. *Appl. Microbiol. Biotechnol.* **2009**, *84*, 1069–1077. [CrossRef] [PubMed]

22. Quillaguaman, J.; Hashim, S.; Bento, F.; Mattiasson, B.; Hatti-Kaul, R. Poly(β-hydroxybutyrate) production by a moderate halophile, *Halomonas boliviensis* LC1 using starch hydrolysate as substrate. *J. Appl. Microbiol.* **2005**, *99*, 151–157. [CrossRef] [PubMed]

23. Huang, T.Y.; Duan, K.J.; Huang, S.Y.; Chen, C.W. Production of polyhydroxyalkanoates from inexpensive extruded rice bran and starch by *Haloferax mediterranei*. *J. Ind. Microbiol. Biotechnol.* **2006**, *33*, 701–706. [CrossRef] [PubMed]

24. Koller, M.; Hesse, P.; Bona, R.; Kutschera, C.; Atlić, A.; Braunegg, G. Potential of various archae-and eubacterial strains as industrial polyhydroxyalkanoate producers from whey. *Macromol. Biosci.* **2007**, *7*, 218–226. [CrossRef] [PubMed]

25. Bhattacharyya, A.; Saha, J.; Haldar, S.; Bhowmic, A.; Mukhopadhyay, U.K.; Mukherjee, J. Production of poly-3-(hydroxybutyrate-*co*-hydroxyvalerate) by *Haloferax mediterranei* using rice-based ethanol stillage with simultaneous recovery and re-use of medium salts. *Extremophiles* **2014**, *18*, 463–470. [CrossRef] [PubMed]

26. Bhattacharyya, A.; Pramanik, A.; Maji, S.K.; Haldar, S.; Mukhopadhyay, U.K.; Mukherjee, J. Utilization of vinasse for production of poly-3-(hydroxybutyrate-*co*-hydroxyvalerate) by *Haloferax mediterranei*. *AMB Express* **2012**, *2*, 34. [CrossRef] [PubMed]

27. Pramanik, A.; Mitra, A.; Arumugam, M.; Bhattacharyya, A.; Sadhukhan, S.; Ray, A.; Mukherjee, J. Utilization of vinasse for the production of polyhydroxybutyrate by *Haloarcula marismortui*. *Folia Microbiol.* **2012**, *57*, 71–79. [CrossRef] [PubMed]

28. Taran, M. Utilization of petrochemical wastewater for the production of poly(3-hydroxybutyrate) by *Haloarcula* sp. IRU1. *J. Hazard. Mater.* **2011**, *188*, 26–28. [CrossRef] [PubMed]

29. Mani, K.; Salgaonkar, B.B.; Bragança, J.M. Culturable halophilic archaea at the initial and final stages of salt production in a natural solar saltern of Goa, India. *Aquat. Biosyst.* **2012**, *8*, 15. [CrossRef] [PubMed]

30. Salgaonkar, B.B.; Bragança, J.M. Biosynthesis of poly(3-hydroxybutyrate-*co*-3-hydroxyvalerate) by *Halogeometricumborinquense* strain E3. *Int. J. Biol. Macromol.* **2015**, *78*, 339–346. [CrossRef] [PubMed]

31. American Public Health Association; American Water Works Association. *Standard Methods for the Examination of Water and Wastewater: Selected Analytical Methods Approved and Cited by the United States Environmental Protection Agency*, 20th ed.; American Public Health Association: Washington, DC, USA, 1981.

32. Raposo, F.; De la Rubia, M.A.; Borja, R.; Alaiz, M. Assessment of a modified and optimised method for determining chemical oxygen demand of solid substrates and solutions with high suspended solid content. *Talanta* **2008**, *76*, 448–453. [CrossRef] [PubMed]

33. Dubois, M.; Gilles, K.A.; Hamilton, J.K.; Rebers, P.A.T.; Smith, F. Colorimetric method for determination of sugars and related substances. *Anal. Chem.* **1956**, *28*, 350–356. [CrossRef]

34. Labconco, C. *A Guide to Kjeldahl Nitrogen Determination Methods and Apparatus*; Labconco Corporation: Houston, TX, USA, 1998.

35. Salgaonkar, B.B.; Mani, K.; Braganca, J.M. Characterization of polyhydroxyalkanoates accumulated by a moderately halophilic salt pan isolate *Bacillus megaterium* strain H16. *J. Appl. Microbiol.* **2013**, *114*, 1347–1356. [CrossRef] [PubMed]

36. Law, J.H.; Slepecky, R.A. Assay of poly-β-hydroxybutyric acid. *J. Bacteriol.* **1961**, *82*, 33–36. [PubMed]

37. Sánchez, R.J.; Schripsema, J.; da Silva, L.F.; Taciro, M.K.; Pradella, J.G.; Gomez, J.G.C. Medium-chain-length polyhydroxyalkanoic acids (PHA mcl) produced by Pseudomonas putida IPT 046 from renewable sources. *Eur. Polym. J.* **2003**, *39*, 1385–1394. [CrossRef]

38. Follonier, S.; Panke, S.; Zinn, M. A reduction in growth rate of *Pseudomonas putida* KT2442 counteracts productivity advances in medium-chain-length polyhydroxyalkanoate production from gluconate. *Microb. Cell Factories* **2011**, *10*, 25. [CrossRef] [PubMed]

39. Castilho, L.R.; Mitchell, D.A.; Freire, D.M. Production of polyhydroxyalkanoates (PHAs) from waste materials and by-products by submerged and solid-state fermentation. *Bioresour. Technol.* **2009**, *100*, 5996–6009. [CrossRef] [PubMed]

40. Da Silva Pinto, C.E.; Arizaga, G.G.C.; Wypych, F.; Ramos, L.P.; Satyanarayana, K.G. Studies of the effect of molding pressure and incorporation of sugarcane bagasse fibers on the structure and properties of poly (hydroxy butyrate). *Compos. Part A Appl. Sci. Manuf.* **2009**, *40*, 573–582. [CrossRef]

41. Vidhate, S.; Innocentini-Mei, L.; D'Souza, N.A. Mechanical and electrical multifunctional poly(3-hydroxybutyrate-*co*-3-hydroxyvalerate)-multiwall carbon nanotube nanocomposites. *Polym. Eng. Sci.* **2012**, *52*, 1367–1374. [CrossRef]

42. Oliveira, L.M.; Araújo, E.S.; Guedes, S.M.L. Gamma irradiation effects on poly(hydroxybutyrate). *Polym. Degrad. Stab.* **2006**, *91*, 2157–2162. [CrossRef]

43. Buzarovska, A.; Grozdanov, A.; Avella, M.; Gentile, G.; Errico, M. Poly(hydroxybutyrate-*co*-hydroxyvalerate)/ titanium dioxide nanocomposites: A degradation study. *J. Appl. Polym. Sci.* **2009**, *114*, 3118–3124. [CrossRef]

44. Sudesh, K. *Polyhydroxyalkanoates from Palm Oil: Biodegradable Plastics*; Springer Science & Business Media: New York, NY, USA, 2012; Volume 76.

45. Hermann-Krauss, C.; Koller, M.; Muhr, A.; Fasl, H.; Stelzer, F.; Braunegg, G. Archaeal production of polyhydroxyalkanoate (PHA) co-and terpolyesters from biodiesel industry-derived by-products. *Archaea* **2013**, *2013*, 129268. [CrossRef] [PubMed]

46. Koller, M.; Hesse, P.; Bona, R.; Kutschera, C.; Atlić, A.; Braunegg, G. Biosynthesis of high quality polyhydroxyalkanoate co-and terpolyesters for potential medical application by the archaeon *Haloferax mediterranei*. *Macromol. Symp.* **2007**, *253*, 33–39. [CrossRef]

47. Cervantes-Uc, J.M.; Catzin, J.; Vargas, I.; Herrera-Kao, W.; Moguel, F.; Ramirez, E.; Lizama-Uc, G. Biosynthesis and characterization of polyhydroxyalkanoates produced by an extreme halophilic bacterium, *Halomonas nitroreducens*, isolated from hypersaline ponds. *J. Appl. Microbiol.* **2014**, *117*, 1056–1065. [CrossRef] [PubMed]

48. Van-Thuoc, D.; Huu-Phong, T.; Thi-Binh, N.; Thi-Tho, N.; Minh-Lam, D.; Quillaguaman, J. Polyester production by halophilic and halotolerant bacterial strains obtained from mangrove soil samples located in Northern Vietnam. *Microbiol. Open* **2012**, *1*, 395–406. [CrossRef] [PubMed]

49. Kulkarni, S.O.; Kanekar, P.P.; Nilegaonkar, S.S.; Sarnaik, S.S.; Jog, J.P. Production and characterization of a biodegradable poly(hydroxybutyrate-*co*-hydroxyvalerate) (PHB-*co*-PHV) copolymer by moderately haloalkalitolerant *Halomonas campisalis* MCM B-1027 isolated from Lonar Lake, India. *Bioresour. Technol.* **2010**, *101*, 9765–9771. [CrossRef] [PubMed]

50. Shrivastav, A.; Mishra, S.K.; Shethia, B.; Pancha, I.; Jain, D.; Mishra, S. Isolation of promising bacterial strains from soil and marine environment for polyhydroxyalkanoates (PHAs) production utilizing Jatropha biodiesel byproduct. *Int. J. Biol. Macromol.* **2010**, *47*, 283–287. [CrossRef] [PubMed]

51. Yu, J.; Stahl, H. Microbial utilization and biopolyester synthesis of bagasse hydrolysates. *Bioresour. Technol.* **2008**, *99*, 8042–8048. [CrossRef] [PubMed]

6

In-Line Monitoring of Polyhydroxyalkanoate (PHA) Production during High-Cell-Density Plant Oil Cultivations using Photon Density Wave Spectroscopy

Björn Gutschmann [1], Thomas Schiewe [2], Manon T.H. Weiske [1], Peter Neubauer [1], Roland Hass [2] and Sebastian L. Riedel [1,*]

[1] Bioprocess Engineering, Department of Biotechnology, Technische Universität Berlin, 13355 Berlin, Germany; bjoern.gutschmann@tu-berlin.de (B.G.); manon.th.weiske@campus.tu-berlin.de (M.T.H.W.); peter.neubauer@tu-berlin.de (P.N.)

[2] innoFSPEC, University of Potsdam, 14476 Potsdam, Germany; tschiewe@uni-potsdam.de (T.S.); rh@pdw-analytics.de (R.H.)

* Correspondence: riedel@tu-berlin.de

Abstract: Polyhydroxyalkanoates (PHAs) are biodegradable plastic-like materials with versatile properties. Plant oils are excellent carbon sources for a cost-effective PHA production, due to their high carbon content, large availability, and comparatively low prices. Additionally, efficient process development and control is required for competitive PHA production, which can be facilitated by *on-line* or *in-line* monitoring devices. To this end, we have evaluated photon density wave (PDW) spectroscopy as a new process analytical technology for *Ralstonia eutropha* (*Cupriavidus necator*) H16 plant oil cultivations producing polyhydroxybutyrate (PHB) as an intracellular polymer. PDW spectroscopy was used for *in-line* recording of the reduced scattering coefficient μ_s' and the absorption coefficient μ_a at 638 nm. A correlation of μ_s' with the cell dry weight (CDW) and μ_a with the residual cell dry weight (RCDW) was observed during growth, PHB accumulation, and PHB degradation phases in batch and pulse feed cultivations. The correlation was used to predict CDW, RCDW, and PHB formation in a high-cell-density fed-batch cultivation with a productivity of 1.65 $g_{PHB} \cdot L^{-1} \cdot h^{-1}$ and a final biomass of 106 $g \cdot L^{-1}$ containing 73 wt% PHB. The new method applied in this study allows *in-line* monitoring of CDW, RCDW, and PHA formation.

Keywords: polyhydroxyalkanoate; PHA; process analytical technologies; PAT; plant oil; high-cell-density fed-batch; photon density wave spectroscopy; PDW; *Ralstonia eutropha*; *Cupriavidus necator*; *on-line*; *in-line*

1. Introduction

When the US Food and Drug Administration (FDA) announced their process analytical technology (PAT) directives, the investigation of PAT became a key research area in bioprocess development. The main objectives are designing, developing, and operating bioprocesses to guarantee a targeted final product quality [1,2]. The focus of this initiative was predominantly on biopharmaceutical processes, while novel PAT tools could be integrated into any bioprocess. Especially, the implementation of PAT for polyhydroxyalkanoate (PHA) production can provide significant benefits to facilitate a consistent and highly efficient production. Techniques such as FTIR, Raman spectroscopy, fluorescence staining associated with flow cytometry, and enzymatic approaches were reported as novel methods for a rapid characterization of PHA production [3–7]. A comprehensive overview of qualitative and quantitative methods for PHA analysis was published by Koller et al. [8]. However, the reported methods have not been applied for *in-line* or *at-line* measurements of the PHA production process so far.

Photon density wave (PDW) spectroscopy is an *in-line* technique, which has been used as an analytical tool for measurements of various highly turbid chemical processes [9–12]. The method is based on the theory of photon migration in multiple light scattering material. If intensity-modulated light is introduced into a strongly light scattering but weakly light absorbing material, a PDW is generated. Absorption and scattering properties of the material influence the amplitude and phase of the PDW. By quantifying these shifts as a function of the emitter fiber and detector fiber distance and of the modulation frequency, the absorption coefficient μ_a and the reduced scattering coefficient μ_s' can be determined independently [9,13,14]. The mentioned features make PDW spectroscopy very attractive for the monitoring of high-cell-density bioprocesses.

Currently, PHA production costs are not compatible with the low-priced production of conventional plastics. The main cost driving factors are the feedstocks for PHA accumulation and the recovery process. Thus, alternative low-cost substrates, e.g., biogenic waste streams, are of high interest to reduce the final production price. Other attempts concentrate on finding more sustainable and price efficient purification strategies [15–21]. *Ralstonia eutropha* (also known as *Cupriavidus necator*) is one of the main species studied for polyhydroxybutyrate (PHB) accumulation and the model organism for PHA accumulation [22]. Growth of *R. eutropha* on oleaginous feedstocks is particularly attractive due to their high carbon contents, high conversion rates to PHA, and low culture dilution in fed-batch processes. Efficient growth on these feedstocks is facilitated by the expression of extracellular lipases, which emulsify the lipids [23–27]. A large biomass accumulation prior to PHA accumulation is very important for a high final product titer. In this context, it has been shown that urea is an inexpensive nitrogen source, which allows excellent growth [24,28]. Despite alternative substrates and downstream approaches, highly efficient bioprocesses are required for an economic feasible PHA production. Recently, high-cell-density cultivations with *R. eutropha* on various renewable feedstocks have been published presenting the production of over 100 $g \cdot L^{-1}$ PHA and space time yields from 1 to 2.5 $g_{PHA} \cdot L^{-1} \cdot h^{-1}$ [21,24,29–31]. However, none of the presented studies describe *in-line* PAT-based monitoring or control strategies for the enhancement of process results.

This work aims to integrate PDW spectroscopy into high-cell-density bioprocesses, for the monitoring of the highly turbid and complex PHB production with *R. eutropha* in plant oil cultivations. As a result, total cell dry weight (CDW) and residual cell dry weight (RCDW, the difference of CDW and the PHB concentration) accumulation could be distinguished with the PDW spectroscopy probe as a new *in-line* tool for bioprocesses.

2. Materials and Methods

2.1. Bacterial Strain

All cultivations were performed with the wild type strain *R. eutropha* H16 (DSM-428, Leibniz Institute DSMZ-German Collection of Microorganisms and Cell Cultures, Germany).

2.2. Growth Media and Preculture Cultivation Conditions

Tryptic soy broth (TSB) media (17 $g \cdot L^{-1}$ tryptone, 5 $g \cdot L^{-1}$ NaCl, 3 $g \cdot L^{-1}$ peptone) was used for the first precultures and with an additional supply of 2% (w·v^{-1}) agar for culture plates. The second precultures and bioreactor cultivations were conducted in mineral salt media (MSM) containing 4.62 $g \cdot L^{-1}$ $NaH_2PO_4 \cdot H_2O$, 5.74 $g \cdot L^{-1}$ $Na_2HPO_4 \cdot 2H_2O$, 0.45 $g \cdot L^{-1}$ K_2SO_4, 0.04 $g \cdot L^{-1}$ NaOH, 0.80 $g \cdot L^{-1}$ $MgSO_4 \cdot 7H_2O$, 0.06 $g \cdot L^{-1}$ $CaCl_2 \cdot 2H_2O$ and 1 $mL \cdot L^{-1}$ trace element solution consisting of 0.48 $g \cdot L^{-1}$ $CuSO_4 \cdot 5H_2O$, 2.4 $g \cdot L^{-1}$ $ZnSO_4 \cdot 7H_2O$, 2.4 $g \cdot L^{-1}$ $MnSO_4 \cdot H_2O$, 15 $g \cdot L^{-1}$ $FeSO_4 \cdot 7H_2O$. All cultivation media and plates contained 10 $mg \cdot L^{-1}$ sterile filtered gentamycin sulfate. Rapeseed oil (Edeka Zentrale AG & Co. KG, Germany) was used as the sole carbon source and urea as the sole nitrogen source in the

MSM. The explicit amounts are described in the text. All chemicals were purchased from Carl Roth GmbH & Co. KG (Germany) unless stated otherwise.

R. eutropha H16 was streaked from a cryoculture on a TSB agar plate and incubated for 3–4 days at 30 °C. A single colony from the plate was used to inoculate the first preculture in 10 mL TSB media in a 125-mL Ultra Yield™ Flask (Thomson Instrument Company, USA) sealed with an AirOtop™ enhanced flask seal (Thomson Instrument Company, USA). After incubating for 16 h, 2.5 mL were used to inoculate the second preculture (250 mL MSM with 3% ($w \cdot v^{-1}$) rapeseed oil and 4.5 $g \cdot L^{-1}$ urea) in a 1-L DURAN® baffled glass flask with a GL45 thread (DWK Life Sciences GmbH, Germany) sealed with an AirOtop membrane. After 24 h of incubation, the complete second preculture was used to inoculate the main bioreactor culture. The precultures were incubated at 30 °C and shaken at 200 rpm (first preculture) or 180 rpm (second preculture) in an orbital shaker (Kühner LT-X incubator, Adolf Kühner AG, Switzerland, 50 mm amplitude).

2.3. Bioreactor Cultivation Conditions

Mineral salts dissolved in deionized (DI) water and rapeseed oil were added prior autoclavation in a 6.6-L stirred tank bioreactor with two six-blade Rushton impellers (BIOSTAT® Aplus, Sartorius AG, Germany). $MgSO_4$, $CaCl_2$, trace elements, gentamycin, and urea were added into the medium after autoclavation from sterile stock solutions. The temperature was maintained at 30 °C and the pH was kept constant at 6.8 ± 0.2 using 2 M NaOH and 1 M H_3PO_4 for pH control. The dissolved oxygen concentration (DO) was kept above 40% using a stirrer cascade ranging from 400 to 1350 rpm. The cultures were aerated with a constant aeration rate of 0.5 vvm throughout the cultivations. Five pairs of cable ties were mounted on the upper part of the stirrer shaft in order to break the foam mechanically and thus preventing overfoaming of the reactor.

2.3.1. Batch Cultivations

For a first evaluation of the PDW spectroscopy signal, three batch cultivations were performed in which the carbon and nitrogen content was varied. The concentrations of rapeseed oil were 3, 4, and 4% ($w \cdot v^{-1}$), and 2.25 (corresponding to 75 mM nitrogen), 4.5, and 2.25 $g \cdot L^{-1}$ for urea, respectively.

2.3.2. Pulse-Based Fed-Batch Cultivation

A cultivation strategy with a pulse feeding was performed in biological duplicates. The cultures initially contained 0.5% ($w \cdot v^{-1}$) rapeseed oil and 4.5 $g \cdot L^{-1}$ urea. Pulses were given whenever the PDW spectroscopy *in-line* signal (μ_s' at 638 nm) indicated a decreased cell activity. After 8.2 h, the first pulse (15 g rapeseed oil) was added, followed by two more rapeseed oil pulses at 14.3 h (30 g) and at 21.1 h (60 g). At 31.7 h a pulse consisting of 110 mL urea solution (122 $g \cdot L^{-1}$), 15.6 mL 0.5 M K_2SO_4, 30 mL 0.042 M $CaCl_2$, 30 mL 0.32 M $MgSO_4$, and 3 mL trace element solution was added to restore the initial media concentrations of the components. The last pulse (120 g rapeseed oil) was added after 48.4 h.

2.3.3. Fed-Batch High-Cell-Density Cultivation

The culture initially contained 4% ($w \cdot v^{-1}$) rapeseed oil and 4.5 $g \cdot L^{-1}$ urea (150 mM nitrogen). Continuous feeding of pure rapeseed oil and a 30% ($w \cdot v^{-1}$) urea solution was started 7 h after inoculation with initial feeding rates of 3.5 $g \cdot h^{-1}$ and 0.39 $mL \cdot h^{-1}$, respectively. Both feeding rates were linearly increased up to 6.58 $mL \cdot h^{-1}$ (urea) at 16 h, after which the urea feed was stopped to cause nitrogen starvation, and 23 $g \cdot h^{-1}$ (rapeseed oil) at 35 h to final concentrations of 480 mM nitrogen and 170 $g \cdot L^{-1}$ rapeseed oil. A single injection of $MgSO_4$, $CaCl_2$, K_2SO_4, and trace elements was performed after 20 h with the amounts as described above (see Section 2.3.2.) to restore the initial concentrations and prevent nutrient depletion. An additional pulse of $CaCl_2$ with the same concentration was added after 32 h.

2.4. Photon Density Wave Spectroscopy

A PDW spectrometer built by the University of Potsdam was used for the measurement of the absorption coefficient μ_a and the reduced scattering coefficient μ_s'. Identical devices are commercially available at PDW Analytics GmbH (Potsdam, Germany). The general set-up of the PDW spectrometer was described by Bressel et al., as follows: "A schematic set-up of the spectrometer is shown in Figure 1. Light from a laser diode with wavelength λ [m] (typ. 400–1000 nm) is sinusoidally intensity modulated by a vector network analyzer (typ. $f = \omega/(2\pi) = 10$–1300 MHz). The light is then coupled into the material via an optical fiber, acting as a point-like light source. A second optical fiber, positioned at a distance r to the emission fiber (typ. r = 5–30 mm), collects light of the PDW and guides it onto an avalanche photodiode (APD) as detector." [13].

To integrate the multifiber PDW spectroscopy *in-line* probe into the system, a DN25 safety Ingold socket (elpotech GmbH & Co. KG, Germany) was welded onto the lid of the bioreactor. The probe was mounted before autoclaving and sterilized with the bioreactor inside the autoclave. The optical fibers of the probe were connected to the PDW spectrometer after autoclaving. μ_s' and μ_a were analyzed at 638 nm with a temporal resolution of 0.8 min^{-1}. A 10-point moving average was used to reduce the signal noise.

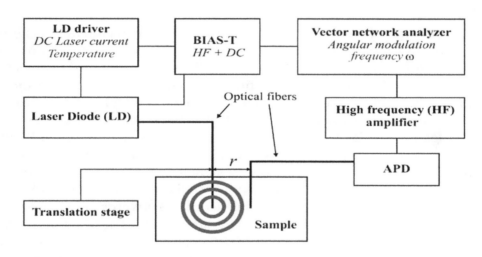

Figure 1. Schematic experimental set-up of a photon density wave (PDW) spectrometer [13].

2.5. Analytical Methods

For each *off-line* reference analysis time point, two aliquots of 10 mL were sampled in preweighed 15-mL polypropylene test tubes. The samples were centrifuged for 15 min at $6000\times g$ and pellets were washed either with a mixture of 5 mL cold deionized (DI) water and 2 mL cold hexane or with 7 mL cold DI water to remove residual lipids. The washed pellets were resuspended in 2–4 mL ice cold DI water, frozen at −80 °C, and dried for 24 h by lyophilization (Gamma 1–20, Martin Christ Gefriertrocknungsanlagen GmbH, Germany). Then the CDW was determined by weighing the test tubes.

The PHB content was determined *off-line* by high-performance liquid chromatography with a diode array detector (HPLC-DAD 1200 series, Agilent Technologies, USA). The method was adapted from Karr et al. [32]. Pure PHB (Sigma-Aldrich Corporation, USA) or 8–15 mg freeze-dried cells were depolymerized by boiling samples with 1 mL concentrated H_2SO_4 to yield crotonic acid. Dilution series of the depolymerized PHB were prepared to yield standards in the range of 0.1–10 mg·mL^{-1}. Samples were diluted with 4 mL 5 mM H_2SO_4, filtered through a 0.2 µm cellulose acetate syringe filter, and subsequently 100 µL were transferred to a HPLC vial containing 900 µL of 5 mM H_2SO_4. HPLC analysis was performed with an injection volume of 20 µL using 5 mM H_2SO_4 as an eluent with an isocratic flow rate of 0.4 mL·min^{-1} for 60 min on a NUCLEOGEL® ION 300 OA column

(Macherey-Nagel, Germany). Crotonic acid was detected at 210 nm. RCDW was determined by subtracting the PHB from the CDW concentration.

The nitrogen content was indirectly determined *off-line* by measuring ammonia, resulting from urea cleavage, in the supernatant using a pipetting robot (Cedex Bio HT Analyzer, Roche Diagnostics International AG, Switzerland) with the NH3 Bio HT test kit (Roche Diagnostics International AG, Switzerland).

During the fed-batch cultivation the *in-line* PDW spectroscopy signals were used for an *on-line* determination of the CDW, RCDW, and PHB content (see Section 3.4).

3. Results

A reduction of the PHA production price is crucial for commercialization [33], which can be facilitated by maximization of the biotechnological process performance using PAT. In this context, our group is interested in developing biotechnological processes for PHA production from renewable resources [24] and biogenic waste streams [17,18].

In this study, PDW spectroscopy was evaluated as a novel *in-line* tool to monitor the PHA production process with *R. eutropha* from plant oil.

3.1. Batch Cultivations

For an initial evaluation of the PDW spectroscopy signals, batch cultivations with different C/N ratios were performed to trigger different PHB and CDW accumulation. The results of the three batch cultivations are shown in Figure 2. Detailed graphs of the cultivations can be found in the in Supplementary Materials in Figures S1–S3. The first batch cultivation was performed as a reference batch containing 3% ($w \cdot v^{-1}$) rapeseed oil and 75 mM nitrogen (2.25 $g \cdot L^{-1}$ urea). Nitrogen limitation was indirectly detected by ammonia quantification, which is released from urea cleavage prior nitrogen uptake [24,28]. Nitrogen was depleted between 9.2 and 14.2 h, which triggered PHB accumulation. Within this period, the CDW increased from 6 to 15 $g \cdot L^{-1}$ and the PHB content increased from 7 to 35 wt% (0.4 to 5.3 $g_{PHB} \cdot L^{-1}$). The PDW spectroscopy signals μ_a and μ_s' did not show any analyzable signals until 6.5 h. Subsequently, both signals exponentially increased until 10.5 h. After this time point, μ_a linearly increased until 16 h and did not further rise afterwards. An increase of the μ_s' signal was detected from 10.5 to 20 h and remained constant afterwards. The maximum CDW (25.5 $g \cdot L^{-1}$) was achieved after 33.2 h containing 65 wt% PHB (16.8 $g_{PHB} \cdot L^{-1}$), which represents an overall yield of 0.56 $g_{PHB} \cdot g_{Oil}^{-1}$. Over the entire cultivation, the urea consumption for biomass accumulation was 0.26 $g_{Urea} \cdot g_{RCDW}^{-1}$.

The purpose of Batch 2 was to decrease the C/N ratio compared to the reference batch for an increased accumulation of active biomass (RCDW) and decreased PHB content. Nitrogen was depleted between 11.6 and 19.8 h. Within this period, the CDW and PHB content increased from 12.3 to 38.3 $g \cdot L^{-1}$ and 7 to 62 wt% (0.8 to 23.7 $g_{PHB} \cdot L^{-1}$), respectively. The PDW spectroscopy signals μ_a and μ_s' simultaneously increased from the beginning of the cultivation until 12 h in an exponential manner. No significant changes of μ_a were detected after that time point, whereas μ_s' increased until 18 h and subsequently remained constant. After 23.8 h the maximum CDW of 40 $g \cdot L^{-1}$ containing 63 wt% (26.2 $g_{PHB} \cdot L^{-1}$) was achieved. Overall, a PHB yield of 0.66 $g_{PHB} \cdot g_{Oil}^{-1}$ and urea consumption of 0.29 $g_{Urea} \cdot g_{RCDW}^{-1}$ was achieved.

The C/N ratio in the third batch was increased compared to the reference batch by keeping the initial nitrogen concentration constant (75 mM) but increasing the rapeseed oil content to 4% ($w \cdot v^{-1}$). Depletion of nitrogen occurred between 10.3 and 11.3 h. Within this period the CDW increased from 7.7 $g \cdot L^{-1}$ with 8 wt% PHB (0.6 $g_{PHB} \cdot L^{-1}$) to 9.6 $g \cdot L^{-1}$ with 20 wt% PHB (1.9 $g_{PHB} \cdot L^{-1}$). PDW signals were detectable after 2 h. A comparable increase of μ_a and μ_s' was detected until 10 h. An attenuated increase of μ_a until 17.5 h was detected, whereas a diminished increase of μ_s' was detected from 10 to 22.5 h. The CDW further increased to 33.7 $g \cdot L^{-1}$ containing 73 wt% PHB (24.5 $g_{PHB} \cdot L^{-1}$). Overall, a PHB yield of 0.61 $g_{PHB} \cdot g_{Oil}^{-1}$ and urea consumption of 0.25 $g_{Urea} \cdot g_{RCDW}^{-1}$ was achieved.

To summarize, the C/N ratio influenced the yield coefficients for PHB accumulation and urea usage within these three batch cultivations. During the batch cultivations, μ_s' and μ_a simultaneously increased until nitrogen depletion. Subsequently, μ_a leveled off and μ_s' increased further until maximum PHB accumulation.

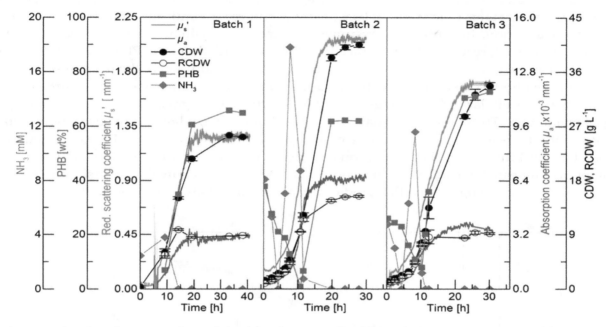

Figure 2. Batch cultivations for polyhydroxybutyrate (PHB) production by *R. eutropha* H16. Batch 1 contained 3% (w·v^{-1}) rapeseed oil and 2.25 g·L^{-1} urea (75 mM nitrogen), Batch 2 contained 4% (w·v^{-1}) rapeseed oil and 4.5 g·L^{-1} urea, and Batch 3 contained 4% (w·v^{-1}) rapeseed oil and 2.25 g·L^{-1} urea. Ammonia content (grey diamonds, mM), PHB content (green squares, wt%), cell dry weight (CDW) (filled circles, g·L^{-1}), residual cell dry weight (RCDW) (empty circles, g·L^{-1}), reduced scattering coefficient μ_s' (red line, mm^{-1}), and absorption coefficient μ_a (blue line, ×10^{-3} mm^{-1}) at 638 nm are shown. Error bars indicate minimum and maximum values of technical duplicates.

3.2. Pulse-Based Fed-Batch Cultivation

While the batch cultivations aimed to initially evaluate the relationship of the PDW spectroscopy signals with process relevant characteristics, a pulse-based fed-batch cultivation was conducted to: (i) show the feasibility to control the process by monitoring the process with PDW spectroscopy; (ii) confirm signal relationships; and (iii) validate the reproducibility during biological duplicate cultivations with independent seed trains. The process intention was implemented without difficulty: a pulse of either rapeseed oil or a nutrient bolus (see dashed lines in Figure 3 and details in the legend), respectively, was added to the culture when the μ_s' signal showed no further changes and indicated either carbon or nitrogen limitation. It is worth mentioning, that no signal deflections were observed whenever oil pulses were added to the bioreactor, which would account for an effect of the added oil on μ_s' or μ_a.

Within the first 8 h, the intracellular PHB content decreased from 36 to 14 wt% (Figure 3). An increase of the CDW, RCDW, μ_s' and μ_a was observed in the first 6 h of the cultivation. Subsequently, only minor increases of the CDW and RCDW of about 0.7 g·L^{-1} were detected until 8 h, whereas the PDW spectroscopy signals did not further increase during this period. After addition of the first pulse (0.5% (w·v^{-1}) rapeseed oil), μ_s' and μ_a resumed to increase until reaching constant levels at 11.5 h until the next pulse addition (1% (w·v^{-1}) rapeseed oil). The increase of μ_a stopped at 16.5 h, whereas μ_s' increased further until 18 h. Nitrogen depletion was detected at 17.5 h (Figure S4, Supplementary Materials). The PHB content had already increased to 30 wt% at this time point and further increased to 38 wt% before addition of the next pulse. The CDW increased up to 23 g·L^{-1}, whereas the RCDW stopped at a value of 14 g·L^{-1} at 20.5 h. Addition of the next rapeseed oil pulse (2% (w·v^{-1})) at 21 h did not trigger a significant change of the μ_s' signal and of the RCDW. In contrast, a sharp increase of μ_s' resulted from the rapeseed oil addition and the CDW increased up to 43 g·L^{-1} at 31 h containing 66 wt% PHB. At 31.6 h, a bolus containing urea as a nitrogen source was supplemented to the culture. The addition resulted in a dilution of the culture, which was seen in a step decrease of both PDW spectroscopy signals at this time point. Subsequently, μ_s' decreased until 42 h and stayed constant until the next pulse addition. In contrast, μ_a resumed to increase until 45 h. The CDW decreased during this period to 24 g·L^{-1}, resulting from intracellular PHB degradation. The PHB content decreased to 30 wt%. At the same time, the RCDW increased to 17 g·L^{-1}. The addition of the next pulse (4% (w·v^{-1}) rapeseed oil) resulted in resumed growth on rapeseed oil as the primary carbon source instead of degrading the intracellular carbon storage. The PDW spectroscopy signals increased after the rapeseed oil supplementation. The scattering signal μ_s' increased until 70 h, whereas the absorption coefficient μ_a only slightly increased after 52 h. At 52 h, nitrogen was depleted again. The cells accumulated 66 wt% PHB until the end of the cultivation and the CDW increased to 53 g·L^{-1}. When the whole cultivation was repeated, the PDW spectroscopy signals showed an equivalent course and a final CDW of 52 g·L^{-1} CDW containing 64 wt% PHB, indicating a high robustness, i.e., repeatability, of the process.

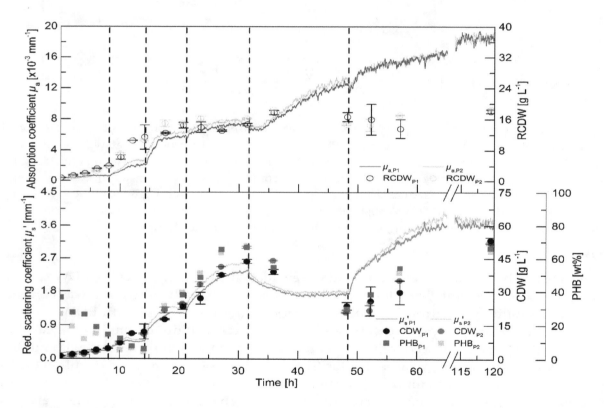

Figure 3. Pulse feeding cultivations for PHB production by *R. eutropha* H16. The cultures were initially started with 0.5% (w·v^{-1}) rapeseed oil and 4.5 g·L^{-1} urea (150 mM nitrogen). The dashed vertical lines represent time points of pulse additions: 0.5% (w·v^{-1}) rapeseed oil at 8.2 h, 1% (w·v^{-1}) rapeseed oil at 14.3 h, and 2% (w·v^{-1}) rapeseed oil at 21.1 h. At 31.7 h a bolus consisting of 110 mL urea solution (122 g·L^{-1}), 15.6 mL 0.5 M K$_2$SO$_4$, 30 mL 0.32 M MgSO$_4$, 30 mL 0.042 mM CaCl$_2$, and 3 mL trace element solution was added to the bioreactor. A final bolus of 4% (w·v^{-1}) rapeseed oil was added at 48.4 h. Data from the reference experiment is shown in color (indexed P1) and the biological duplicate (with an independent seed train) in transparent grey (indexed P2). Absorption coefficient μ_a (upper graph, solid line, ×10^{-3} mm^{-1}), RCDW (upper graph, empty circles, g·L^{-1}), PHB content (bottom graph, squares, wt%), CDW (bottom graph, filled circles, g·L^{-1}), and reduced scattering coefficient μ_s' (bottom graph, solid line, mm^{-1}) are shown. Error bars indicate minimum and maximum values of technical duplicates.

The pulse-based fed-batch experiment showed the possibility to control the rapeseed oil-based cultivation for PHB production using the *in-line* PDW spectroscopy probe. The highly reproducible course of the PDW spectroscopy signals strongly imply the connection of biological events with this measurement technique. The coefficients μ_s' and μ_a show the same trend as the CDW and RCDW, respectively. Nevertheless, a more significant change was observed for the μ_a signal after addition of the nitrogen pulse than the *off-line* determined RCDW. Regardless of this discrepancy, the hypothesis was that the μ_s' and μ_a correspond and can be correlated with the CDW and RCDW, respectively.

3.3. PDW Spectroscopy Signal Correlation

For an analysis of the correlation between the PDW spectroscopy signals μ_s' and μ_a at 638 nm, respectively, with process relevant characteristics, the experiments described above were analyzed. The reduced scattering coefficient μ_s' followed the course of the CDW, while the absorption coefficient μ_a increased with the rise of RCDW. The correlation of the respective values from all five cultivations (i.e., batch and pulse-based fed-batch cultures) are shown in Figure 4. A root mean squared error of 0.96 for the linear correlation of μ_s' and the CDW was obtained, whereas the linear correlation of μ_a and the RCDW resulted in a R^2 of 0.90. Equations (A1)–(A3) in Appendix A show the obtained formulas for calculating the CDW, RCDW, and subsequently the PHB concentration using the linear relationships of the *in-line* PDW spectroscopy signals with the corresponding *off-line* values. Due to disproportionally large PDW spectroscopy signal intensification after 48 h of the pulse feeding experiments (Figure 3), the last four data points of each replicate were not included in the correlation analysis.

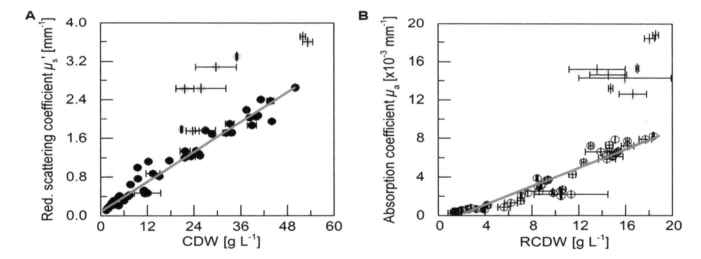

Figure 4. Correlation analysis of PDW spectroscopy signals with *off-line* values. (**A**) Reduced scattering coefficient μ_s' at 638 nm is correlated with CDW. (**B**) Absorption coefficient μ_a at 638 nm is correlated with RCDW. Data points are from five cultivations with different rapeseed oil and urea contents (cf. Figures 2 and 3). The gray data points comprise the last four samples of the two pulse-based fed-batch experiments (cf. Figure 3), which differ significantly from the other samples and were therefore not considered for the linear fit with the experimental data. A squared correlation coefficient R^2 of 0.96 was obtained for μ_s' and CDW and R^2 of 0.90 for μ_a and RCDW. Error bars indicate minimum and maximum values of technical duplicates (CDW, RCDW). STDEV of the 10-point average is shown for μ_s' and μ_a.

3.4. High-Cell-Density Fed-Batch Cultivation

High-cell-density cultivations are essential for a competitive production of PHA biopolymers. Therefore, it was aimed to perform such a cultivation to evaluate the PDW spectroscopy probe performance at industrial relevant biomass concentrations. The correlation factors obtained from the first five cultivations (Equations (A1)–(A3) in Appendix A) were used to calculate the CDW, RCDW, and PHB content from the *in-line* PDW spectroscopy signals (Figure 5).

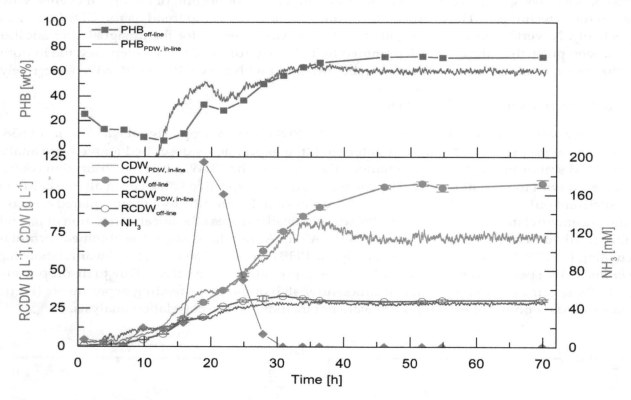

Figure 5. Fed-batch cultivation of *R. eutropha* H16 for PHB production. The culture was started with 4% (w·v^{-1}) rapeseed oil and 4.5 g·L^{-1} urea (150 mM nitrogen). Rapeseed oil feeding linearly increased from 7 h with an initial feeding rate of 3.5 g·h^{-1} to 35 h with a final feeding rate of 23 g·h^{-1} up to a total concentration of 17% (w·v^{-1}). Urea (30% (w·v^{-1})) feeding was linearly increased from 7 h with an initial feeding rate of 0.39 mL·h^{-1} up to a final feeding rate of 6.58 mL·h^{-1} at 16 h to a total concentration of 14.4 g·L^{-1}. *In-line* PHB content (upper graph, green line, wt%) by PDW spectroscopy and *off-line* PHB content (upper graph, green squares, wt%), estimated *in-line* CDW by PDW spectroscopy (bottom graph, red line, g·L^{-1}), *off-line* CDW (bottom graph, red filled circles, g·L^{-1}), estimated *in-line* RCDW by PDW spectroscopy (bottom graph, blue line, g·L^{-1}), *off-line* RCDW (bottom graph, blue empty circles, g·L^{-1}), and ammonia content (bottom graph, grey diamonds, mM) are shown. Error bars indicate minimum and maximum values of technical duplicates.

In the first 13 h, the *off-line* CDW increased to 8.5 g·L^{-1} and the RCDW to 8.2 g·L^{-1}. In the same period, the PHB content decreased to 4 wt%. Within this time frame, the *in-line* signals overestimated CDW and RCDW to 13.9 and 12.3 g·L^{-1} at 13 h, respectively. The estimated RCDW was higher than the estimated CDW until 12 h, which yielded a negative calculated PHB content. An accumulation of ammonia up to 195 mM at 19 h was detected, which subsequently decreased and was depleted after 28 h. The CDW and RCDW increased up to 63 and 32 g·L^{-1}, respectively, during this period. Contrary to expectation, PHB accumulation was detected from 13 to 19 h up to 33 wt%. Subsequently, the PHB content decreased again to 28 wt% during ammonia consumption and further accumulation started after 25 h. Subsequently, the PHB content increased to 72 wt% at 46 h. The preliminary accumulation and degradation of PHB until 25 h was also detected by the calculated *in-line* PHB signal. Subsequently, the *in-line* PHB content increased to 61 wt% at 46 h and ceased afterwards. At 46 h, the *off-line* CDW and RCDW had increased to 106 and 30 g·L^{-1}, respectively. In total, 76 g·L^{-1} PHB accumulated in 46 h, which represents a space time yield of 1.65 g$_{PHB}$·L^{-1}·h^{-1} and a yield coefficient of 0.43 g$_{PHB}$·g$_{Oil}^{-1}$. The overall urea usage was 0.48 g$_{Urea}$·g$_{RCDW}^{-1}$. The *in-line* RCDW signal indicated an accumulation until 29 h to a final RCDW of 29 g·L^{-1}. The *in-line* CDW signal increased simultaneously with the *off-line* CDW until 36 h to 82 g·L^{-1}. Unexpectedly, the *in-line* CDW signal decreased afterwards until 46 h and stayed constant at a value of 71 g·L^{-1}.

The high-cell-density cultivation showed that PDW spectroscopy is capable of a qualitative tracking of the CDW, RCDW, and PHB content. Nevertheless, the quantitative accuracy was not precise during the first 12 h of the cultivation at low cell densities. However, the estimated *in-line* CDW showed a good representation until 36 h. Also, the *in-line* RCDW was estimated from the μ_a with a very good accuracy for the rest of the cultivation. In contrast, a drop of the μ_s' signal after 36 h did not correlate with CDW during PHB production.

4. Discussion

The purpose of this study was to evaluate the potential of PDW spectroscopy for monitoring plant oil-based *R. eutropha* cultivations. The batch (Figure 2) and pulse-based fed-batch (Figure 3) cultivations showed that the reduced scattering coefficient μ_s' correlates strongly with the CDW and the absorption coefficient μ_a with the RCDW (Figure 4). These results demonstrate that PDW spectroscopy is a valuable tool for *in-line* monitoring of the CDW, RCDW, and PHB accumulation. To the best of our knowledge, the results of this study are the first data showing *in-line* quantification of PHB. The lack of such *in-line* or *on-line* monitoring devices for an adaptive control of the production process was recently emphasized by Koller et al. [3]. Previously, Cruz et al. reported the possibility to use a NIR transflectance probe for an *in-line* quantification of PHB. However, the authors showed only *at-line* data quantifying the PHB and CDW concentration (up to 9.3 g·L^{-1} and 13.7 g·L^{-1}, respectively) during batch cultivations [34].

During the initial batch cultivations, PHB yields of 0.56–0.66 g$_{PHB}$·g$_{Oil}$$^{-1}$ were obtained (Figure 2), which are similar to previously reported yields for *R. eutropha* H16 growth on palm oil [35]. During the high-cell-density fed-batch cultivation 106 g·L^{-1} CDW (72 wt% PHB) with a space time yield of 1.65 g$_{PHB}$·L^{-1}·h^{-1} were reached, which is comparable to other published high-cell-density plant oil cultivations [21,24,25,29,36,37].

In the pulse experiment, a nitrogen bolus was added (32 h) after the PHB production phase to trigger PHB degradation, as described previously [38]. The PDW spectroscopy signal μ_s' decreased with the declining CDW while μ_a increased with an increasing RCDW (Figure 3). Currently, we do not understand why the strength of the signal was not proportional with the determined *off-line* value changes after that time point (32 h). For this reason, these measurement points were not used for the linear correlation. A potential hypothesis could be an unequal distribution of PHA granules during PHA mobilization, as it was reported for *Pseudomonas putida* [39], which might have an effect on scattering and absorption coefficients during PHA degradation.

Atypical PHB formation before nitrogen depletion was detected during the high-cell-density cultivation (Figure 5). This preliminary formation of PHB could explain the low yield coefficient of 0.43 g$_{PHB}$·g$_{Oil}$$^{-1}$, which is significant lower than the typical yield of PHB in *R. eutropha* plant oil cultivations [37]. The formation of PHB without nutrient starvation could indicate a stress response triggered by the high urea levels. Stress responses typically involve the formation of the alarmone (p)ppGpp. For *R. eutropha* it is known that formation of this alarmone triggers PHB formation [38,40], but (p)ppGpp formation due to excess urea or ammonia availability was not studied so far. Additionally, it was reported that controlled induction of stress could also been used for an enhanced PHB formation [41]. Such stress responses should be thoroughly considered during the scale-up of a *R. eutropha* PHA production process, as zones of high or low substrate availability occur in large scale bioreactors. The high impact of such substrate gradients on a reduction of biomass and product yields were intensively studied for *Escherichia coli* [42]. Adapting the feeding strategy during the PHA production process by using *in-line* monitoring devices could be a potential scenario for avoiding such negative impacts on the process.

A reduction of the μ_s' signal after 35 h was observed during the high-cell-density fed-batch cultivation, whereas μ_a stayed constant during that period (Figure 5). The decrease of μ_s' instead of a leveling off of the signal contradicts that a signal saturation effect was observed. Additionally, scattering coefficients in suspensions with particle contents of up to 40% (v·v^{-1}) were measured successfully with this technology [9,10]. Heavy foaming, which occurred after 35 h, could be the reason for the observed signal reductions. The surplus of foam was constantly forced into the liquid phase, which increased the overall gas hold-up and total reaction volume in the system. This additional gas volume results in a dilution of the system, which could explain the μ_s' decrease (35–45 h) even though the culture continued to accumulate PHB (Figure 5). The foaming occurred after the end of the continuous rapeseed oil feed at 35 h. Before 35 h, the added oil functioned as a natural antifoam agent by decreasing the surface tension of the culture broth.

During plant oil cultivations foaming occurs through the emulsification process. *R. eutropha* emulsifies plant oils before uptake, which is catalyzed by extracellular lipases. The lipases cleave the triacylglycerols in diacylglycerols, monoacylglycerols, glycerol, and free fatty acids (FFAs) [23,43,44], which causes heavy foaming during aerated bioreactor cultivations. Nevertheless, μ_s' stayed constant after the CDW did not further increase at 45 h, which could indicate the perfect time point for harvesting in an industrial process. A reliable *in-line* quantification of CDW, RCDW, and subsequently PHA concentration was reached until a CDW of 84 g·L^{-1}. To increase the robustness of the method further at higher cell densities, the calculated gas-hold up in the bioreactor [45] could be integrated in the correlation of the PDW spectroscopy signals. In order to quantify the direct impact of the PHA concentration on the optical coefficients, further studies referencing cell counts and sizes by flow cytometry and microscopy need to be conducted.

A wavelength of 638 nm was used to evaluate the PDW spectroscopy signals during this study, which did not show any correlations with the oil addition or emulsification (Figure 3). The emulsification process is very important for an efficient growth. It was previously shown that an overexpression of lipases results in a reduced lag phase and subsequently a more efficient process [43]. In previous studies, it was shown that PDW spectroscopy was used to measure emulsions [46,47]. An *in-line* determination of the oil content or the emulsion formation would be a very valuable additional information for bioprocess development and process control. This information might result from integration of additional wavelengths into the PDW spectroscopy set-up.

To summarize, PDW spectroscopy allows *in-line* estimation of the CDW, RCDW, and PHB content in real-time. In contrast, *off-line* analysis is typically carried out to determine the PHB content, which includes drying the cells and polymer derivatization before time-consuming HPLC or GC analysis. By using PDW spectroscopy, process development and scale-up could be accelerated. In addition, this technology could be used at a large scale for process monitoring and control of *R. eutropha* cultivations. Specifically, the real-time adjustment of feeding strategies according to the PHA production rates—determined by PDW spectroscopy signals—holds great potential.

5. Conclusions

Here, we show that *in-line* PDW spectroscopy is a powerful PAT tool for monitoring *R. eutropha*-based PHB production. The reduced scattering coefficient μ_s' and absorption coefficient μ_a showed very reproducible signals during different biological cultivations. The new method described in this study allows *in-line* monitoring of CDW, RCDW, and PHB concentrations in *R. eutropha* cultivations up to a CDW of 84 g·L^{-1}. PDW spectroscopy could contribute to improving the scaling-up process and thus to performing PHA production processes in an economical efficient way with the ultimate goal to commercialize a green sustainable plastic.

Author Contributions: Conceptualization, B.G., T.S., R.H., S.L.R.; Methodology, B.G., T.S., R.H., S.L.R.; Validation, B.G., T.S.; Formal analysis, B.G., T.S.; Investigation, B.G., T.S., M.T.H.W.; Resources, P.N., R.H., S.L.R.; Data curation, B.G., T.S.; Writing—original draft preparation, B.G., S.L.R.; Writing—review and editing, B.G., T.S., M.T.H.W., P.N., R.H., S.L.R.; Visualization, B.G.; Supervision, R.H., S.L.R.; Project administration, S.L.R.; Funding acquisition, R.H., S.L.R., P.N.

Acknowledgments: We thank Roche CustomBiotech (Mannheim, Germany) for the supply of the Cedex Bio HT Analyzer. We thank Thomas Högl for helping with the installation of the PDW spectroscopy probe. We acknowledge support by the Open Access Publication Funds of TU Berlin.

Appendix A

$$\text{CDW } [\text{g·L}^{-1}] = 19.012 \, [\text{mm·g·L}^{-1}] \, \mu_s' \, [\text{mm}^{-1}] - 1.2404 \, [\text{g·L}^{-1}] \quad R^2 = 0.96 \tag{A1}$$

$$\text{RCDW } [\text{g·L}^{-1}] = 1872 \, [\text{mm·g·L}^{-1}] \, \mu_a \, [\text{mm}^{-1}] + 2.5722 \, [\text{g·L}^{-1}] \quad R^2 = 0.90 \tag{A2}$$

$$\text{PHB } [\text{wt\%}] = 100 \, [\text{wt\%}] \, (\text{CDW } [\text{g·L}^{-1}] - \text{RCDW } [\text{g·L}^{-1}])/\text{CDW } [\text{g·L}^{-1}] \tag{A3}$$

References

1. FDA Guidance for Industry PAT: A Framework for Innovative Pharmaceutical Development, Manufacuring, and Quality Assurance. FDA Off. Doc. 2004. Available online: https://www.fda.gov/regulatory-information/search-fda-guidance-documents/pat-framework-innovative-pharmaceutical-development-manufacturing-and-quality-assurance (accessed on 18 September 2019).
2. Gomes, J.; Chopda, V.R.; Rathore, A.S. Integrating systems analysis and control for implementing process analytical technology in bioprocess development. *J. Chem. Technol. Biotechnol.* **2015**, *90*, 583–589. [CrossRef]
3. Karmann, S.; Follonier, S.; Bassas-Galia, M.; Panke, S.; Zinn, M. Robust *at-line* quantification of poly(3-hydroxyalkanoate) biosynthesis by flow cytometry using a BODIPY 493/503-SYTO 62 double-staining. *J. Microbiol. Methods* **2016**, *131*, 166–171. [CrossRef] [PubMed]
4. Lee, J.H.; Lee, S.H.; Yim, S.S.; Kang, K.; Lee, S.Y.; Park, S.J.; Jeong, K.J. Quantified High-Throughput Screening of *Escherichia coli* Producing Poly (3-hydroxybutyrate) Based on FACS. *Appl. Biochem. Biotechnol.* **2013**, *170*, 1767–1779. [CrossRef] [PubMed]
5. Samek, O.; Obruča, S.; Šiler, M.; Sedláček, P.; Benešová, P.; Kučera, D.; Márova, I.; Ježek, J.; Bernatová, S.; Zemánek, P. Quantitative Raman Spectroscopy Analysis of Polyhydroxyalkanoates Produced by *Cupriavidus necator* H16. *Sensors* **2016**, *16*, 1808. [CrossRef] [PubMed]
6. Porras, M.A.; Cubitto, M.A.; Villar, M.A. A new way of quantifying the production of poly(hydroxyalkanoate)s using FTIR. *J. Chem. Technol. Biotechnol.* **2016**, *91*, 1240–1249. [CrossRef]
7. Hesselmann, R.P.X.; Fleischmann, T.; Hany, R.; Zehnder, A.J.B. Determination of polyhydroxyalkanoates in activated sludge by ion chromatographic and enzymatic methods. *J. Microbiol. Methods* **1999**, *35*, 111–119. [CrossRef]
8. Koller, M.; Rodríguez-Contreras, A. Techniques for tracing PHA-producing organisms and for qualitative and quantitative analysis of intra- and extracellular PHA. *Eng. Life Sci.* **2015**, *15*, 558–581. [CrossRef]
9. Hass, R.; Munzke, D.; Vargas Ruiz, S.; Tippmann, J.; Reich, O. Optical monitoring of chemical processes in turbid biogenic liquid dispersions by Photon Density Wave spectroscopy. *Anal. Bioanal. Chem.* **2015**, *407*, 2791–2802. [CrossRef]
10. Münzberg, M.; Hass, R.; Dinh Duc Khanh, N.; Reich, O. Limitations of turbidity process probes and formazine as their calibration standard. *Anal. Bioanal. Chem.* **2017**, *409*, 719–728. [CrossRef]
11. Häne, J.; Brühwiler, D.; Ecker, A.; Hass, R. Real-time inline monitoring of zeolite synthesis by Photon Density Wave spectroscopy. *Microporous Mesoporous Mater.* **2019**, *288*, 109580. [CrossRef]
12. Hass, R.; Münzberg, M.; Bressel, L.; Reich, O. Industrial applications of photon density wave spectroscopy for *in-line* particle sizing. *Appl. Opt.* **2013**, *52*, 1429–1431. [CrossRef] [PubMed]
13. Bressel, L.; Hass, R.; Reich, O. Particle sizing in highly turbid dispersions by Photon Density Wave spectroscopy. *J. Quant. Spectrosc. Radiat. Transf.* **2013**, *126*, 122–129. [CrossRef]
14. Hass, R.; Reich, O. Photon density wave spectroscopy for dilution-free sizing of highly concentrated

nanoparticles during starved-feed polymerization. *ChemPhysChem* **2011**, *12*, 2572–2575. [CrossRef] [PubMed]

15. Koller, M.; Shahzad, K.; Braunegg, G. Waste streams of the animal-processing industry as feedstocks to produce polyhydroxyalkanoate biopolyesters. *Appl. Food Biotechnol.* **2018**, *5*, 193–203.

16. Kourmentza, C.; Plácido, J.; Venetsaneas, N.; Burniol-Figols, A.; Varrone, C.; Gavala, H.N.; Reis, M.A.M. Recent Advances and Challenges towards Sustainable Polyhydroxyalkanoate (PHA) Production. *Bioengineering* **2017**, *4*, 55. [CrossRef] [PubMed]

17. Riedel, S.L.; Jahns, S.; Koenig, S.; Bock, M.C.E.; Brigham, C.J.; Bader, J.; Stahl, U. Polyhydroxyalkanoates production with *Ralstonia eutropha* from low quality waste animal fats. *J. Biotechnol.* **2015**, *214*, 119–127. [CrossRef] [PubMed]

18. Brigham, C.J.; Riedel, S.L. The Potential of Polyhydroxyalkanoate Production from Food Wastes. *Appl. Food Biotechnol.* **2018**, *6*, 7–18.

19. Ong, S.Y.; Kho, H.P.; Riedel, S.L.; Kim, S.W.; Gan, C.Y.; Taylor, T.D.; Sudesh, K. An integrative study on biologically recovered polyhydroxyalkanoates (PHAs) and simultaneous assessment of gut microbiome in yellow mealworm. *J. Biotechnol.* **2018**, *265*, 31–39. [CrossRef]

20. Riedel, S.L.; Brigham, C.J.; Budde, C.F.; Bader, J.; Rha, C.; Stahl, U.; Sinskey, A.J. Recovery of poly(3-hydroxybutyrate-*co*-3-hydroxyhexanoate) from *Ralstonia eutropha* cultures with non-halogenated solvents. *Biotechnol. Bioeng.* **2013**, *110*, 461–470. [CrossRef]

21. Obruca, S.; Marova, I.; Snajdar, O.; Mravcova, L.; Svoboda, Z. Production of poly(3-hydroxybutyrate-*co*-3-hydroxyvalerate) by *Cupriavidus necator* from waste rapeseed oil using propanol as a precursor of 3-hydroxyvalerate. *Biotechnol. Lett.* **2010**, *32*, 1925–1932. [CrossRef]

22. Reinecke, F.; Steinbüchel, A. *Ralstonia eutropha* strain H16 as model organism for PHA metabolism and for biotechnological production of technically interesting biopolymers. *J. Mol. Microbiol. Biotechnol.* **2008**, *16*, 91–108. [CrossRef] [PubMed]

23. Brigham, C.J.; Budde, C.F.; Holder, J.W.; Zeng, Q.; Mahan, A.E.; Rha, C.K.; Sinskey, A.J. Elucidation of β-oxidation pathways in *Ralstonia eutropha* H16 by examination of global gene expression. *J. Bacteriol.* **2010**, *192*, 5454–5464. [CrossRef] [PubMed]

24. Riedel, S.L.; Bader, J.; Brigham, C.J.; Budde, C.F.; Yusof, Z.A.M.; Rha, C.; Sinskey, A.J. Production of poly(3-hydroxybutyrate-*co*-3-hydroxyhexanoate) by *Ralstonia eutropha* in high cell density palm oil fermentations. *Biotechnol. Bioeng.* **2012**, *109*, 74–83. [CrossRef]

25. Kahar, P.; Tsuge, T.; Taguchi, K.; Doi, Y. High yield production of polyhydroxyalkanoates from soybean oil by *Ralstonia eutropha* and its recombinant strain. *Polym. Degrad. Stab.* **2004**, *83*, 79–86. [CrossRef]

26. Rehm, B.H.A. Polyester synthases: Natural catalysts for plastics. *Biochem. J.* **2003**, *376*, 15–33. [CrossRef] [PubMed]

27. Pohlmann, A.; Fricke, W.F.; Reinecke, F.; Kusian, B.; Liesegang, H.; Cramm, R.; Eitinger, T.; Ewering, C.; Pötter, M.; Schwartz, E.; et al. Genome sequence of the bioplastic-producing "Knallgas" bacterium *Ralstonia eutropha* H16. *Nat. Biotechnol.* **2006**, *24*, 1257–1262. [CrossRef] [PubMed]

28. Ng, K.S.; Ooi, W.Y.; Goh, L.K.; Shenbagarathai, R.; Sudesh, K. Evaluation of jatropha oil to produce poly(3-hydroxybutyrate) by *Cupriavidus necator* H16. *Polym. Degrad. Stab.* **2010**, *95*, 1365–1369. [CrossRef]

29. Arikawa, H.; Matsumoto, K. Evaluation of gene expression cassettes and production of poly(3-hydroxybutyrate-*co*-3-hydroxyhexanoate) with a fine modulated monomer composition by using it in *Cupriavidus necator*. *Microb. Cell Fact.* **2016**, *15*, 1–11. [CrossRef]

30. Arikawa, H.; Matsumoto, K.; Fujiki, T. Polyhydroxyalkanoate production from sucrose by *Cupriavidus necator* strains harboring csc genes from *Escherichia coli* W. *Appl. Microbiol. Biotechnol.* **2017**, *101*, 7497–7507. [CrossRef]

31. Sato, S.; Maruyama, H.; Fujiki, T.; Matsumoto, K. Regulation of 3-hydroxyhexanoate composition in PHBH synthesized by recombinant *Cupriavidus necator* H16 from plant oil by using butyrate as a co-substrate. *J. Biosci. Bioeng.* **2015**, *120*, 246–251. [CrossRef]

32. Karr, D.B.; Waters, J.K.; Emerich, D.W. Analysis of poly-β-hydroxybutyrate in *Rhizobium japonicum* bacteroids by ion-exclusion high-pressure liquid chromatography and UV detection. *Appl. Environ. Microbiol.* **1983**, *46*, 1339–1344. [PubMed]

33. Koller, M. Advances in Polyhydroxyalkanoate (PHA) Production. *Bioengineering* **2017**, *4*, 88. [CrossRef] [PubMed]

34. Cruz, M.V.; Sarraguça, M.C.; Freitas, F.; Lopes, J.A.; Reis, M.A.M. Online monitoring of P(3HB) produced from used cooking oil with near-infrared spectroscopy. *J. Biotechnol.* **2015**, *194*, 1–9. [CrossRef] [PubMed]

35. Budde, C.F.; Riedel, S.L.; Hübner, F.; Risch, S.; Popović, M.K.; Rha, C.; Sinskey, A.J. Growth and polyhydroxybutyrate production by *Ralstonia eutropha* in emulsified plant oil medium. *Appl. Microbiol. Biotechnol.* **2011**, *89*, 1611–1619. [CrossRef] [PubMed]

36. Sato, S.; Fujiki, T.; Matsumoto, K. Construction of a stable plasmid vector for industrial production of poly(3-hydroxybutyrate-*co*-3-hydroxyhexanoate) by a recombinant *Cupriavidus necator* H16 strain. *J. Biosci. Bioeng.* **2013**, *116*, 677–681. [CrossRef] [PubMed]

37. Fadzil, F.I.B.M.; Tsuge, T. Bioproduction of polyhydroxyalkanoate from plant oils. In *Microbial Applications*; Springer: Berlin, Germany, 2017; Volume 2, pp. 231–260. ISBN 9783319526690.

38. Juengert, J.R.; Borisova, M.; Mayer, C.; Wolz, C.; Brigham, C.J.; Sinskey, A.J.; Jendrossek, D. Absence of ppGpp leads to increased mobilization of intermediately accumulated poly(3-hydroxybutyrate) in *Ralstonia eutropha* H16. *Appl. Environ. Microbiol.* **2017**, *83*, 1–16. [CrossRef] [PubMed]

39. Karmann, S.; Panke, S.; Zinn, M. The Bistable Behaviour of *Pseudomonas putida* KT2440 during PHA Depolymerization under Carbon Limitation. *Bioengineering* **2017**, *4*, 58. [CrossRef] [PubMed]

40. Brigham, C.J.; Speth, D.R.; Rha, C.K.; Sinskey, A.J. Whole-genome microarray and gene deletion studies reveal regulation of the polyhydroxyalkanoate production cycle by the stringent response in *Ralstonia eutropha* H16. *Appl. Environ. Microbiol.* **2012**, *78*, 8033–8044. [CrossRef] [PubMed]

41. Obruca, S.; Marova, I.; Svoboda, Z.; Mikulikova, R. Use of controlled exogenous stress for improvement of poly(3-hydroxybutyrate) production in *Cupriavidus necator*. *Folia Microbiol.* **2010**, *55*, 17–22. [CrossRef]

42. Enfors, S.O.; Jahic, M.; Rozkov, A.; Xu, B.; Hecker, M.; Jürgen, B.; Krüger, E.; Schweder, T.; Hamer, G.; O'Beirne, D.; et al. Physiological responses to mixing in large scale bioreactors. *J. Biotechnol.* **2001**, *85*, 175–185. [CrossRef]

43. Lu, J.; Brigham, C.J.; Rha, C.; Sinskey, A.J. Characterization of an extracellular lipase and its chaperone from *Ralstonia eutropha* H16. *Appl. Microbiol. Biotechnol.* **2013**, *97*, 2443–2454. [CrossRef] [PubMed]

44. Riedel, S.L.; Lu, J.; Stahl, U.; Brigham, C.J. Lipid and fatty acid metabolism in *Ralstonia eutropha*: Relevance for the biotechnological production of value-added products. *Appl. Microbiol. Biotechnol.* **2014**, *98*, 1469–1483. [CrossRef] [PubMed]

45. Sieblist, C.; Lübbert, A. Gas Holdup in Bioreactors. In *Encyclopedia of Industrial Biotechnology*; John Wiley & Sons: New York, NY, USA, 2010; pp. 1–8.

46. Reich, O.; Bressel, L.; Hass, R. Sensing emulsification processes by Photon Density Wave spectroscopy. In *the 21st International Conference on Optical Fiber Sensors*; The Institute of Electrical and Electronics Engineers: New York, NY, USA, 2011; Volume 7753, p. 77532J.

47. Münzberg, M.; Hass, R.; Reich, O. *In-line* characterization of phase inversion temperature emulsification by photon density wave spectroscopy. *SOFW J.* **2013**, *4*, 38–46.

7

Prospecting for Marine Bacteria for Polyhydroxyalkanoate Production on Low-Cost Substrates

Rodrigo Yoji Uwamori Takahashi, Nathalia Aparecida Santos Castilho, Marcus Adonai Castro da Silva, Maria Cecilia Miotto and André Oliveira de Souza Lima *

Centro de Ciências Tecnológicas da Terra e do Mar, Universidade do Vale do Itajaí, R. Uruguai 458, Itajaí-SC 88302-202, Brazil; rodrigo.aquicultura@gmail.com (R.Y.U.T.); nathi_zuca@hotmail.com (N.A.S.C.); marcus.silva@univali.br (M.A.C.S.); cecilia.miotto@gmail.com (M.C.M.)
* Correspondence: andreolima@gmail.com

Academic Editor: Martin Koller

Abstract: Polyhydroxyalkanoates (PHAs) are a class of biopolymers with numerous applications, but the high cost of production has prevented their use. To reduce this cost, there is a prospect for strains with a high PHA production and the ability to grow in low-cost by-products. In this context, the objective of this work was to evaluate marine bacteria capable of producing PHA. Using Nile red, 30 organisms among 155 were identified as PHA producers in the medium containing starch, and 27, 33, 22 and 10 strains were found to be positive in media supplemented with carboxymethyl cellulose, glycerol, glucose and Tween 80, respectively. Among the organisms studied, two isolates, LAMA 677 and LAMA 685, showed strong potential to produce PHA with the use of glycerol as the carbon source, and were selected for further studies. In the experiment used to characterize the growth kinetics, LAMA 677 presented a higher maximum specific growth rate ($\mu max = 0.087 \ h^{-1}$) than LAMA 685 ($\mu max = 0.049 \ h^{-1}$). LAMA 677 also reached a D-3-hydroxybutyrate (P(3HB)) content of 78.63% (dry biomass), which was 3.5 times higher than that of LAMA 685. In the assay of the production of P(3HB) from low-cost substrates (seawater and biodiesel waste glycerol), LAMA 677 reached a polymer content of 31.7%, while LAMA 685 reached 53.6%. Therefore, it is possible to conclude that the selected marine strains have the potential to produce PHA, and seawater and waste glycerol may be alternative substrates for the production of this polymer.

Keywords: biopolymer; seawater; waste glycerol; deep sea

1. Introduction

PHAs (polyhydroxyalkanoates) are a class of polyesters produced by prokaryotic microorganisms that are accumulated inside cells as carbon and energy reserves [1,2]. These biopolymers have drawn great interest due to their biodegradability, biocompatibility, the possibility of biosynthesis from renewable resources, and similar physical and chemical characteristics to the main petrochemical polymers [3,4]. Despite the environmental benefits and the potential of using this raw material in several areas, the production costs are relatively high compared to those of conventional polymers [5].

There are bacteria known to be PHA producers, such as *Cupriavidus necator*, *Azohydromonas lata* and *Azotobacter vinelandii*. However, the other organisms capable of using low-cost substrates, accumulating a high PHA content, presenting high productivity, producing copolymers from single carbon sources, and producing other PHA monomer compositions have high economic importance [3,4,6,7]. According to Quillaguamán et al. [8], the marine environment has been poorly explored in terms of prospecting for PHA-producing organisms. However, recent research on halophiles indicates a strong potential for biotechnological production of PHAs, based on a study of the bacterium *Halomonas hydrothermalis*, which was able to accumulate a polyhydroxybutyrate (P(3HB)) content of 75.8% when cultivated in residual glycerol as the only source of carbon [9]. Two other *Halomonas* spp. have been isolated and showed great potential for low-cost PHA production. *Halomonas* sp. TD01 grew rapidly to over 80 g/L cell dry weight (CDW) in a lab fermentor and accumulated a P(3HB) content of over 80% on a glucose–salt medium [10]. *H. campaniensis* LS21 was able to grow in artificial seawater and kitchen-waste-like mixed substrates consisting of cellulose, proteins, fats, fatty acids and starch [11].

There are low-cost ways to synthesize PHAs with the use of halophilic bacteria [12] whose salinity requirements may inhibit the growth of non-halophilic microorganisms, allowing for growth in non-sterile conditions [8,12]. This result was noted in a study by Tan et al. [10], who found evidence of the production of P(3HB) using a non-sterile fermentation process in cultures of *Halomonas* TD01. Moreover, a study by Kawata and Aiba [13] reported the growth of *Halomonas* sp. KM-1 in unsterilized medium cultures. Additionally, seawater can be used as a source of minerals and low-cost nutrients for cultivation, according to a study conducted by Pandian et al. [14] which illustrated the production of PHAs in cultures of *Bacillus megaterium* SRKP-3 (an organism isolated from the marine environment) from seawater, milk residues and rice bran. According to Yin et al. [15], due to their characteristics, using halophiles to produce PHAs can reduce the costs of the fermentation and recovery processes, making them a promising alternative for PHA production.

Therefore, this study describes a search for marine bacteria for PHA production using low-cost substrates, as well as the growth, productivity, and PHA characteristics of the selected producers.

2. Materials and Methods

2.1. Marine Bacteria

In the present study, marine bacteria maintained in the culture collection of the Laboratory of Applied Microbiology (LAMA) of the University of Vale do Itajaí (UNIVALI) were used. The bacteria were obtained through the South Atlantic MAR-ECO Patterns and Processes of the Ecosystems of The Southern Mid-Atlantic projects. The organisms were isolated from sediment and water samples collected between the surface and a depth of 5500 m from the Mid-South Atlantic ridge, the Rio Grande Rise and the Walvis Ridge.

2.2. Qualitative Screening of PHA Producers

All isolates were evaluated based on their ability to synthesize PHAs according to the method described by Spiekermann et al. [16]. Thus, the organisms were cultivated in: (a) Zobell Marine Agar 2216 (AM); (b) AM with added glucose (0.5% w/v); and (c) a mineral medium (MM) with added starch (0.5%), carboxymethyl cellulose (CMC), glycerol, glucose, or Tween 80. The MM composition described by Baumann et al. [17] for 1 L of the medium with a pH of 7.5 was 11.7 g of NaCl, 12.32 g of $MgSO_4 \cdot 7H_2O$, 0.745 g of KCl, 1.47 g of $CaCl_2 \cdot 2H_2O$, 6.05 g of Tris (hydroxymethyl) aminomethane ($C_4H_{11}NO_3$), 6.65 g of NH_4Cl, 0.062 g of $K_2HPO_4 \cdot 3H_2O$, and 0.026 g of $FeSO_4 \cdot 7H_2O$. In order to evaluate the potential of the bacteria to produce PHA, all media (except the control) were supplemented with Nile red (final concentration: 0.5 μg/mL), using a stock solution (0.025% m/v in dimethyl sulfoxide). Methods using Nile red are fast and can detect PHA inside intact cells [18,19]; however, the fluorophore can also stain other lipophilic compounds [20]. Organisms were cultivated in an incubator at 28 °C until evident growth occurred (1–9 days). After growth, the colonies were inspected by direct exposure to ultraviolet light (λ = 312 nm; transilluminator UV-TRANS, model UVT-312) and photographed with a digital camera (Canon, EOS Rebel model, Canon Inc., Tokyo, Japan). The fluorescence intensity of isolates is proportional to the PHA content in the cells, as reported by Degelau et al. [21] and Reddy et al. [22]. In this context, the fluorescence intensity was used (Image-Pro Plus, version 6.0, Media Cybernetics Inc., Rockville, MD, USA) as an indicator of polymer content, and the colony area was associated with the growth capacity. The fluorescence intensity and colony area data were statistically analyzed, and the organisms in each test were compared using a Kruskal-Wallis test. In cases of significance ($p < 0.05$), Dunn's test was applied. Statistical analyses were performed using BioEstat (version 5.3, Instituto Mamirauá; Tefé, Brasil) and Statistica (version 8.0, StatSoft Inc., Oklahoma, OK, USA) software.

2.3. Production of PHA in Semi-Solid and Liquid Mediums

The preculture was prepared by inoculating isolated bacterial colonies in Zobell Marine Broth 2216 (CM) or the MM supplemented with glycerol (5%), and cultivating (28 °C; 150 rpm) for 24–48 h. After growth, the preinoculum (10% v/v) was precipitated via centrifugation, and resuspended to assay the PHA production media. Then, the broth was divided into three aliquots (each 100 mL), and cultivated (68 h; 28 °C; 150 rpm; in 250 mL Erlenmeyer flasks) to assay the PHA production. In the first evaluation round, the bacteria were cultured in a MM supplemented with one of the following carbon sources: (a) starch (2% w/v), (b) CMC (2% w/v), (c) glycerol (5% v/v), (d) glucose (5% w/v), or (e) Tween 80 (5% v/v). To evaluate the growth kinetics, cells were cultivated (28 °C; 150 rpm; in triplicate) for 143 h in a MM supplemented with glycerol (5% v/v). Periodically, samples were collected to determine the biomass using gravimetry.

When evaluating the PHA synthesis from the low-cost carbon sources, different media formulations were used (Table 1). For those experiments, residual glycerol from biodiesel production was used, which was provided by a Brazilian company from Rio Grande do Sul (not identified). Selected bacteria were cultivated (28 °C; 150 rpm; 69 h), and the produced biomass was recovered by centrifugation, washed (with distilled water), lyophilized, weighed, and frozen for further analysis by gas chromatography (GC). The PHA quantification analysis was performed using a complex sample that constituted of equal amounts of each replicate. Thus, the PHA content results represent the average of the three cultures.

Assays were also conducted in semi-solid media. Organisms were inoculated (streak) and incubated (28 °C; 6 days) on the MM with 1.8 g· L^{-1} of agar, and separately supplemented with the following carbon sources: starch (2% w/v), CMC (2% w/v), glycerol (5% v/v), glucose (5% v/v) and Tween 80 (5% v/v). After culturing, cells were resuspended in a saline solution, recovered, washed (with distilled water), lyophilized, weighed, and frozen for further analysis by GC (Shimadzu, modelo 17A, Kyoto, Japan).

Table 1. Culture media formulations used in PHA synthesis assays from different sources of low-cost carbon, minerals and nutrients (residual glycerol and seawater).

Culture Media	Composition
1	90% MM medium * + 5% (v/v) glycerol + 5% distilled water
2	90% MM medium + 10% (v/v) glycerol
3	90% seawater + 5% (v/v) glycerol + 5% distilled water
4	90% seawater + 5% (v/v) residual glycerol + 5% distilled water
5	90% seawater + 10% (v/v) residual glycerol
6	90% MM medium + 5% (v/v) residual glycerol + 5% distilled water

* Marine mineral medium.

2.4. Quantification of PHA by Gas Chromatography

For PHA quantification, approximately 50 mg of freeze-dried cells were weighed and transferred to a screw-cap tube. Then, to extract the polymer and submit it to the methanolysis reaction to form monomers of methyl esters, 2 mL of H_2SO_4/methanol (5:95) was added, along with 2 mL of chloroform and 250 μL of internal standard solution, which was composed of 20 mg of benzoic acid in 1 mL of methanol. Due to the reduced biomass in some assays, 1 mL of chloroform was added during methanolysis. The tubes were then heated (100 °C) in a dry block for 3 h with occasional stirring. After the reaction, the tubes were cooled to room temperature, 1 mL of distilled water was added, and the tubes were vortexed for 30 s and left for phase separation. The chloroform phase was transferred to chromatography vials for analysis. A Shimadzu gas chromatograph (model 17A) equipped with a flame ionization detector (FID) adjusted to 280 °C was used. The temperature program was set to an initial temperature of 50 °C for 2 min, then increased from 50 °C to 110 °C at a rate of 20 °C/min, and finally increased to 250 °C at a rate of 20 °C/min. The injector was maintained at 250 °C, and the oven was maintained at 120 °C. The column used was a VB-WAX (VICI) column, 30 m long, 0.25 mm in diameter, and with 0.25 mm film thickness. The volume injected was 1 µL, with the helium flow set at 1 mL/min, with a total run time of 12 min. The standard curve was made using P(3HB) (Sigma-Aldrich, St. Louis, MO, USA) as an external standard, with a mass ranging from 0.005 to 0.020 g. The standards were submitted to the methanolysis process, as described previously for the bacterial samples.

3. Results and Discussion

3.1. Screening for PHA-Producing Marine Bacteria

Among the 155 isolates evaluated, 40.6% (63 isolates) presented fluorescence indicative of PHA production in at least one of the growth conditions tested. The results were distributed as follows: 19.4% (30 isolates) were active in one culture condition, 5.8% (9 isolates) were active in two mediums, 5.2% (8 isolates) were active in three mediums, 7.1% (11 isolates) were active in four mediums, 0.6% (1 isolate) was active in five mediums, 1.9% (3 isolates) were active in six mediums, and 0.6% (1 isolate) presented fluorescence in every medium tested (Table 2). Regarding the substrates added in the MM, the number of isolates capable of producing PHA was higher when starch was added, with 30 representatives, followed by 27, 23, 22 and 10 organisms classified as producers in assays supplied with CMC, glycerol, glucose and Tween 80, respectively. In the MM supplied with glycerol, 23 positive isolates (14.7%) were identified as polymer producers, a percentage similar to that observed by Shrivastav et al. [9], who selected PHA producers from the marine environment using residual glycerol as the carbon source, and identified 14% positives using the Nile red method.

Table 2. Results of the qualitative analysis of the growth and PHA production of marine bacteria isolates when exposed to two culture media (marine agar and mineral marine) with different formulations.

Isolate LAMA [3]	MA [1]		MM [2]					Isolate LAMA	MA		MM				
	NS	GU	ST	CM	GL	GU	TW		NS	GU	ST	CM	GL	GU	TW
570	−	−	wg	wg	wg	wg	wg	671	−	−	−	+	−	−	−
571	−	−	wg	wg	wg	−	wg	672	−	−	−	−	−	−	−
572	−	−	+	wg	−	−	−	673	−	−	+	+	+	+	−
573	−	−	wg	wg	wg	−	wg	674	+	−	+	+	+	+	+
574	−	−	wg	wg	wg	wg	−	675	−	−	wg	wg	wg	wg	−
575	−	−	wg	wg	wg	wg	wg	677	−	−	+	+	+	−	+
576	−	−	wg	wg	wg	wg	wg	679	+	+	+	+	+	+	+
577	−	−	wg	wg	−	−	wg	680	−	−	−	−	−	−	−
580	−	−	wg	−	−	−	wg	681	−	−	wg	−	−	−	wg
582	−	−	wg	wg	−	−	wg	683	−	−	wg	wg	wg	wg	+
583	−	−	−	wg	−	−	−	684	−	−	wg	wg	wg	wg	−
584	+	−	wg	wg	wg	wg	wg	685	−	−	−	+	+	+	−
585	−	−	wg	wg	−	−	wg	687	−	−	wg	wg	wg	wg	wg
587	−	−	−	−	−	−	−	688	−	−	wg	wg	wg	wg	wg
592	+	−	−	−	−	−	−	689	−	−	−	−	wg	−	−
593	−	−	wg	wg	wg	wg	wg	690	−	−	wg	−	wg	−	−
594	−	−	+	+	+	−	−	691	−	−	wg	wg	wg	wg	−
595	−	−	wg	−	wg	wg	wg	692	−	−	wg	wg	wg	wg	wg
597	−	−	wg	wg	−	wg	wg	693	−	−	−	−	−	−	−
598	−	−	wg	wg	wg	−	−	694	−	−	wg	wg	wg	wg	wg
599	−	−	+	+	−	−	−	695	−	−	−	−	−	−	−
600	−	−	−	−	−	−	−	696	−	−	wg	wg	wg	wg	wg
601	−	−	+	+	−	−	−	697	+	+	−	−	−	+	−
604	−	−	+	−	wg	−	−	698	−	−	wg	wg	wg	wg	wg
606	−	−	wg	wg	wg	wg	wg	699	−	−	wg	wg	wg	−	−

Table 2. *Cont.*

Isolate LAMA [3]	MA [1]		MM [2]					Isolate LAMA	MA		MM				
	NS	GU	ST	CM	GL	GU	TW		NS	GU	ST	CM	GL	GU	TW
607	−	−	wg	wg	wg	−	−	700	−	−	−	wg	−	−	−
608	−	−	wg	wg	wg	wg	wg	701	−	−	wg	wg	−	wg	wg
610	−	−	wg	−	−	−	wg	702	−	−	+	+	+	+	−
611	−	−	wg	wg	wg	wg	−	703	−	−	+	−	−	−	−
612	−	−	+	+	+	+	−	704	−	−	+	+	−	−	−
613	−	−	wg	wg	wg	wg	wg	705	−	−	−	wg	−	−	−
614	−	−	wg	wg	+	−	wg	706	−	−	wg	wg	wg	wg	+
615	−	−	−	−	−	−	−	707	+	+	−	−	−	−	wg
616	−	−	wg	wg	wg	−	−	708	−	−	wg	−	−	−	wg
617	−	−	wg	wg	wg	wg	−	709	−	−	wg	wg	−	wg	wg
618	−	+	wg	wg	wg	wg	wg	710	−	−	+	wg	−	−	−
619	−	−	wg	wg	wg	−	−	711	+	+	+	+	+	+	−
644	−	+	−	−	−	−	−	712	−	−	wg	wg	wg	wg	−
647	−	−	−	−	−	−	−	713	−	−	wg	wg	wg	wg	wg
650	−	−	wg	−	−	+	−	715	−	−	wg	wg	wg	−	wg
653	+	−	−	−	−	−	+	716	−	−	wg	wg	wg	wg	wg
659	−	−	+	+	+	+	−	717	−	+	−	wg	wg	wg	wg
667	−	−	−	wg	−	−	−	718	−	−	wg	wg	−	−	wg
669	−	−	−	−	−	+	−	719	−	−	+	−	−	−	−
720	−	−	wg	wg	wg	wg	wg	759	−	−	wg	wg	wg	wg	wg
722	−	−	wg	+	−	+	−	760	−	−	+	+	+	−	−
723	−	−	−	−	−	−	−	761	+	+	−	+	+	+	+
725	−	−	−	wg	wg	−	wg	762	−	−	wg	wg	−	−	−
726	+	−	−	wg	−	+	+	763	+	+	wg	wg	wg	−	−
727	−	−	−	wg	wg	−	wg	764	−	−	wg	wg	wg	wg	wg
728	−	−	wg	wg	−	−	wg	765	−	−	+	+	+	+	−
729	−	+	+	+	+	+	−	766	−	−	+	wg	−	−	−
730	−	−	wg	wg	−	−	−	767	−	−	−	−	−	−	−
731	−	−	−	−	−	−	wg	768	−	−	wg	wg	wg	wg	wg
732	+	−	−	wg	wg	−	−	769	−	−	wg	wg	wg	wg	wg
733	−	−	−	−	−	−	−	773	−	−	+	+	+	+	−
734	−	+	−	wg	wg	−	−	775	−	−	wg	wg	−	−	wg
735	−	−	wg	wg	−	+	−	778	−	−	+	−	−	−	−
736	+	−	−	wg	+	+	−	779	−	−	−	−	wg	−	−
737	−	−	+	+	+	+	−	781	−	−	wg	−	−	−	−
738	−	−	wg	wg	−	−	−	782	−	−	−	wg	wg	wg	wg
739	−	−	wg	wg	wg	wg	wg	786	−	−	−	−	−	−	−
741	−	−	wg	wg	wg	−	−	790	−	−	+	−	−	−	−
742	−	−	wg	−	wg	−	−	791	−	−	+	−	−	−	−
743	−	−	wg	wg	−	−	wg	M 112	−	−	wg	wg	wg	wg	wg
744	−	+	wg	wg	wg	−	−	M 135	+	+	wg	wg	wg	wg	wg
746	−	−	wg	wg	−	−	−	M 151	+	−	wg	wg	−	−	−
747	−	−	−	−	−	−	+	M 169	−	−	wg	wg	wg	−	wg
748	−	−	+	+	+	−	−	M 171	+	−	wg	wg	wg	−	−
749	−	−	wg	wg	wg	−	wg	M 173	−	+	wg	wg	−	wg	wg
750	−	−	wg	−	+	+	−	M 180	+	+	+	+	−	−	−
751	−	−	−	−	wg	−	wg	M 189	−	−	−	−	−	−	−
753	−	−	wg	wg	wg	−	wg	M 198	−	−	wg	wg	−	−	wg
754	−	−	+	+	+	−	−	M 199	−	−	+	+	+	−	+
755	−	−	wg	wg	wg	wg	−	M 211	−	−	−	+	−	−	−
756	−	−	wg	wg	wg	−	−	M 84	+	−	wg	wg	wg	wg	wg
757	−	−	wg	−	−	−	wg	M 97	−	−	wg	−	−	wg	wg
758	+	−	−	+	+	+	−								

(+) PHA producer; (−) non-PHA producer; (wg) without growth; [1] MA—commercial marine agar medium (NS—not supplemented, and GU—supplemented with glucose); [2] MM—mineral medium (ST—supplemented with starch, CM—supplemented with carboxymethylcellulose, GL—supplemented with glycerol, GU—supplemented with glucose, and TW—supplemented with Tween 80). [3] Marine bacteria of the Laboratory of Applied Microbiology of the University of Vale do Itajaí.

When considering the use of agro-industrial by-products as substrates to produce PHA, the great potential of some bacteria can be recognized, particularly the isolates LAMA 674, LAMA 677,

LAMA 679 and M 199. The statistical analyses of the colony area and fluorescence on semi-solid media were not significantly different among organisms (data not shown). However, it was possible to identify the organisms with higher fluorescence and growth (area) in each medium, and these were selected for further evaluation. LAMA 679 and LAMA 732 growing in the AM medium were among the highest producers. On the other hand, LAMA 711 and LAMA 644 performed better in the AM medium supplemented with glucose, and LAMA 748 and LAMA 737 performed well in the MM with added starch. In the MM with CMC, LAMA 748 and LAMA 674 performed the best; in the MM with added glycerol, LAMA 685 and LAMA 677 were also sufficient. Finally, when cultivated in the MM with added Tween 80, LAMA 726 and M 199 presented the highest indices.

According to the data provided by the MAR-ECO project, 49.7% of the isolates were obtained from water samples, and 50.3% were obtained from sediments. When considering the sea zonation defined by Hedgpeth [23], 76.6% of the bacteria were collected in the epipelagic zone, 16.9% in the mesopelagic zone, and 6.5% in the bathypelagic zone. In the benthic domain, 29.5% of bacteria were collected from the bathyal zone, and 70.5% were obtained from the abyssal zone. Regarding the sampling locations, 49.7% were from the Rio Grande Rise, 38.7% were from the Walvis Ridge, and 11.6% were from the Mid-South Atlantic ridge. Given this information, the most promising organisms for PHA production were taken from the epipelagic zone, with six organisms, while one isolate was taken from the mesopelagic zone. It is also important to emphasize that 93.5% of the studied bacteria were from the epi- and meso-pelagic zones. To produce PHA, microorganisms need substrates with excess carbon sources, as can be observed in the epi- and meso-pelagic zones [24]. This fact may explain why most of the organisms capable of accumulating PHA were from these zones.

3.2. PHA Production in Different Substrates

Among the bacteria screened in the MM assay, those with higher intensities and colony areas were chosen to be evaluated for PHA production in liquid media. As shown in Table 3, production of PHA was not detected in the assays where the media were supplemented with starch, CMC or Tween. However, it is believed that most of these samples were composed of the carbon source itself, so the GC did not detect PHA production. On the other hand, when evaluated in the medium supplemented with glycerol, LAMA 677 had a productivity of 0.0058 g· L^{-1}·h^{-1}, reaching 1.41 ± 0.18 g· L^{-1} of biomass with 28.28% of a polymer identified by GC as P(3HB). Additionally, LAMA 685 presented better indices with a biomass of 2.03 ± 0.27 g· L^{-1} (32.79% P(3HB)) with 0.0098 g· L^{-1}·h^{-1} productivity. A higher P(3HB) production was observed when the media was supplemented with glycerol as a carbon source. This condition was even better than when the media were supplemented with glucose, which is a readily assimilated source. For example, LAMA 685 accumulated 17% more biomass (2.03 g· L^{-1}) and 3.4 times more P(3HB) (32.79%) when growing in glycerol compared to glucose as the carbon source. Similar results have been reported previously. Chien et al. [25] evaluated the use of bacteria isolated from mangrove sediments to produce PHA and reported the best PHA content with the use of glycerol in comparison to glucose. According to Chien et al. [25], the organism named M11 produced 30.2% PHA in cells in glycerol culture and 8.1% in the supplied glucose culture. Mahishi et al. [26] used recombinant *Escherichia coli* cultures and reported PHA contents of up to 60% in relation to dry-weight PHA with the glycerol supply, and 38% when supplied with glucose as the carbon source. When evaluating different carbon sources, Mohandas et al. [27] reported that glycerol supported the maximum biomass yield and P(3HB) production, followed by glucose and fructose. The biomass yield and P(3HB) production in the presence of glycerol were 3 g· L^{-1} and 68% (w/w), respectively.

Due to the presence of insoluble substrates in some of the treatments tested, it was difficult to detect PHA production, as the biomass analyzed by GC consisted of both the microbial biomass and the residual substrate. Taking this into consideration, a new set of assays were carried out in a semi-solid medium, as an alternative to obtain cells free from the substrates. For instance, when using media supplemented with CMC, LAMA 674 and LAMA 748 produced PHA as 22.26% and 27.64% of biomass, respectively. Although the substrate used in this study was a source of carbon that is more easily assimilated by microorganisms compared to sources used in other studies, these results are comparable to a study by Van-Thuoc et al. [28]. Van-Thuoc et al. [28] used xylose as the carbon source for PHA production by *Halomonas boliviensis* LC1, and obtained polymer yields ranging from 23.1% to 33.8% of biomass. Bertrand et al. [29] reported a 19.6% accumulation of PHA in cultures of *Pseudomonas pseudoflava* supplied with xylose. Silva et al. [30] reported a PHA content of between 35% and 58.2% when using xylose, and contents of 15.39% to 23.22% when using hydrolyzed sugarcane. Additionally, the PHA accumulation results from this study where Tween 80 was added to the medium were interesting, as isolated LAMA 726 reached a biomass content of 55.24% PHA. In a study by Fernández et al. [31], who used cultures of *Pseudomonas aeruginosa* NCIB 40045, a PHA content of 29.4% was obtained when residual frying oil was added to the culture. On the other hand, He et al. [32] achieved a yield of 63% polymers by cultivating *Pseudomonas stutzeri* 1317 in soybean oil.

Table 3. Results of the parameters used to analyze the PHA production of the isolates grown in the mineral medium (MM) supplemented with starch, carboxymethylcellulose, glycerol, glucose and Tween 80.

Isolate	Carbon Source	Total Biomass $(g \cdot L^{-1})$	P(3HB) Concentration $(g \cdot L^{-1})$	P(3HB) Content in Total Biomass (%)	P(3HB) Productivity $(g \cdot L^{-1} \cdot h^{-1})$
LAMA 748	Starch	16.80 ± 0.97	0	0	0
LAMA 737	Starch	25.31 ± 1.09	0	0	0
LAMA 748	Carboxymethylcellulose	1.19 ± 0.19	0	0	0
LAMA 674	Carboxymethylcellulose	1.24 ± 0.20	0	0	0
LAMA 677	Glycerol	1.41 ± 0.18	0.4	28.28	0.0058
LAMA 685	Glycerol	2.03 ± 0.27	0.67	32.79	0.0098
LAMA 685	Glucose	1.73 ± 0.22	0.17	9.62	0.0025
LAMA 737	Glucose	0.68 ± 0.10	0.05	7.85	0.0008
LAMA 726	Tween 80	0.91 ± 0.08	0	0	0
M 199 A	Tween 80	0.56 ± 0.05	0	0	0

3.3. Growth Kinetics and P(3HB) Production

As the best productivity was obtained when using glycerol as the carbon source, this condition was further analyzed during bacterial growth. However, the biomass production was much lower than in the previous experiments. For example, LAMA 677 produced 0.16 g· L^{-1} of biomass, of which 22.74% was polymer, compared to the 28.28% of polymer produced in the previous liquid test. This fact can be explained by the longer cultivation time at this stage, allowing for the consumption of the polymer. Moreover, LAMA 685 in the same condition had a 78.63% yield of P(3HB) in the biomass. This value is higher than reported by Cavalheiro et al. [33], who reported a PHA content of 62% when evaluating the production of *Cupriavidus necator* DSM 545 (the main organism used in the industrial production of PHA) under controlled conditions and high cellular concentrations, with a supply of pure glycerol. Mothes et al. [34] cultured the bacteria *Cupriavidus necator* JMP 134 at low cell concentrations using pure glycerol and obtained a 70% accumulation. When compared to other studies, the PHA content of LAMA 685 was similar. Zhu et al. [35] reported a content of 81.9% in cultures of *Burkholderia cepacia* ATCC 17759 grown in a shaker using residual glycerol as the carbonic substrate. Kangsadan et al. [36] cultivated *Cupriavidus necator* ATCC 17699 with pure and residual glycerol and reached contents of 83.23% and 78.26%, respectively. Considering these results, the polymer content obtained in this experiment with LAMA 685 is relevant because it is comparable to that of the organisms used in the production of the polymer on a commercial scale.

Compared to the literature, the maximum specific growth rates obtained in this study were lower, possibly because the medium culture used was poorer in nutrients than those employed in other studies. The values of maximum specific growth rates (μmax) obtained were μmax = 0.087 h^{-1} for LAMA 677 and μmax = 0.049 h^{-1} for LAMA 685. Piccoli et al. [37] studied PHA-producing lines using pure glycerol and found specific speeds ranging from 0.155 to 0.222 h^{-1}. Rodrigues [38] reported values of 0.27, 0.24 and 0.23 h^{-1} in glucose, apple cake and starch residues, respectively, in cultures of *Cupriavidus necator* DSM 545. Nascimento et al. [39] cultured *Burkholderia sacchari* LFM 101 on glucose, sucrose and glycerol, but did not observe differences in the maximum specific growth rates obtained with glucose and sucrose (an average of 0.539 h^{-1}). However, the highest PHA productivity (0.054 $g \cdot L^{-1} \cdot h^{-1}$) was seen with glucose at 35 °C.

3.4. Production of P(3HB) in Seawater and Residual Biodiesel Glycerol

An increase in the concentration of pure glycerol in the MM led to a reduction in the dry biomass as well as the PHA content in the cells, as seen in Table 4. When 5% pure glycerol was used in the culture, LAMA 677 produced 1.11 ± 0.04 g· L^{-1} biomass, with 64.28% P(3HB) content, and LAMA 685 produced 0.95 ± 0.06 g· L^{-1} biomass including 43.64% of the biopolymer. When the carbon source increased to 10%, the biomass of LAMA 677 decreased to 0.85 ± 0.05 g· L^{-1}, corresponding to a 23.4% reduction in total biomass. In terms of the P(3HB) content, the numbers showed a small decrease when comparing the two concentrations of the carbon source (64.28% at 5% glycerol and 64.04% at 10% glycerol), corresponding to a reduction of 0.24% in the total biomass. The biomass produced by LAMA 685 in the 10% glycerol assay resulted in 0.36 ± 0.07 g· L^{-1}, presenting a 62% reduction, and 28.08% P(3HB), presenting a reduction of 15.56%. The negative effect of increasing the concentration of pure glycerol on the growth of organisms has already been reported in other studies. Zhu et al. [35] cultured *Burkholderia cepacia* ATCC 17759 and obtained lower biomass values in the assay of 9% pure glycerol (1.3 g· L^{-1}) compared to the assay of 3% of the same carbon source, which yielded 2.8 g· L^{-1}.

Table 4. Results of the parameters used to analyze the P(3HB) production of the isolates LAMA 677 and LAMA 685, in seawater and residual biodiesel glycerol, with different formulations.

Isolate	Culture Medium Composition	Total Biomass ($g \cdot L^{-1}$)	P(3HB) Concentration ($g \cdot L^{-1}$)	P(3HB) Content in Total Biomass (%)	P(3HB) Productivity ($g \cdot L^{-1} \cdot h^{-1}$)
LAMA 677	90% mineral medium + 5% glycerol + 5% distilled water	1.11 ± 0.04	0.71	64.28	0.0103
	90% mineral medium + 10% glycerol	0.85 ± 0.05	0.55	64.04	0.0079
	90% seawater + 5% glycerol + 5% distilled water	0.10 ± 0.03	0.03	35.04	0.0005
	90% seawater + 5% residual glycerol + 5% distilled water	0.06 ± 0.01	0.02	31.70	0.0003
	90% seawater + 10% residual glycerol	*	*	*	*
	90% mineral medium + 5% residual glycerol + 5% distilled water	1.39 ± 0.05	0.74	52.94	0.0107
LAMA 685	90% mineral medium + 5% glycerol + 5% distilled water	0.95 ± 0.06	0.42	43.64	0.0060
	90% mineral medium + 10% glycerol	0.36 ± 0.07	0.10	28.08	0.0015
	90% seawater + 5% glycerol + 5% distilled water	0.27 ± 0.03	0.13	48.26	0.0019
	90% seawater + 5% residual glycerol + 5% distilled water	0.32 ± 0.01	0.17	53.60	0.0024
	90% seawater + 10% residual glycerol	0.10 ± 0.02	0.01	10.97	0.0002
	90% mineral medium + 5% residual glycerol + 5% distilled water	2.71 ± 0.96	1.22	44.95	0.0177

* Unobserved growth.

To evaluate the production of P(3HB) from low-cost minerals and nutrients, a medium composed of seawater and pure glycerol (5%) was used. When comparing this condition to a treatment where the MM was used instead of seawater, a drastic reduction in growth was evident. For instance, the biomass production levels in LAMA 677 and LAMA 685 were approximately 90% and 70% lower, respectively, when using seawater. The seawater used in the medium culture was collected on the Brazilian continental shelf, where the currents have low nutrient concentrations, as reported by Yoneda [40]. These low nutrient concentrations may have led to less growth when compared to the MM. On the other hand, when considering the PHA content in the biomass, the differences were much smaller.

For example, LAMA 677 produced approximately 21% less P(3HB) in the seawater cultures. In fact, LAMA 685 exhibited a slight increase when using seawater, reaching 48.26% P(3HB), or a gain of 10%. It is possible that this difference occurred because the collected seawater was poor in nutrients, and this limiting condition favored the accumulation of P(3HB) in LAMA 685.

A low-cost medium composed of seawater as a source of nutrients and minerals, and residual biodiesel glycerol as the carbon source was also tested as a medium for bacterial PHA production. In this case, the medium was supplied with two concentrations of the carbon source, 5% and 10%. The first assay resulted in a biomass and P(3HB) content of 0.06 ± 0.01 g· L^{-1} and 31.7% for the LAMA 677 isolate, and 0.32 ± 0.01 g· L^{-1} and 53.6% for LAMA 685. When using the 10% residual glycerol culture media, LAMA 677 showed no growth, and LAMA 685 presented lower values in relation to the other treatment (5% glycerol). This result was also observed by Zhu et al. [35] in *Burkholderia cepacia* ATCC 17759 cultures with residual glycerol in concentrations ranging from 3% to 9%. The authors reported that the increase in the carbon source concentration resulted in a gradual reduction of biomass and PHA content in the cells. Shrivastav et al. [9] evaluated the growth potential of organisms isolated from terrestrial and marine environments, using residual glycerol at concentrations of 1%, 2%, 5%, and 10% as the carbon sources, and reported high growth in the experiments with 1% and 2% carbon, and reduced growth in the experiments with 5% and 10%. This difference may have occurred as a result of the impurities (salts, esters, and alcohol) in the residual glycerol, which may have affected the metabolic processes of the microorganism. The use of residual glycerol is being investigated as an opportunity to reduce the costs of bacterial PHA production. For example, the new *Pannonibacter phragmitetus* ERC8 isolate was found to be capable of producing PHA (0.43 g· L^{-1}) from residual glycerol as the sole carbon source. The maximum PHA production was 1.36 g· L^{-1} when a low concentration (0.80%) of residual glycerol was applied [41]. Naranjo et al. [42] demonstrated the valorization of glycerol for P(3HB) production when working with *Bacillus megaterium*. The study successfully produced 4.8 g· L^{-1} of P(3HB) using 2% (w/v) purified glycerol under controlled conditions. Similarly, Jincy et al. [43] performed statistical optimization for P(3HB) production (0.60 g· L^{-1}) using 2% (v/v) residual glycerol by *Bacillus firmus* NII 0830. A study by Hermann-Krauss et al. [6] compared residual glycerol to pure glycerol and did not reveal any negative effects in terms of productivity or polyester properties in *Haloferax mediterranei*. This finding demonstrated that expensive carbon sources for archaic PHA production can be replaced with low residual glycerol phase surplus products from the biodiesel production process.

Conversely, in work carried out by Rodríguez-Contreras et al. [44], the authors obtained high cell dry mass and growth rates when glycerol was used together with glucose in a fermentation with *Cupriavidus necator*. When analyzing the biomass and the growth of *Burkholderia sacchari* using glycerol as a carbon source, the strain properly synthesized P(3HB); however, the biopolymers obtained from both fermentations with glycerol showed low molecular masses.

The impurities present in the residual glycerol vary according to the raw material and the biodiesel production process, as seen from the characterizations. Onwudili and Williams [45] described the composition of a residual glycerol containing 20.8% methanol, 33.1% esters, 1.52% moisture and 2.28% ash. Mothes et al. [34] analyzed residual glycerol from several companies and reported compositions of 5.3% to 14.2% moisture, 0.01% to 1.7% methanol, and 1% to 6% salts. Thompson and He [46] reported extremely variable methanol concentrations, ranging from 23.4% to 37.5%. According to the results obtained in the test with the MM and 5% pure or residual glycerol, the best biomass production was identified with the use of residual glycerol. Under this condition, LAMA 677 reached 1.39 ± 0.05 g· L^{-1}, representing an increase of 0.28 g· L^{-1} or 25.2%, compared to the use of pure glycerol. LAMA 685 reached a value of 2.71 ± 0.96 g· L^{-1}, with the biomass 2.85 times greater than that recorded in the test with pure glycerol. Thus, the best growth rates occurred when residual glycerol was added to the culture medium. It is possible that this result was obtained as a function of the nutrients contained in the carbon source, as Thompson and He [46] found varying contents of calcium (11.0 to 163.3 mg·L^{-1}), potassium (216.7 mg·L^{-1}), magnesium (0.4 to 126.7 mg·L^{-1}), phosphorus (12.0 to 134.7 mg·L^{-1}), sulfur (14.0 to 128.0 mg·L^{-1}), and sodium (1.07 to 1.4 mg·L^{-1}) when analyzing residual glycerol from several companies.

LAMA 677 had a lower biomass P(3HB) value of 11.34% (absolute value) when grown in residual glycerol (5%) than when grown in a medium with pure glycerol. Other studies have also shown the negative effect of residual glycerol on the accumulation of PHA compared to pure glycerol when used in the same concentrations. This difference also occurred in a study by Kawata and Aiba [13], which used residual and pure glycerol at a concentration of 5% in cultures of *Halomonas* sp. KM-1. The study reported the lowest PHA content in the medium with added biodiesel by-product, compared to purified glycerol. When *Cupriavidus necator* bacteria was cultured in pure and residual glycerol, Posada et al. [47] measured 57.1 g· L^{-1} of PHA in the purified substrate and 27.8 g· L^{-1} in the crude substrate. Kangsadan et al. [36] reported a PHA yield of 83.23% of the biomass with pure glycerol, and 78.26% in the crude form. AndreeBen et al. [48] reported contents of 11.85% of polymer with the use of pure glycerol and 5.24% of PHA with the addition of crude glycerol in cultures of recombinant *Escherichia coli*. This fact can be explained by the influence of impurities in the residual glycerol on PHA synthesis.

In a recent study by de Paula et al. [49], *Pandoraea* sp. MA03 showed strong potential to produce P(3HB) from crude glycerol. Experiments were performed for a 10–50 g· L^{-1} carbon source, and the best values for P(3HB) production were shown in crude glycerol cultivations, compared to pure glycerol, with a polymer accumulation ranging from 49.0% to 63.6% cell dry weight. Based on the P(3HB) production parameters of the evaluated organisms, it is possible to conclude that LAMA 685 has a higher capacity to tolerate impurities in crude glycerol, even though lower growth rates occurred in the medium formulated with seawater and residual glycerol (10%), especially considering that LAMA 677 did not show any growth under the same conditions. Additionally, the cultivation of LAMA 685 in the MM with residual glycerol (5%) resulted in superior biomass, 1.76 g· L^{-1}, when compared to the MM with pure glycerol (5%). Moreover, LAMA 677 showed an increase of 0.28 g· L^{-1} in total biomass when using residual glycerol, although the biomass results obtained from both organisms in the MM supplemented with 5% pure glycerol were similar. In addition, LAMA 685 was found to be very similar in P(3HB) content in the assays of the MM with 5% pure or residual glycerol, whereas LAMA 677 achieved a lower P(3HB) content in the residual glycerol cultivation.

The isolates cultured in low-cost media (5% seawater and residual glycerol) expressed a P(3HB) content of 0.02 g· L^{-1} in LAMA 677 and 0.17 g· L^{-1} in LAMA 685. When compared to the results of Pandian et al. [14], who evaluated the PHA production of *Bacillus megaterium* SRKP-3 (isolated organism from the marine environment) in a medium formulated with low-value inputs including seawater, rice bran and dairy residues, the low PHA content is verified, as they obtained results ranging from 0.196 to 6.376 g· L^{-1} of PHA. However, in the study by Pandian et al. [14], the culture medium used for the polymer synthesis could be considered more nutritionally rich, as, according to Silva [50], the effluents of the dairy industry are characterized by high amounts of organic matter, vitamins and minerals. Therefore, the medium culture used by Pandian et al. [14] was composed of higher concentrations of nutrients compared to the medium used in this study, thus possibly explaining the better polymer contents reported.

4. Conclusions

It is possible to conclude that marine bacteria have great potential for PHA production. Specifically, the use of marine bacteria from the epipelagic zone, which are exposed to a myriad of substrates, increases the opportunity to accumulate the biopolymer as an energy reserve. This ability was verified in the laboratory when isolates were able to use various carbon sources simulating agro-industrial residues (starch, CMC, glycerol, etc.). Two isolates that were efficient in producing PHA in a high concentration from pure glycerol were further investigated. In addition, the potential of those selected bacteria to synthesize P(3HB) using seawater and residual glycerol from biodiesel as the culture media was revealed. These bacteria will be further characterized in order to optimize their production and evaluate their performance in other low-cost substrates, as well to conduct their molecular identification. These results open a new avenue to explore marine bacteria as efficient converters of by-products into biomass rich in PHA, thus reducing the production costs.

Acknowledgments: ICGEB/CNPq (Brazil, Process 577915/2008-8) and CNPq/INCT-Mar COI (Brazil, Process 565062/2010-7) supported this work. We are also thankful to Jose Angel Alvarez Perez, coordinator of South Atlantic MAR-ECO Patterns and Processes of the Ecosystems, and Andre Silva Barreto for providing samples. We also thank CNPq for the scholarship provided to Andre Oliveira De Souza Lima (Process 311010/2015-6) and Maria Cecilia Miotto, as well to Santa Catarina State Govern for Nathalia Aparecida Santos Castilho's scholarship.

Author Contributions: Rodrigo Yoji Uwamori Takahashi and Andre Oliveira De Souza Lima conceived and designed the study and experiments. Rodrigo Yoji Uwamori Takahashi, Nathalia Aparecida Santos Castilho and Marcus Adonai Castro Da Silva performed the experiments under Andre Oliveira De Souza Lima's supervision. Rodrigo Yoji Uwamori Takahashi and Maria Cecilia Miotto contributed to the analysis. Rodrigo Yoji Uwamori Takahashi and Maria Cecilia Miotto wrote the paper with the suggestions/corrections of Andre Oliveira De Souza Lima.

References

1. Rehm, B.H.A.; Steinbüche, A. Biochemical and genetic analysis of PHA synthases and other proteins required for PHA synthesis. *Int. J. Biol. Macromol.* **1999**, *25*, 3–19. [CrossRef]

2. Sudesh, K.; Abe, H.; Doi, Y. Synthesis, structure and properties of polyhydroxyalkanoates: Biological polyesters. *Prog. Polym. Sci.* **2000**, *25*, 1503–1555. [CrossRef]

3. Sheu, D.S.; Chen, W.M.; Yang, J.Y.; Chang, R.C. Thermophilic bacterium *Caldimonas taiwanensis* produces poly(3-hydroxybutyrate-co-3-hydroxyvalerate) from starch and valerate as carbon sources. *Enzym. Microb. Technol.* **2009**, *44*, 289–294. [CrossRef]

4. Tay, B.Y.; Lokesh, B.E.; Lee, C.Y.; Sudesh, K. Polyhydroxyalkanoate (PHA) accumulating bacteria from the gut of higher termite *Macrotermes carbonarius* (Blattodea: Termitidae). *World J. Microbiol. Biotechnol.* **2010**, *26*, 1015–1024. [CrossRef]

5. Valentin, H.E.; Broyles, D.L.; Casagrande, L.A.; Colburn, S.M.; Creely, W.L.; Delaquil, P.A.; Felton, H.M.; Gon-zalez, K.A.; Houmiel, K.L.; Lutke, K.; et al. PHA production, from bacteria to plants. *Int. J. Biol. Macromol.* **1999**, *25*, 303–306. [CrossRef]

6. Hermann-Krauss, C.; Koller, M.; Muhr, A.; Fasl, H.; Stelzer, F.; Braunergg, G. Archaeal production of polyhydroxyalkanoate (PHA) co-and terpolyesters from biodiesel industry-derived by-products. *Archaea* **2013**. [CrossRef] [PubMed]

7. Koller, M.; Maršálek, L.; Dias, M.M.S.; Braunegg, G. Producing microbial polyhydroxyalkanoate (PHA) biopolyesters in a sustainable manner. *New Biotechnol.* **2017**, *37*, 24–38. [CrossRef] [PubMed]

8. Quillaguamán, J.; Guzmán, H.; Van-Thuoc, D.; Hatti-Kaul, R. Synthesis and production of polyhydroxyalkanoates by halophiles: Current potential and future prospects. *Appl. Microbiol. Biotechnol.* **2010**, *85*, 1687–1696. [CrossRef] [PubMed]

9. Shrivastav, A.; Mishra, S.K.; Shethia, B.; Pancha, I.; Jain, D.; Mishra, S. Isolation of promising bacterial strains from soil and marine environment for polyhydroxyalkanoates (PHAs) production utilizing *Jatropha* biodiesel byproduct. *Int. J. Biol. Macromol.* **2010**, *47*, 283–287. [CrossRef] [PubMed]

10. Tan, D.; Xue, Y.S.; Aibaidula, G.; Chen, G.Q. Unsterile and continuous production of polyhydroxybutyrate by *Halomonas* TD01. *Bioresour. Technol.* **2011**, *102*, 8130–8136. [CrossRef] [PubMed]

11. Yue, H.; Ling, C.; Yang, T.; Chen, X.; Chen, Y.; Deng, H.; Wu, Q.; Chen, J.; Chen, G.-Q. A seawater-based open and continuous process for polyhydroxyalkanoates production by recombinant *Halomonas campaniensis* LS21 grown in mixed substrates. *Biotechnol. Biofuels* **2014**, *7*, 108. [CrossRef]

12. Margesin, R.; Schinner, F. Potential of halotolerant and halophilic microorganisms for biotechnology. *Extremophiles* **2001**, *5*, 73–83. [CrossRef] [PubMed]

13. Kawata, Y.; Aiba, S. Poly(3-hydroxybutyrate) production by isolated *Halomonas* sp. KM-1 using waste glycerol. *Biosci. Biotechnol. Biochem.* **2010**, *74*, 175–177. [CrossRef] [PubMed]

14. Pandian, S.R.; Venkatraman, D.; Kalishwaralal, K.; Rameshkumar, N.; Jeraraj, M.; Gurunathan, S. Optimization and fed-batch production of PHB utilizing dairy waste and sea water as nutrient sources by *Bacillus megaterium* SRKP-3. *Bioresour. Technol.* **2010**, *101*, 705–711. [CrossRef] [PubMed]

15. Yin, J.; Chen, J.C.; Wu, Q.; Chen, G.Q. Halophiles, coming stars for industrial biotechnology. *Biotechnol. Adv.* **2015**, *33*, 1433–1442. [CrossRef] [PubMed]

16. Spiekermann, P.; Rehm, B.H.A.; Kalscheuer, R.; Baumeister, D.; Steinbüchel, A. A sensitive, viable-colony staining method using Nile red for direct screening of bacteria that accumulate polyhydroxyalkanoic acids and other lipid storage compounds. *Arch. Microbiol.* **1999**, *171*, 73–80. [CrossRef] [PubMed]

17. Baumann, P.; Baumann, L.; Mandel, M. Taxonomy of Marine Bacteria: The Genus *Beneckea*. *J. Bacteriol.* **1971**, *107*, 268–294. [PubMed]

18. Alves, L.P.; Almeida, A.T.; Cruz, L.M.; Pedrosa, F.O.; De Souza, E.M.; Chubatsu, L.S.; Müller-Santos, M.; Valdameri, G. A simple and efficient method for poly-3-hydroxybutyrate quantification in diazotrophic bacteria within 5 minutes using flow cytometry. *Braz. J. Med. Biol. Res.* **2017**, *50*, e5492. [CrossRef] [PubMed]

19. Zuriani, R.; Vigneswari, S.; Azizan, M.N.M.; Majid, M.I.A.; Amirul, A.A.A. High throughput Nile red fluorescence method for rapid quantification of intracellular bacterial polyhydroxyalkanoates. *Biotechnol. Bioprocess Eng.* **2013**, *18*, 472–478. [CrossRef]

20. Arikawa, H.; Sato, S.; Fujiki, T.; Matsumoto, K. Simple and rapid method for isolation and quantitation of polyhydroxyalkanoate by SDS-sonication treatment. *J. Biosci. Bioeng.* **2017**, *S1389–S1723*, 30664–30668. [CrossRef] [PubMed]

21. Degelau, A.; Scheper, T.; Bailey, J.E.; Guske, C. Fluorometric measurement of poly-β hydroxybutyrate in *Alcaligeneseutrophus* by flow cytometry and spectrofluorometry. *Appl. Microbiol. Biotechnol.* **1995**, *42*, 653–657. [CrossRef]

22. Reddy, C.S.K.; Ghai, R.; Rashmi; Kalia, V.C. Polyhydroxyalkanoates: An overview. *Bioresour. Technol.* **2003**, *87*, 137–146. [CrossRef]

23. Hedgpeth, J. *Classification of Marine Environments*; Reseck, J., Jr., Ed.; Marine Biology: Englewood Cliffs, NJ, USA, 1957; pp. 18–27.

24. Longhurst, R.A.; Harrison, G.W. Vertical nitrogen flux from the oceanic photic zone by diel migrant zooplankton and nekton. *Deep Sea Res. Part A Oceanogr. Res. Pap.* **1988**, *35*, 881–889. [CrossRef]

25. Chien, C.C.; Chen, C.C.; Choi, M.H.; Kung, S.S.; Wei, Y.H. Production of poly-ß-hydroxybutyrate (PHB) by *Vibrio* spp. isolated from marine environment. *J. Biotechnol.* **2007**, *132*, 259–263. [CrossRef] [PubMed]

26. Mahishi, L.H.; Tripathi, G.; Rawal, S.K. Poly(3-hydroxybutyrate) (PHB) synthesis by recombinant *Escherichia coli* harbouring *Streptomyces aureofaciens* PHB biosynthesis genes: Effect of various carbon and nitrogen sources. *Microbiol. Res.* **2003**, *158*, 19–27. [CrossRef] [PubMed]

27. Mohandas, S.P.; Balan, L.; Lekshmi, N.; Cubelio, S.S.; Philip, R.; Sing, I.S.B. Production and characterization of Polyhydroxybutyrate from *Vibrio Harveyi* MCCB 284 utilizing glycerol as carbon source. *J. Appl. Microbiol.* **2016**, *122*, 698–707. [CrossRef] [PubMed]

28. Van-Thuoc, D.; Quillaguamán, J.; Mamo, G.; Matiason, B. Utilization of agricultural residues for poly(3-hydroxybutyrate) production by *Halomonasboliviensis* LC1. *J. Appl. Microbiol.* **2008**, *104*, 420–428. [PubMed]

29. Bertrand, J.L.; Ramsay, B.A.; Ramsay, J.A.; Chavarie, C. Biosynthesis of poly-β-hydroxyalkanoates from pentoses by *Pseudomonas pseudoflava*. *Appl. Environ. Microbiol.* **1990**, *56*, 3133–3138. [PubMed]

30. Silva, L.F.; Taciro, M.K.; Ramos, M.E.M.; Carter, J.M.; Pradella, J.G.C.; Gomez, J.G.C. Poly-3-hydroxybutyrate (P3HB) production by bacteria from xylose, glucose and sugarcane bagasse hydrolysate. *J. Ind. Microbiol. Biotechnol.* **2004**, *31*, 245–254. [CrossRef] [PubMed]

31. Fernández, D.; Rodríguez, E.; Bassas, M.; Viñas Solanas, A.M.; Liorens, J.; Marquéz, A.M.; Manresa, A. Agro-industrial oily wastes as substrates for PHA production by the new strain *Pseudomonas aeruginosa* NCIB 40045: Effect of culture conditions. *Biochem. Eng. J.* **2005**, *26*, 159–167. [CrossRef]

32. He, W.; Tian, W.; Zhang, G.; Chen, G.-Q.; Zhang, Z. Production of novel polyhydroxyalkanoates by *Pseudomonas stutzeri* 1317 from glucose and soybean oil. *FEMS Microbiol. Lett.* **1998**, *169*, 45–49. [CrossRef]

33. Cavalheiro, J.M.B.T.; Almeida, M.C.M.D.; Grandfils, C.; Fonseca, M.M.R. Poly(3-hydroxybutyrate) production by *Cupriavidus necator* using waste glycerol. *Process Biochem.* **2009**, *44*, 509–515. [CrossRef]

34. Mothes, G.; Schnorpfeil, C.; Ackermann, J.U. Production of PHB from crude glycerol. *Eng. Life Sci.* **2007**, *7*, 475–479. [CrossRef]

35. Zhu, C.; Nomura, C.T.; Perrotta, J.A.; Stipanovic, A.J.; Nakas, J.P. Production and characterization of poly-3-hydroxybutyrate from biodiesel-glycerol by *Burkholderia cepacia* ATCC 17759. *Biotechnol. Prog.* **2010**, *26*, 424–430. [PubMed]

36. Kangsadan, T.; Swadchaipon, N.; Kongruang, S. Value-added utilization of crude glycerol from biodiesel production by microbial synthesis of polyhydroxybutyrate-valerate. *Curr. Opin. Biotechnol.* **2011**, *22*, S1–S35. [CrossRef]

37. Piccoli, R.A.M.; Silva, E.S.; Taciro, M.K.; Maiorano, A.E.; Ribeiro, C.M.S.; Rodrigues, M.F.A. Produção de polihidroxibutirato a partir de glicerol resíduo da produção de biodiesel. In *Simpósio Nacional de Bioprocessos*; 17, 2011, Caxias do Sul. Anais...; Associação Brasileira de Engenharia Química: Caxias do Sul, Brazil, 2011; pp. 1–6.

38. Rodrigues, R.C. Condições de Cultura Para a Produção de Poli(3-hidroxibutirato) por *Ralstoniaeu tropha* a partir de Resíduos de Indústrias de Alimento. Master's Thesis, Universidade Federal de Santa Catarina, Trindade, Florianópolis, Brasil, 2005.

39. Nascimento, V.M.; Silva, L.F.; Gomez, J.G.C.; Fonseca, G.G. Growth of *Burkholderia sacchari* LFM 101 cultivated in glucose, sucrose and glycerol at different temperatures. *Sci. Agricola.* **2016**, *73*, 429–433. [CrossRef]

40. Yoneda, N.T. Área Temática: Plâncton, 1999. Centro de Estudos do Mar, Universidade Federal do Paraná. Available online: http://www.brasil-rounds.gov.br/round7/arquivos_r7/PERFURACAO_R7/refere/pl%E2ncton.pdf (accessed on 23 December 2011).

41. Ray, S.; Prajapati, V.; Patel, K; Triedi, U. Optimization and characterization of PHA from isolate *Pannonibacter phragmitetus* ERC8 using glycerol waste. *Int. J. Biol. Macromol.* **2016**, *86*, 741–749. [CrossRef] [PubMed]

42. Naranjo, J.M.; Posada, J.A.; Higuita, J.C.; Cardona, C.A. Valorization of glycerol through the production of biopolymers: The PHB case using *Bacillus megaterium*. *Bioresour. Technol.* **2013**, *133*, 38–44. [CrossRef] [PubMed]

43. Jincy, M.; Sindhu, R.; Pandey, A.; Binod, P. Bioprocess development for utilizing biodiesel industry generated crude glycerol for production of poly-3-hydroxybutyrate. *J. Sci. Ind. Res.* **2013**, *72*, 596–602.

44. Rodríguez-Contreras, A.; Koller, M.; Dias, M.M.S.; Calaffel-Monfort, M.; Braunegg, G.; Marqués-Calvo, M.S. Influence of glycerol on poly(3-hydroxybutyrate) production by *Cupriavidus necator* and *Burkholderia sacchari*. *Biochem. Eng. J.* **2015**, *94*, 50–57. [CrossRef]

45. Onwudili, J.A.; Williams, P.T. Hydrothermal reforming of bio-diesel plant waste: Products distribution and characterization. *Fuel* **2010**, *89*, 501–509. [CrossRef]

46. Thompson, J.C.; He, B.B. Characterization of crude glycerol from biodiesel production from multiple feedstocks. *Appl. Eng. Agric.* **2006**, *22*, 261–265. [CrossRef]

47. Posada, J.A.; Naranjo, J.M.; López, J.A.; Higuita, J.C.; Cardona, C.A. Design and analysis of poly-3-hydroxybutyrate production processes from crude glycerol. *Process Biochem.* **2011**, *46*, 310–317. [CrossRef]

48. Andreeßen, B.; Lange, A.B.; Robenek, H.; Steinbüchel, A. Conversion of glycerol to poly(3-Hydroxypropionate) in recombinant *Escherichia coli*. *Appl. Environ. Microbiol.* **2010**, *76*, 622–626. [CrossRef] [PubMed]

49. De Paula, F.C.; Kakazu, S.; de Paula, C.B.C.; Contiero, J. Polyhydroxyalkanoate production from crude glycerol by newly isolated *Pandoraea* sp. *J. King Saud Univ. Sci.* **2017**, *29*, 166–173. [CrossRef]

50. Silva, A.M.X.P. Degradação de Efluentes Lácteo sem Reactores UASB Com Recirculação. Master's Thesis, Universidade de Aveiro, Aveiro, Portugal, 2008.

8

Biomedical Processing of Polyhydroxyalkanoates

Dario Puppi *, Gianni Pecorini and Federica Chiellini *

Department of Chemistry and Industrial Chemistry, University of Pisa, UdR INSTM – Pisa, Via G. Moruzzi 13, 56124 Pisa, Italy; gianni.pecorini21@gmail.com
* Correspondence: dario.puppi@unipi.it (D.P.); federica.chiellini@unipi.it (F.C.)

Abstract: The rapidly growing interest on polyhydroxyalkanoates (PHA) processing for biomedical purposes is justified by the unique combinations of characteristics of this class of polymers in terms of biocompatibility, biodegradability, processing properties, and mechanical behavior, as well as by their great potential for sustainable production. This article aims at overviewing the most exploited processing approaches employed in the biomedical area to fabricate devices and other medical products based on PHA for experimental and commercial applications. For this purpose, physical and processing properties of PHA are discussed in relationship to the requirements of conventionally-employed processing techniques (e.g., solvent casting and melt-spinning), as well as more advanced fabrication approaches (i.e., electrospinning and additive manufacturing). Key scientific investigations published in literature regarding different aspects involved in the processing of PHA homo- and copolymers, such as poly(3-hydroxybutyrate), poly(3-hydroxybutyrate-*co*-3-hydroxyvalerate), and poly(3-hydroxybutyrate-*co*-3-hydroxyhexanoate), are critically reviewed.

Keywords: polyhydroxyalkanoates processing; electrospinning; additive manufacturing; selective laser sintering; fused deposition modeling; computer-aided wet-spinning

1. Introduction

The always increasing interest on polyhydroxyalkanoates (PHA) for biomedical applications stems from their well-ascertained biocompatibility and biodegradability in physiological environments [1]. In addition, their microbial synthesis by means of sustainable processes with potential for large-scale industrial production [2], together with a better processing versatility and superior mechanical properties in comparison with other polymers from natural resources, make PHA unique polymer candidates for advanced research and development approaches.

From a chemical point of view, PHA are aliphatic polyesters with a variable number of carbon atoms in the monomeric unit (Figure 1). They are generally classified as short-chain length (SCL)-PHA when they consist of monomers C3-C5 in length, and medium-chain length (MCL)-PHA when they consist of monomers C6–C14 in length [3]. SCL-PHA consist of monomeric units of 3-hydroxybutyrate (3HB), 4-hydroxybutyrate (4HB), or 3-hydroxyvalerate (HV). MCL-PHA consist of monomeric units of 3-hydroxyhexanoate (HHx), 3-hydroxyoctanoate (HO), 3-hydroxydecanoate (HD), 3-hydroxydodecanoate (HDD), 3-hydroxytetradecanoate (HTD), or even longer-chain comonomer units [4]. In general, SCL-PHA have high crystallinity degree and behave as a stiff and brittle material, while MCL-PHA have reduced crystallinity and increased flexibility showing elastomeric properties. The length of the pendant groups of the monomer units plays a key role in the resulting polymer physical properties, so that SCL-PHA copolymers with ethyl side groups can show elongation at break values varying in the range 5–50% depending in comonomers units ratio [5]. In addition, copolymers consisting of both SCL- and MCL-subunits can have properties between those of the two states. While most bacteria accumulate PHA granules of only one type, i.e., SCL or MCL, bacteria accumulating SCL-MCL-PHA copolymers were also isolated [6].

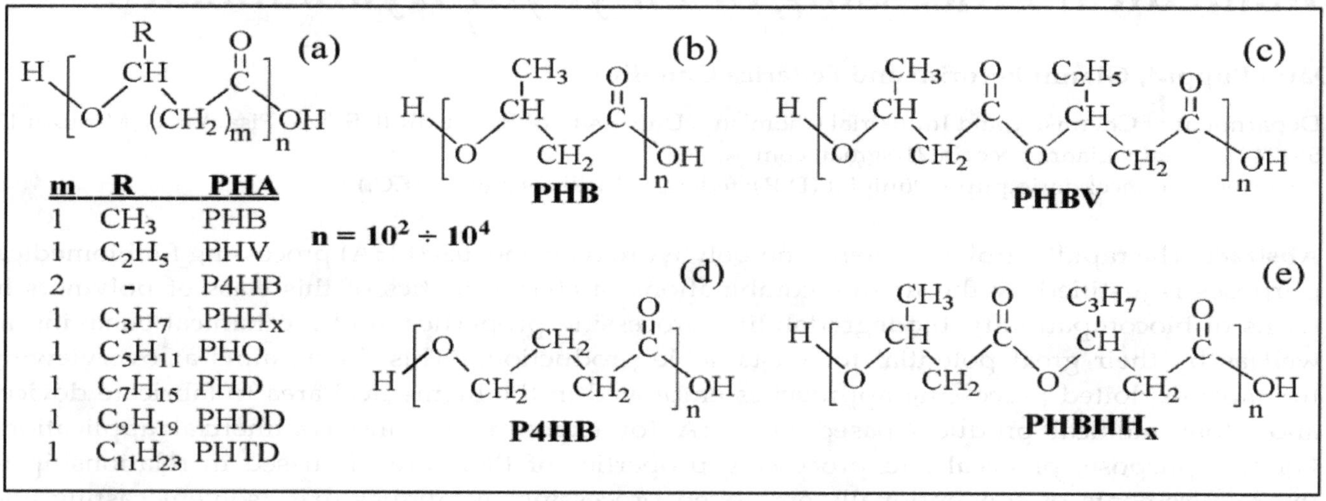

Figure 1. (**a**) General chemical structure of polyhydroxyalkanoates (PHA); chemical structure of (**b**) poly(3-hydroxybutyrate) (PHB), (**c**) poly(3-hydroxybutyrate-*co*-3-hydroxyvalerate) (PHBV), (**d**) poly(4-hydroxybutyrate) (P4HB), and (**e**) poly(3-hydroxybuyrate-*co*-3-hydroxyexanoate) (PHBHHx).

Poly(3-hydroxybutyrate) (PHB), which was the first discovered PHA in the 1920s [7], together with its copolymers poly(3-hydroxybutyrate-*co*-3-hydroxyvalerate) (PHBV), represent the most investigated microbial polyesters in the biomedical area thanks to the thermoplastic behavior, mechanical properties suitable for load-bearing applications, and versatile synthesis methods [8,9]. Novel microbial synthesis procedures have allowed the biomedical investigation of PHA with a wide range of molecular structures, in terms of molecular weight, length of alkyl side group, and ratio of comonomer units, as a means to develop materials with physical properties tailored to specific applications [10,11]. A widely investigated example in this context is represented by poly(3-hydroxybuyrate-*co*-3-hydroxyexanoate) (PHBHHx), which shows tunable elasticity by varying HHx percentage, exploitable for different applications, such as engineering tissues with much different stiffness (e.g., bone, cartilage, nerve, and blood vessel) [12]. Thanks to its low crystallinity, poly(4-hydroxybutyrate) (P4HB) has higher flexibility, ductility, and processing properties in comparison to PHB, which have been exploited to develop biomedical devices for soft tissue repair [13]. P4HB products currently on the clinical market include, among others, GalaFLEX®, a surgical mesh of knitted fibers [14], TephaFLEX®sutures, meshes, tubes, and thin films [15], MonoMax®sutures [16], Phantom Fiber™ sutures and BioFiber®Surgical Mesh [17].

PHA versatility in terms of processing approaches and conditions has allowed the investigation and application of a wide range of fabrication techniques relevant to biomedical research and industrial application. As it will be discussed in detail in the following sections, techniques based on different working principles have been successfully employed to process PHA, either in the form of a melt or a solution, and shape them at different scale length levels. Techniques with an old industrial history, such as melt spinning and blow extrusion, are currently employed to produce implantable medical devices made from PHA that are available on the market. Other approaches industrially employed in the case of commodity use polymers, such as solvent casting and injection molding, or under development for porous polymer structures fabrication, such as freeze drying and phase separation methods, have been also optimized for PHA processing. In this context, this review article is aimed at summarizing PHA processing properties in relationship to the requirements of the different techniques developed so far for their processing, as well as at critically overviewing key literature on this topic. Emphasis is dedicated to cutting edge advancement reported in literature on electrospinning and additive manufacturing application to PHA processing.

2. Physical and Processing Properties of PHA

As previously mentioned, the side group length significantly affects PHA crystallinity, mechanical behavior, as well as processing properties. PHB has a glass transition temperature (T_g) ~0 °C, melting temperature (T_m) ~180 °C, crystallinity degree in the range 60–80%, and degradation temperature (T_{deg}) ~220 °C. The narrow window between T_m and T_{deg}, typically requires the use of plasticizers for PHB melt processing in order to prevent polymer decomposition, which can occur at temperatures above 150 °C, as well as to enhance melt strength and elasticity [13,18]. By increasing the molar percentage of HV in copolymers, T_m can be decreased down to 130 °C and crystallinity to 35%, without marked effect on T_g and T_{deg}, thus widening processing temperature window and enhancing melt processability. Analogously, in the case of PHBHHx, T_m can be decreased down to 54 °C and crystallinity to 15%. Values of Young modulus and elongation at break reported in literature for PHBV (0.5–3.5 GPa, 5–50%) and PHBHHx (0.1–0.5 GPa, 5–850%) are generally different than those reported for PHB (0.9–4.0 GPa, 5–20%). As a consequence of the absence of alkyl side groups along the macromolecular chain, P4HB has much lower T_g, (~ −50°C), T_m (55–70 °C), and crystallinity (<40%) than PHB, resulting in enhanced melt processability, lower stiffness, and much larger elongation at break. These differences justify the successful application of different melt processing techniques for the fabrication of biomedical products currently available on the market, as discussed in the next section. Copolymers of 3HB and 4HB monomeric units, poly(3-hydroxybutyrate-co-4-hydroxybutyrate) (P3HB-co-4HB), show thermal and morphological parameters, as well as mechanical and processing properties, in between those of the two relevant homopolymers [19].

SCL-PHA are soluble only in a few organic solvents, including chloroform, dichloromethane, dimethyl formamide, tetrahydrofuran, and dioxane. In addition, their solubilization can require high temperatures or sonication to form a homogeneous solution at concentrations suitable for processing techniques commonly used in the biomedical field (e.g., solvent casting, phase separation, and electrospinning) [20]. This aspect is particularly significant in the case of PHB often resulting in suspensions rather than homogeneous solutions when mixed with organic solvents.

The theoretical and practical aspects of the most exploited processing techniques in the biomedical field, as well as of advanced fabrication approaches based on electrospinning [21] and additive manufacturing [22], are overviewed in the following section. Schematic representations of various techniques commonly employed for PHA biomedical processing are reported in Figure 2. They include techniques to fabricate 3D molded objects (e.g., injection molding), films (e.g., blow extrusion), continuous fibers (e.g., wet-spinning), nanoparticles (e.g., double emulsion), as well as a number of processing approaches to obtain a porous structure. The latter are due to a widespread interest raised within the scientific and clinical communities for the development of biodegradable scaffolds with a porous architecture tailored to tissue regeneration strategies.

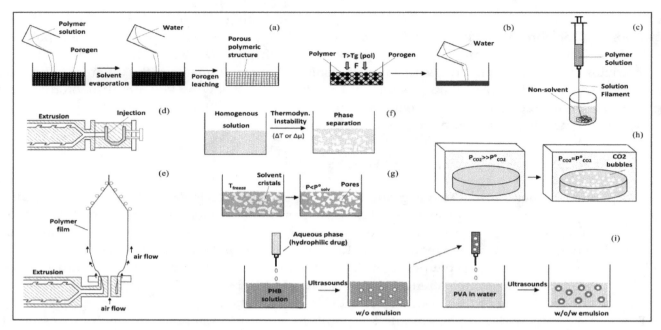

Figure 2. Schematic representation of techniques commonly employed for PHA processing in the biomedical area (modified from [23]). (**a**) Solvent casting-particle leaching process: a polymer solution is cast into a mold filled with porogen particles, the solvent is allowed to evaporate and the porogen is finally water-leached out; (**b**) melt molding-particulate leaching: a powder mixture of polymer and porogen is placed in a mold and heated above the polymer T_g while a pressure (F) is applied, the porogen is then water-leached out; (**c**) representative fiber spinning technique, i.e., wet-spinning: a polymeric solution is extruded directly into a coagulation bath leading to the formation of a continuous polymer fiber by non-solvent-induced phase inversion; (**d**) injection molding: a polymeric material is melt-extruded and injected into a mold; (**e**) film extrusion: a polymeric material is melt-extruded in the form of a tubular film by using a circular die and air pressure; (**f**) phase separation: a thermodynamic instability is established in a homogeneous polymer solution that separates into a polymer-rich and a polymer-poor phase; (**g**) freeze drying: a polymer solution is cooled down leading to the formation of solvent ice crystals, then a pressure lower than the equilibrium vapor pressure of the solvent (P_{solv}) is applied; (**h**) gas foaming: a chemical or physical blowing agent is mixed with the polymeric material, typically during extrusion (in the case depicted in figure a polymeric sample is exposed to high pressure CO_2 allowing saturation of the gas in the polymer and the subsequent gas pressure reduction causes the nucleation of CO_2 bubbles); (**i**) double emulsion for particles preparation: an aqueous phase containing a drug is added to a polymer solution, a water-in-oil (w/o) emulsion is obtained through sonication and added to a second aqueous phase, to form a w/o/w emulsion, particles are then separated by centrifugation, after organic solvent evaporation under stirring, and possibly resuspended in an aqueous phase.

3. Biomedical Processing of PHA

Given the thermoplastic behavior and solubility in organic solvents of this class of polyesters, various approaches have been investigated for processing PHA into systems with different potential biomedical applications. Representative tailored processing strategies to obtain PHA-based constructs with a morphology engineered at different scale levels are described in the following, as seen in Figure 3. A particular focus is given to electrospinning and additive manufacturing, which represents the most advanced processing approaches with great potential for biomedical industry translation. Electrospinning is the technique of election for fabricating PHA nanofibers organized into 3D assemblies with structural features mimicking those of the native tissues' extracellular matrix. This aspect together with other inherent advantages, including high surface to volume ratio of ultrafine fiber systems and processing versatility for drug-loading, make electrospinning of PHA suitable for a wide array of biomedical applications, as discussed more in depth in Section 3.1. Moreover, combining the sustainable production potential of PHA with the high technological level of additive manufacturing, in terms of reproducibility, automation degree, and control on composition and structure at different length scales, is inspiring a growing body of current literature that can have a tremendous impact on the biomedical industry.

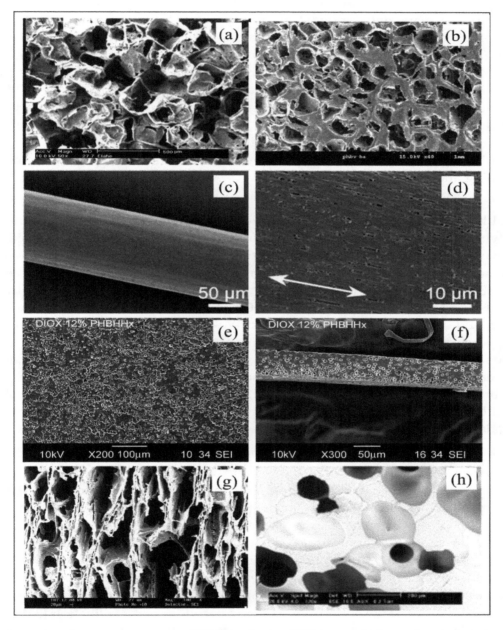

Figure 3. Scanning electron microscopy (SEM) analysis of PHA morphology after processing: (**a**) PHB porous structure by solvent casting/particulate leaching (scale bar 500 µm; reproduced from [24]); (**b**) PHBV/hydroxyapatite (HA) porous structure by melt molding/particulate leaching (scale bar 1 mm; reproduced from [25]); (**c**) low and (**d**) high magnification analysis of PHBV fiber fabricated by melt-spinning, and subjected to isothermal crystallization and 10 times one-step-drawing at room temperature (reproduced from [26]); (**e**) top view and (**f**) cross-section of PHBHHx film prepared by phase inversion of a 1,4 dioxane solution through immersion in water (reproduced from [27]); (**g**) PHBV scaffold by means of emulsion freezing/freeze drying technique (scale bar 20 µm; reproduced from [28]); (**h**) PHBV porous morphology obtained by gas foaming (scale bar 200 µm; reproduced from [29]).

Solvent casting represents one of the most straightforward approaches to process PHA into two-dimensional (2D) membranes. For instance, Basnett et al. [24] developed films based on poly(3-hydroxyoctanoate) (PHO) blended with PHB by dissolving the two polymers in chloroform at different weight ratios (80:20, 50:50 or 20:80) for a total concentration of 5% wt. The solutions were mixed by sonication, cast into a glass petri dish, and then air dried for one week to obtain films of 180–220 μm thickness. Drying in an atmosphere saturated with the solvent is often employed to achieve slow solvent evaporation and avoid internal stress formation. Combination of solvent casting with salt leaching is an effective means to obtain a PHA porous structure. As an example, Masaeli et al. [30] added NaCl particles (200–250 μm) to chloroform solutions of PHB and, after solvent evaporation and vacuum drying, submitted the solid to extensive water washing for five days. The resulting salt-leached membranes have a thickness of around 500 μm and a porosity of around 90% (Figure 3a). Advantages of this processing approach is the ease of fabrication, the possibility to vary the pore's size over a large range, as well as to control pore size and porosity independently. The main limitation is that only small thicknesses can be achieved due to difficulty in removing salt particles along thick sections. In addition, membrane shape is given by the mold, and obtaining customized geometries requires designing and fabricating ad-hoc molds.

Melt molding can be alternatively employed to fabricate thin PHA membranes, possibly in combination with salt leaching for developing porous architectures. After filling a mold with polymer and porogen particles, the system is heated above polymer T_g, while applying pressure to the powder. Once the polymer particles are fused together, the mold is removed and the porogen is leached out. As an example, PHBV/hydroxyapatite (HA) powder (9:1 w/w) was mixed with NaCl particles (100~300 μm) at a 1:17 weight ratio and then cast in a mold at 180 °C [25]. After leaching out salt particles, through water washing, and drying it under a vacuum, PHBV scaffolds exhibited an interconnected, porous network with pore sizes ranging from several microns to around 400 μm (Figure 3b). This processing approach holds some advantages and disadvantages of solvent casting-based techniques, such as the independent control of pore shape and porosity, the ease of fabrication, the limited design freedom in terms of membrane shape and thickness. Moreover, while it avoids the use of organic solvents that can be harmful for biological systems, it requires high temperature processing with the related risks of thermal degradation and energy costs.

Fiber spinning techniques, i.e., melt-, dry-, and wet-spinning, were recently investigated to process PHBV into single fibers or tridimensional (3D) fibrous macroporous scaffolds. They involve the extrusion of a polymer as a melt, in the case of melt-spinning, or dissolved in a solvent and then extruded in air or directly into a coagulation bath, in the case of dry- or wet-spinning, respectively. Depending on the technique employed, the final applications, and other product requirements, the fibers are submitted to different post-processing treatments (e.g., drying, washing, and drawing) or assembled into 3D fibrous systems. In the case of melt spinning, different methods have been investigated to overcome PHA processing shortcomings, such as adhesion, high brittleness, and low melt strength, related to slow crystallization rate, large spherulite size, and secondary crystallization [31]. Blending with organic or inorganic particles acting as nucleating agents, graft copolymerization, or fiber stretching are effective means to control and optimize the crystal structure and crystallization behavior of PHBV. For instance, PHBV fibers with around 1 GPa tensile strength were prepared by quenching during melt spinning, followed by isothermal crystallization near the T_g, and one-step-drawing at room temperature (Figure 3c) [26]. As previously mentioned, melt spinning is commercially employed for producing monofilaments or multifilaments made of P4HB (T_m ~60 °C) that are used as sutures or further processed using conventional textile processes, such as braiding, knitting, and weaving, to produce scaffolds and surgical meshes [14,15]. In particular, P4HB monofilament sutures show superior tensile strength characteristics than polydioxanone and polypropylene sutures. Wet-spinning technique has been investigated to overcome the aforementioned shortcomings related to thermal processing of PHA. For instance, Alagoz et al. [32] extruded a chloroform solution of PHBV into a coagulation bath of methanol. The fibers produced were kept in methanol overnight at −4 °C for solidification, then placed into a cylindrical Teflon mold, and dried in a vacuum oven. The resulting scaffolds had a diameter of 4 mm, height of 2 mm, interconnected porosity of 75%, average fiber diameter of 90 μm, and pore size of 250 μm.

Injection molding and film extrusion are also used to process PHA into 2D or 3D objects with potential application in the biomedical industry. These techniques involve processing the polymer in a screw extruder, pumping the melt through a die. The melt is either injected in a mold, or axially drawn and radially expanded in the form of a thin-walled tube to obtain a continuous film. Injection molding can be also combined with the particulate leaching strategy, or integrated with blowing agents that are blended with the raw polymer and activated upon heating to form a porous structure [33]. A range of melt extrusion grade formulations based on PHB or PHBV blended with additives, other polymers, and/or inorganic fillers, have been developed to enhance the material toughens processability, as well as to reduce costs. They are currently available on the market for applications other than medical ones. Examples are injection molding and film blowing grade PHA formulations approved for food contact that are marketed by Telles and TianAn [34]. Although the employment of these melt processing approaches to biomedical research is limited, the trademark TephaFLEX®by Tepha Inc. includes, besides the previously cited sutures produced by melt spinning, P4HB surgical tubes and films made by injection molding and blow extrusion, respectively [15].

Phase separation approaches are widely investigated for the preparation of porous PHA systems. They generally rely on establishing a thermodynamic instability in a polymer solution, through changes of physical conditions (e.g., temperature) or chemical composition (e.g., non-solvent addition), to induce a separation into two phases at different composition. Li et al. [35] obtained a nanofibrous network through phase separation by lowering the temperature of a PHB/chloroform/dioxane ternary mixture, with the resulting formation of a gel, which was then water-washed and freeze dried. This method was suitable also for the preparation of nanofibrous systems made of PHB blended with either PHBHHx or P4HB, whose tensile modulus, strength, and elongation at break could be modulated by varying the blend composition. Similarly, Tsujimoto et al. [36] obtained a microporous PHBHHx architecture by quenching a homogeneous polymer/DMSO solution that was prepared at 85 °C. Injectable formulations based on PHA can be prepared by dissolving the polymer in an organic solvent considered as not-toxic. This strategy is based on polymer film formation upon solution injection as a consequence of solvent dilution by the aqueous body fluids. Dai et al. [27] injected in the intra-abdominal position of rats formulations of PHBHHx dissolved in different solvents, i.e., N-methyl pyrrolidone, dimethylacetamide, 1,4-dioxane, dimethyl sulfoxide, and 1,4-butanolide. In particular, they found that PHBHHx films with a porous structure were formed when the solution came into contact with aqueous fluids because of a non-solvent-induced phase inversion process (Figure 3d). The wet-spinning methods described in this article also rely on a phase separation process induced by immersion into a polymer non-solvent [37].

Freeze drying is another processing approach investigated to fabricate porous PHA systems starting from a polymeric solution. As demonstrated by Sultana and Wang [28,38,39], porous scaffolds based on PHBV alone or in blends with poly(l-lactic acid) (PLLA), possibly loaded with HA, can be fabricated through an emulsion freezing/freeze drying process. In detail, the process involves adding an acetic acid aqueous phase to a polymer solution in order to obtain an emulsion that is then frozen and lyophilized. After sublimation, solvent (e.g., chloroform) and water phase crystals leave behind an anisotropic highly porous structure (Figure 3e). In the case of composite development, HA particles are added to the water phase before emulsion formation.

Foaming of PHA can be achieved by means of the employment of physical or chemical blowing agents, typically during melt extrusion. In the case of employment of physical agents, such as supercritical CO_2, pressurized machinery with a more complex technology for gas pumping, screw extrusion, and pressure profile control is required [29]. Either exothermic (e.g., azodicarbonamide [40]) or endothermic chemical blowing agents (e.g., sodium bicarbonate and citric acid [41]) can be employed, the first ones releasing N_2, the others releasing CO_2. Epoxy-functionalized chain extenders and post-extrusion water-quenching have been proposed as effective means to enhance PHA foaming by increasing melt strength and controlling crystallization kinetics [42]. Although the great progress achieved on relevant processing aspects, foaming is not often used in the biomedical field since, despite the technological complexity, pores interconnectivity and surface porosity, which are a key requirement for most applications, are not easily obtained with this approach (Figure 3f).

Formulation of nanoparticles, microspheres, and microcapsules made from PHA has been widely investigated to develop biodegradable systems able to deliver pharmacologically active agents to a specific site of action, at the therapeutically optimal rate and dose regime [43]. Depending on drug hydrophilic/lipophilic behavior and the particle morphological requirements, different formulation methods can be employed, such as polymerization and nanoparticle formation in-situ, modified double emulsion-solvent evaporation, and oil-in-water emulsion-solvent evaporation [44]—as well as the dialysis method [45]. For instance, standard double-emulsion protocols for PHB nanoparticles preparation involve i) adding an aqueous phase, containing a drug and possibly an emulsifier, to an organic polymer solution under vigorous stirring or sonication ii) adding the obtained water-in-oil (w/o) emulsion to a second aqueous phase containing a hydrophilic polymer, e.g., poly(vinyl alcohol) (PVA), to form a w/o/w emulsion, iii) stirring until complete organic solvent evaporation, centrifugation, and resuspension in an aqueous phase. Folate-conjugated PHB nanoparticles loaded with an anti-cancer drug were recently prepared by following this method [46].

3.1. Electrospinning

Electrospinning is the most employed technique for the production, on a lab and industrial scale, of polymeric nanofibers and nanofibrous meshes suitable for different applications, such as patches for tissue engineering and wound repair, nanostructured systems for drug release, filtration membranes, and protective and high-tech clothes [47]. This technique is based on an electrostatically-driven process that involves feeding a polymeric solution through a capillary into a high voltage electric field. The liquid drop is deformed, under the action of electrostatic forces and surface tension, assuming a shape similar to that of a cone. At a critical value of the applied voltage, a thin fluid jet is ejected at the apex of the cone and accelerated towards a grounded or oppositely-charged electrode, typically a flat metallic plate (Figure 4a). The stretching forces acting on the jet and the contemporary solvent evaporation, amplified by the violent whipping and splitting the jet undergoes during its travel, lead to the formation of fibers with a diameter in the range of a few micrometers down to tens of nanometers. The electrospun fibers can be collected in the form of nonwoven, yarn, 3D assemblies, and patterned structures, depending on the electrode/counter-electrode configuration [48].

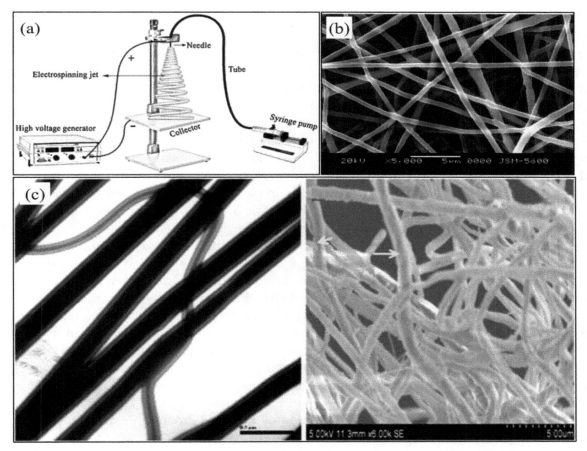

Figure 4. (**a**) Schematic representation of electrospinning set up (reproduced from [49]); (**b**) SEM micrograph of electrospun PHBHHx/ poly(d,l-lactic acid) (PDLLA) blend fiber mesh (reproduced from [50]); (**c**) TEM (left) and SEM (right) micrographs of PHB/gelatin core/sheath coaxial fibers (scale bars 0.5 μm and 5 μm, respectively; yellow arrows indicate the polymeric core/sheath structure; reproduced from [51]).

The great interest by the biomedical science and engineering community on electrospinning is justified by its tremendous potential for the development of nanostructured systems designed for advanced tissue engineering and drug release applications. Indeed, electrospun nanofiber assemblies highly mimic the nanostructure of native extracellular matrix, thus providing cells with a 3D nanofibrous environment which allows them to better maintain their phenotypic shape and establish natural behavior patterns, in comparison to what observed in 2D cell culture and 3D macroporous architectures [52]. In addition, the simplicity and inexpensive nature of the fabrication setup making possible its scale up, and the high design freedom of fibers assembly architecture and composition, together with the versatility in the development of tailored drug-loading methods for functionalizing polymeric nanofibers with a wide variety of therapeutics, have led to a fast growing amount of literature published on electrospinning for drug release [53].

One of the first articles on electrospinning of PHA was published in 2006 and described an investigation of the relationship between processing parameters and electrospun fiber assembly morphology, as observed by means of SEM [54]. The study resulted in the development of a set of scaffolds made of PHB, PHBV, or their blend (75:25, 50:50, or 25:75 weight ratio) with average fiber diameter of few microns. High magnification SEM analysis showed that PHB/PHBV blend fibers had a rough surface, which was explained by the authors with a phase inversion process related to the rapid evaporation of chloroform, employed as a solvent. These scaffolds sustained the growth in vitro of mouse fibroblasts and human osteoblasts, at higher levels than analogous cast-films [55]. Electrospun PHB meshes were also recently shown to be a suitable substrate for human mesenchymal stem cells (MSCs) adhesion, proliferation, and differentiation [56,57]. As systematically investigated by Zhu et al. [58], the variation of PHBV concentration in the starting solution significantly influenced the morphology of the resulting fibers, likely because of an effect on chain entanglement during electrospinning [59]. They were able to change the electrospun structure by gradually increasing polymer concentration, from beaded to string-on-beads morphology, and then to uniform fibrous mesh, with a relevant increase of surface hydrophobicity, as a consequence of the increased roughness.

As reviewed by Sanhueza et al. [21], a current research trend is devoted to electrospinning PHA blended with other synthetic or natural polymers, with the aim of tuning the properties of the resulting fibers or endowing them with intrinsic bioactivity. Cheng et al. [50] processed PHBHHx and poly(d,l-lactic acid) (PDLLA) dissolved in chloroform mixed with dimethylformamide (DMF) (80/20 w/w) to increase the solution electrical conductivity. In particular, they showed that by increasing the PDLLA weight percentage from 25 to 50 or 75%, the tensile modulus was decreased and the elongation at break increased, while the biodegradation rate was higher, due to the more amorphous morphology of PDLLA in comparison to the semicrystalline nature of PHBHHx. The possibility of tuning PHBV meshes mechanical properties through blending with MCL-PHA was also recently shown. Indeed, electrospun meshes crystallinity, tensile strength, and modulus decreased, while the elongation at break increased, by blending PHBV with poly(3-hydroxyoctanoate-co-3-hydroxyhexanoate) (PHOHHx) (25% wt.), which is an amorphous MCL-PHA with elastomeric properties at room temperature. PHB blending with poly(l-lactide-co-ε-caprolactone) led to fiber diameter decreasing and hydrophobicity increasing, without any resulting effect on electrospun mesh mechanical properties [60]. An acetyl triethyl citrate/poly(vinyl acetate) blend was employed as a plasticizer and compatibilizer to improve the miscibility between PHB and poly(propylene carbonate) [61]. The resulting blend was electrospun into meshes with decreased crystallinity and T_m in comparison to PHB meshes. Nagiah et al. [51, 62] investigated the modulation of the properties of PHB/gelatin membranes designed for skin tissue engineering, by adopting different electrospinning strategies. Indeed, by simultaneously electrospinning the two polymers with two separated syringes, processing a blend of the two polymers, or employing a coaxial electrospinning approach, they developed membranes composed by single gelatin fibers and PHB fibers, PHB/gelatin blend fibers, or biphasic fibers composed by a PHB core and a gelatin sheath, respectively (Figure 4c). The different fibers' composition and architecture resulted in significant differences in mechanical properties, wettability, and proliferation of human dermal fibroblasts and keratinocytes cultured in vitro on the membranes. Zhinjiang et al. [63] demonstrated that a variation of PHB/cellulose ratio in a starting chloroform/DMF mixture significantly affected the biodegradation rate, as well as the wettability and mechanical properties of the resulting electrospun blend nanofibers. Blend fibers made of PHB/chitosan blends with different weight ratio were also electrospun by using trifluoroacetic acid, as a common solvent [64]. The addition of chitosan resulted in increased wettability and biodegradation rate, as well as decreased tensile strength. The possibility of tuning in a wide range of the tensile strength and elongation at break of zein/P3HB-co-4HB blend meshes by electrospinning was also recently shown by varying the weight ratio between the two polymers. Different chemical modification strategies have been also adopted to improve cell adhesion onto electrospun PHA meshes, e.g., epoxy functionalization [65] and polysaccharide-grafting [66,67]. Other PHA fibers functionalization approaches include combination with antibacterial particles (e.g., silver [68] and zinc oxide [69] nanoparticles) or electrosprayed osteoconductive ceramics (e.g., HA nanoparticles [70]), as well as grafting with carbon nanotubes mechanical-reinforcing fillers [71].

A large body of literature has been dedicated to investigate and modulate electrospun PHA fibers organization and topography, as a means to control cell behavior and mesh mechanical properties. Yiu et al. [72] carried out a significant comparative study on the influence of topographic morphology of PHBHHx membranes fabricated by compression-molding, solvent-casting or electrospinning, on human MSCs adhesion, proliferation, and differentiation in vitro. Differently to what was observed in the two other kinds of membrane, MSCs showed a specific orientation on the electrospun fibrous meshes, exploitable for guided tissue regeneration and co-culturing of cells with orientation specificity (e.g., nerve, muscle and ligament cells). Aligned PHBV fibers systems can be fabricated by employing a rotating cylinder as fibers collector and auxiliary electrode [73]. Various studies have shown that fibers' alignment can significantly influence physical-chemical, mechanical, and biological properties of the resulting membrane. For instance, an article reported enhanced wettability for PHBV aligned fibers in comparison to PHBV randomly-oriented fibers [74]. In addition, tensile testing revealed that the aligned PHBV fibers membranes were stronger in the longitudinal direction, but weaker in the transverse direction, in comparison to non-woven PHBV meshes showing instead an isotropic behavior. Fibers alignment resulted also in a different morphology of human osteosarcoma SaOS-2 cells that elongated when cultured in vitro. Similarly, Wang et al. [75] observed higher tensile modulus and strength, as well as increased MSCs elongation and differentiation, when electrospun PHBHHx fibers were aligned along their axes (Figure 5). The relationship between fibers' alignment and mechanical properties was further investigated by a recent study reporting on electrospinning of PHB, P3HB-*co*-4HB, PHBV, and PHBHHx [76]. In all cases, fibers' alignment resulted in enhanced tensile mechanical properties with an overall effect on surface properties. The employment of a rotating fiber collector has been also widely investigated for the production of non-woven meshes with a tubular geometry investigated, as suitable nerve conduits [77] or blood vessel scaffolds [78].

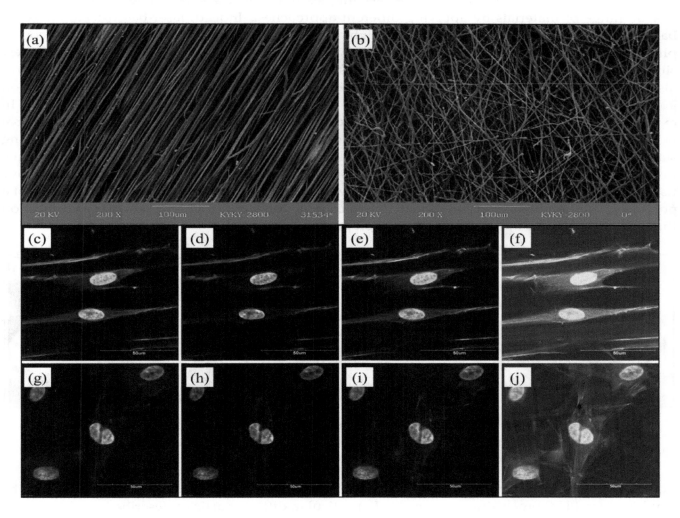

Figure 5. SEM micrographs of electrospun PHBHHx meshes composed by (**a**) aligned fibers or (**b**) randomly-oriented fibers (scale bar: 100 µm); confocal laser scanning microscopy images of MSCs cultured for three days on PHBHHx meshes composed by (**c–f**) aligned fibers or (**g–j**) randomly-oriented fibers (adhesion complexes (vinculin) in green, actin in red and nuclei in blue; scale bar in (**a**) and (**b**) 100 µm, scale bar in (**c–j**) 50 µm; reproduced from [75]).

3.2. Additive Manufacturing

As described in this section, various additive manufacturing approaches have been successfully applied to process different PHA, mainly into 3D porous scaffolds (Figure 6). Additive manufacturing was defined by ASTM as "the process of joining materials to make objects from 3D model data, usually layer upon layer" [79]. Additive manufacturing techniques are based on a computer-controlled design and fabrication process involving a sequential delivery of materials and/or energy to build up 3D layered objects. The geometrical and dimensional details, as well as other product specifications, such as density and composition gradients, are defined in a digital file which is then converted into a numerical control programming language, which specifies the motion of automated manufacturing tools. This approach enables advanced control over composition, shape, and dimensions of the object, in terms of design freedom and resolution. An advantage of additive manufacturing peculiar to the biomedical field is the possibility of deriving the 3D model data from medical imaging techniques commonly used for diagnostic purposes, such as computer tomography and magnetic resonance imaging. In this way, the anatomical features of biological tissues and organs can be reproduced through the fabrication process [80].

The various additive manufacturing techniques developed so far enable the processing of a wide range of materials by applying different approaches. Indeed, laser-based techniques are based on directing a beam or projection of light either to a photosensitive resin that is selectively photopolymerized, like in the case of stereolithography [81], or to a powder bed that is selectively sintered or fused, in the case of selective laser sintering (SLS) [82]. In the case of binder jetting, also referred to as 3D printing, a liquid binder, typically a polymer solvent, is deposited on a powder bed, which is selectively dissolved and fused upon solvent evaporation [83]. In extrusion-based techniques, a polymer in the form of a melt [84] or a solution/suspension [85] is extruded under controlled environmental conditions and selectively deposited onto a building stage. Fused deposition modeling (FDM) is a widely investigated example of melt-extrusion technique involving the extrusion and controlled deposition of a polymeric filament at a temperature above its T_g. Computer-aided wet-spinning (CAWS) involves the controlled extrusion and deposition of a polymeric solution or suspension directly into a coagulation bath to achieve polymer solidification through a non-solvent-induced phase-inversion process.

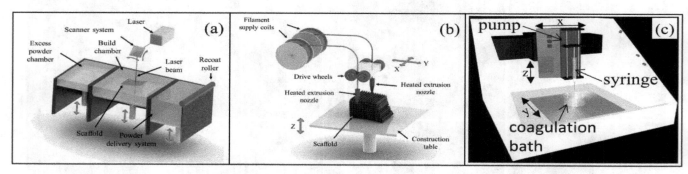

Figure 6. Schematic representation of additive manufacturing techniques applied to PHA processing: (**a**) selective laser sintering (SLS) and (**b**) fused deposition modeling (FDM) (reproduced from [80]); (**c**) computer-aided wet-spinning (CAWS) (reproduced from [86]).

Selective laser sintering (SLS) was the first reported additive manufacturing technique to be employed for PHA processing (Figure 6a). In particular, different articles described the fabrication of 2D and 3D constructs with a predefined shape and interconnected porous architecture, through automated sintering of PHB particles, without any significant change in polymer chemical composition and thermal properties [87–89]. Duan et al. [90–92] published a few articles showing that PHBV scaffolds loaded with calcium phosphate (Ca–P) nanoparticles could be fabricated by processing nanocomposite microspheres previously prepared by means of an emulsion-solvent evaporation method (Figure 7a). They also showed that the scaffolds could be functionalized with the recombinant human bone morphogenetic protein-2 growth factor that was bound to a heparin coating applied on the surface. Although, this set of PHA scaffolds showed high fidelity on the macroscopic scale to the virtual model geometrical features, a low topographical resolution, was observed under SEM analysis (Figure 7c–f). This morphological feature, distinctive of SLS, is a consequence of an incomplete coalescence of polymeric particles during sintering due to the low laser energy employed to prevent thermal decomposition.

Figure 7. Additive manufactured PHA scaffolds: photograph of (**a**) PHBV and (**b**) CaP-loaded PHBV scaffolds by SLS, and relevant SEM micrographs of (**c,d**) PHBV and (e, f) CaP-loaded PHBV scaffolds (scale bars in (**c**) and (**e**) 200 μm, in (**d**) and (**f**) 100 μm; reproduced from [90]); SEM micrographs of (**g**) top view and (**h**) cross-section PHBV/PCL scaffolds by FDM (reproduced from [93]); SEM micrographs of (**i**) top-view and (**j**) cross-section of PHBHHx scaffolds by CAWS (inserts are high magnification micrographs, reproduced from [94]).

Fused deposition modeling (FDM) feasibility for processing PHA and exploiting their thermoplastic behavior has been assessed by a number of research activities (Figure 6b). However, the employment of FDM is still delayed in comparison to what achieved with other aliphatic polyesters, (e.g., PLLA), as a consequence of the limited thermal stability of PHA as well as their low melt elasticity and strength. For this reason, blending with other polymers or plasticizers is needed for successful PHA processing by FDM. Kosorn et al. [93] developed scaffolds made of poly(ε-caprolactone) (PCL)/PHBV blends with different weight ratios between the two polymers (75:25, 50:50, and 25:75) (Figure 7g,h). They demonstrated that the synergistic effect of increase in PHBV ratio and surface low-pressure plasma treatment significantly increased the proliferation and chondrogenic differentiation of porcine chondrocytes cultured in vitro in combination with the scaffolds. Blending with commercial monomeric plasticizers based on esters of citric acid (Citroflex®) was recently shown to be an effective means for FDM processing of PHB/PDLLA/plasticizer blends (60/25/15 wt.) into dog bone shape samples for tensile test [95]. Selection of the optimal plasticizer allowed a remarkable increase in elongation at break (from 5 to 187%) as well as a reduction of warping effects in the printed part upon solidification. The development of filaments made of maleic anhydride-grafted PHA loaded with palm fibers [96], wood flour [97], or multi-walled carbon nanotubes [98] was also recently reported in literature.

Computer-aided wet-spinning (CAWS) represents a successful example of hybrid additive manufacturing technique applied for the processing of PHA. CAWS approach involves the extrusion of a chloroform or tetrahydrofuran suspension of PHBHHx directly into a non-solvent bath (e.g., ethanol) to fabricate a device with a layer-layer process (Figure 6c). Under optimal conditions, the phase inversion process governing polymer solidification leads to the formation of a microporosity in the polymeric matrix integrated with the macroporous network created by the controlled deposition process (Figure 7i,j) [85,99]. PHBHHx scaffolds with different shape and porous architecture, resulting in varied mechanical response, can be fabricated by changing the design model and the processing conditions [94]. Indeed, anatomical PHBHHx scaffolds with the shape and dimensions of a critical size segment of a New Zealand rabbit's radius model, and endowed with a longitudinal macrochannel for optimal bone regeneration conditions, were recently developed [100]. This kind of PHBHHx scaffolds, possibly in the form of a blend with PCL [101], were demonstrated to sustain the in vitro adhesion and proliferation of MC3T3-E1 murine preosteoblast cells. CAWS approach was also recently implemented with a rotating mandrel, as a fiber collector during polymer coagulation, to fabricate small-caliber biodegradable stents made of either PHBHHx or PCL [102]. Tubular constructs with different porous architectures were developed by controlling the synchronized motion of the deposition needle and the rotating mandrel.

4. Conclusions and Future Perspectives

Thanks to their thermoplastic behavior and suitable rheological properties when dissolved or suspended in proper organic solvents, processing of PHA has been investigated by adopting a number of tailored technological approaches. Indeed, the wide processing versatility given by the great variety, in terms of macromolecular structure and morphology, offered by PHA, makes this class of biodegradable polymers one of the most investigated in the biomedical field, especially when a load-bearing role is required. These aspects, together with the promising perspectives for PHA sustainable development [103], are propelling a fast growing research on relevant processing approaches to specific biomedical requirements. Both techniques with a long-history of industrial use, such as melt-spinning and foaming, and emerging techniques currently in the phase of being industrially-implemented, i.e., electrospinning and additive manufacturing, are applied for biomedical processing of PHA. The next frontier could be represented by the combination of different processing approaches to integrate in a single PHA device the high resolution down to the micro/nanoscale given by techniques like electrospinning and electrospraying, with the advanced control at the macroscale guaranteed by automated additive manufacturing approaches.

Acknowledgments: The financial support of the University of Pisa PRA-2018-23 project entitled "Functional Materials" is gratefully acknowledged.

References

1. Puppi, D.; Chiellini, F.; Dash, M.; Chiellini, E. Biodegradable polymers for biomedical applications. In *Biodegradable Polymers: Processing, Degradation & Applications*; Felton, G.P., Ed.; Nova Science Publishers: New York, NY, USA, 2011; pp. 545–560.
2. Koller, M. Advances in polyhydroxyalkanoate (PHA) production. *Bioengineering* **2017**, *4*, 88. [CrossRef] [PubMed]
3. Nomura, C.T.; Tanaka, T.; Gan, Z.; Kuwabara, K.; Abe, H.; Takase, K.; Taguchi, K.; Doi, Y. Effective enhancement of short-chain-length–medium-chain-length polyhydroxyalkanoate copolymer production by coexpression of genetically engineered 3-ketoacyl-acyl-carrier-protein synthase iii (fabh) and polyhydroxyalkanoate synthesis genes. *Biomacromolecules* **2004**, *5*, 1457–1464. [CrossRef] [PubMed]
4. Chen, S.; Liu, Q.; Wang, H.; Zhu, B.; Yu, F.; Chen, G.-Q.; Inoue, Y. Polymorphic crystallization of fractionated microbial medium-chain-length polyhydroxyalkanoates. *Polymer* **2009**, *50*, 4378–4388. [CrossRef]
5. Yang, Q.; Wang, J.; Zhang, S.; Tang, X.; Shang, G.; Peng, Q.; Wang, R.; Cai, X. The properties of poly(3-hydroxybutyrate-co-3-hydroxyhexanoate) and its applications in tissue engineering. *Curr. Stem Cell Res. Ther.* **2014**, *9*, 215–222. [CrossRef] [PubMed]
6. Yoshie, N.; Inoue, Y. Chemical composition distribution of bacterial copolyesters. *Int. J. Biol. Macromol.* **1999**, *25*, 193–200. [CrossRef]
7. Lemoigne, M. Produits de deshydration et de polymerisation de l'acide b-oxobutyrique. *Bull. Soc. Chem. Biol.* **1926**, *8*, 770–782.
8. Nigmatullin, R.; Thomas, P.; Lukasiewicz, B.; Puthussery, H.; Roy, I. Polyhydroxyalkanoates, a family of natural polymers, and their applications in drug delivery. *J. Chem. Technol. Biotechnol.* **2015**, *90*, 1209–1221. [CrossRef]
9. Rodriguez-Contreras, A. Recent advances in the use of polyhydroyalkanoates in biomedicine. *Bioengineering* **2019**, *6*, 82. [CrossRef]
10. Morelli, A.; Puppi, D.; Chiellini, F. Polymers from renewable resources. *J. Renew. Mater.* **2013**, *1*, 83–112. [CrossRef]
11. Koller, M. Biodegradable and biocompatible polyhydroxy-alkanoates (PHA): Auspicious microbial macromolecules for pharmaceutical and therapeutic applications. *Molecules* **2018**, *23*, 362. [CrossRef]
12. Chen, G.-Q.; Wu, Q. The application of polyhydroxyalkanoates as tissue engineering materials. *Biomaterials* **2005**, *26*, 6565–6578. [CrossRef] [PubMed]
13. Manavitehrani, I.; Fathi, A.; Badr, H.; Daly, S.; Negahi Shirazi, A.; Dehghani, F. Biomedical applications of biodegradable polyesters. *Polymers* **2016**, *8*, 20. [CrossRef] [PubMed]
14. Galatea Surgical Scaffolds. Available online: https://www.galateasurgical.com/ (accessed on 30 October 2019).
15. Tepha Medical Devices. Available online: https://www.tepha.com (accessed on 30 October 2019).
16. B. Braun Italia. Available online: https://www.bbraun.it (accessed on 30 October 2019).
17. Wright Medical Group N.V. Available online: http://www.wright.com (accessed on 30 October 2019).
18. Leroy, E.; Petit, I.; Audic, J.L.; Colomines, G.; Deterre, R. Rheological characterization of a thermally unstable bioplastic in injection molding conditions. *Polym. Degrad. Stab.* **2012**, *97*, 1915–1921. [CrossRef]
19. Miranda De Sousa Dias, M.; Koller, M.; Puppi, D.; Morelli, A.; Chiellini, F.; Braunegg, G. Fed-batch synthesis of poly(3-hydroxybutyrate) and poly(3-hydroxybutyrate-co-4-hydroxybutyrate) from sucrose and 4-hydroxybutyrate precursors by burkholderia sacchari strain dsm 17165. *Bioengineering* **2017**, *4*, 36. [CrossRef]
20. Madkour, M.H.; Heinrich, D.; Alghamdi, M.A.; Shabbaj, I.I.; Steinbüchel, A. PHA recovery from biomass. *Biomacromolecules* **2013**, *14*, 2963–2972. [CrossRef]
21. Sanhueza, C.; Acevedo, F.; Rocha, S.; Villegas, P.; Seeger, M.; Navia, R. Polyhydroxyalkanoates as biomaterial for electrospun scaffolds. *Int. J. Biol. Macromol.* **2019**, *124*, 102–110. [CrossRef]
22. Puppi, D.; Chiellini, F. Additive manufacturing of PHA. In *Handbook of Polyhydroxyalkanoates*; Koller, M., Ed.; CRC Press: Boca Raton, FL, USA, 2020.
23. Puppi, D.; Chiellini, F.; Piras, A.M.; Chiellini, E. Polymeric materials for bone and cartilage repair. *Prog. Polym. Sci.* **2010**, *35*, 403–440. [CrossRef]

24. Basnett, P.; Ching, K.Y.; Stolz, M.; Knowles, J.C.; Boccaccini, A.R.; Smith, C.; Locke, I.C.; Keshavarz, T.; Roy, I. Novel poly(3-hydroxyoctanoate)/poly(3-hydroxybutyrate) blends for medical applications. *React. Funct. Polym.* **2013**, *73*, 1340–1348. [CrossRef]

25. Baek, J.-Y.; Xing, Z.-C.; Kwak, G.; Yoon, K.-B.; Park, S.-Y.; Park, L.S.; Kang, I.-K. Fabrication and characterization of collagen-immobilized porous PHBV/HA nanocomposite scaffolds for bone tissue engineering. *J. Nanomater.* **2012**, *2012*, 171804. [CrossRef]

26. Tanaka, T.; Fujita, M.; Takeuchi, A.; Suzuki, Y.; Uesugi, K.; Ito, K.; Fujisawa, T.; Doi, Y.; Iwata, T. Formation of highly ordered structure in poly[(r)-3-hydroxybutyrate-co-(r)-3-hydroxyvalerate] high-strength fibers. *Macromolecules* **2006**, *39*, 2940–2946. [CrossRef]

27. Dai, Z.-W.; Zou, X.-H.; Chen, G.-Q. Poly(3-hydroxybutyrate-co-3-hydroxyhexanoate) as an injectable implant system for prevention of post-surgical tissue adhesion. *Biomaterials* **2009**, *30*, 3075–3083. [CrossRef] [PubMed]

28. Sultana, N.; Khan, T.H. In vitro degradation of PHBV scaffolds and nHA/PHBV composite scaffolds containing hydroxyapatite nanoparticles for bone tissue engineering. *J. Nanomater.* **2012**, *2012*, 190950. [CrossRef]

29. Le Moigne, N.; Sauceau, M.; Benyakhlef, M.; Jemai, R.; Benezet, J.-C.; Rodier, E.; Lopez-Cuesta, J.-M.; Fages, J. Foaming of poly(3-hydroxybutyrate-co-3-hydroxyvalerate)/organo-clays nano-biocomposites by a continuous supercritical co2 assisted extrusion process. *Eur. Polym. J.* **2014**, *61*, 157–171. [CrossRef]

30. Masaeli, E.; Morshed, M.; Rasekhian, P.; Karbasi, S.; Karbalaie, K.; Karamali, F.; Abedi, D.; Razavi, S.; Jafarian-Dehkordi, A.; Nasr-Esfahani, M.H.; et al. Does the tissue engineering architecture of poly(3-hydroxybutyrate) scaffold affects cell-material interactions? *J. Biomed. Mater. Res. Part A* **2012**, *100*, 1907–1918. [CrossRef] [PubMed]

31. Xiang, H.; Chen, Z.; Zheng, N.; Zhang, X.; Zhu, L.; Zhou, Z.; Zhu, M. Melt-spun microbial poly(3-hydroxybutyrate-co-3-hydroxyvalerate) fibers with enhanced toughness: Synergistic effect of heterogeneous nucleation, long-chain branching and drawing process. *Int. J. Biol. Macromol.* **2019**, *122*, 1136–1143. [CrossRef]

32. Alagoz, A.S.; Rodriguez-Cabello, J.C.; Hasirci, V. PHBV wet-spun scaffold coated with ELR-REDV improves vascularization for bone tissue engineering. *Biomed. Mater.* **2018**, *13*, 055010. [CrossRef]

33. Raeisdasteh Hokmabad, V.; Davaran, S.; Ramazani, A.; Salehi, R. Design and fabrication of porous biodegradable scaffolds: A strategy for tissue engineering. *J. Biomater. Sci. Polym. Ed.* **2017**, *28*, 1797–1825. [CrossRef]

34. Ashter, S.A. 7-processing biodegradable polymers. In *Introduction to Bioplastics Engineering*; Ashter, S.A., Ed.; William Andrew Publishing: Oxford, UK, 2016; pp. 179–209.

35. Li, X.-T.; Zhang, Y.; Chen, G.-Q. Nanofibrous polyhydroxyalkanoate matrices as cell growth supporting materials. *Biomaterials* **2008**, *29*, 3720–3728. [CrossRef]

36. Tsujimoto, T.; Hosoda, N.; Uyama, H. Fabrication of porous poly(3-hydroxybutyrate-co-3-hydroxyhexanoate) monoliths via thermally induced phase separation. *Polymers* **2016**, *8*, 66. [CrossRef]

37. Puppi, D.; Piras, A.M.; Chiellini, F.; Chiellini, E.; Martins, A.; Leonor, I.B.; Neves, N.; Reis, R. Optimized electro- and wet-spinning techniques for the production of polymeric fibrous scaffolds loaded with bisphosphonate and hydroxyapatite. *J. Tissue Eng. Regen. Med.* **2011**, *5*, 253–263. [CrossRef]

38. Sultana, N.; Wang, M. Fabrication of HA/PHBV composite scaffolds through the emulsion freezing/freeze-drying process and characterisation of the scaffolds. *J. Mater. Sci. Mater. Med.* **2007**, *19*, 2555–2561. [CrossRef] [PubMed]

39. Sultana, N.; Wang, M. PHBV/PLLA-based composite scaffolds fabricated using an emulsion freezing/freeze-drying technique for bone tissue engineering: Surface modification andin vitrobiological evaluation. *Biofabrication* **2012**, *4*, 015003. [CrossRef] [PubMed]

40. Liao, Q.; Tsui, A.; Billington, S.; Frank, C.W. Extruded foams from microbial poly(3-hydroxybutyrate-co-3-hydroxyvalerate) and its blends with cellulose acetate butyrate. *Polym. Eng. Sci.* **2012**, *52*, 1495–1508. [CrossRef]

41. Wright, Z.C.; Frank, C.W. Increasing cell homogeneity of semicrystalline, biodegradable polymer foams with a narrow processing window via rapid quenching. *Polym. Eng. Sci.* **2014**, *54*, 2877–2886. [CrossRef]

42. Ventura, H.; Laguna-Gutiérrez, E.; Rodriguez-Perez, M.A.; Ardanuy, M. Effect of chain extender and water-quenching on the properties of poly(3-hydroxybutyrate-co-4-hydroxybutyrate) foams for its production by extrusion foaming. *Eur. Polym. J.* **2016**, *85*, 14–25. [CrossRef]

43. Shrivastav, A.; Kim, H.-Y.; Kim, Y.-R. Advances in the applications of polyhydroxyalkanoate nanoparticles for novel drug delivery system. *Biomed. Res. Int.* **2013**, *2013*, 581684. [CrossRef]

44. Barouti, G.; Jaffredo, C.G.; Guillaume, S.M. Advances in drug delivery systems based on synthetic poly(hydroxybutyrate) (co)polymers. *Prog. Polym. Sci.* **2017**, *73*, 1–31. [CrossRef]

45. Errico, C.; Bartoli, C.; Chiellini, F.; Chiellini, E. Poly(hydroxyalkanoates)-based polymeric nanoparticles for drug delivery. *J. Biomed. Biotechnol.* **2009**, *2009*, 571702. [CrossRef]

46. Althuri, A.; Mathew, J.; Sindhu, R.; Banerjee, R.; Pandey, A.; Binod, P. Microbial synthesis of poly-3-hydroxybutyrate and its application as targeted drug delivery vehicle. *Bioresour. Technol.* **2013**, *145*, 290–296. [CrossRef]

47. Persano, L.; Camposeo, A.; Tekmen, C.; Pisignano, D. Industrial upscaling of electrospinning and applications of polymer nanofibers: A review. *Macromol. Mater. Eng.* **2013**, *298*, 504–520. [CrossRef]

48. Teo, W.E.; Ramakrishna, S. A review on electrospinning design and nanofibre assemblies. *Nanotechnology* **2006**, *17*, R89–R106. [CrossRef] [PubMed]

49. Almuhamed, S.; Bonne, M.; Khenoussi, N.; Brendle, J.; Schacher, L.; Lebeau, B.; Adolphe, D.C. Electrospinning composite nanofibers of polyacrylonitrile/synthetic na-montmorillonite. *J. Ind. Eng. Chem.* **2016**, *35*, 146–152. [CrossRef]

50. Cheng, M.-L.; Chen, P.-Y.; Lan, C.-H.; Sun, Y.-M. Structure, mechanical properties and degradation behaviors of the electrospun fibrous blends of phbhhx/pdlla. *Polymer* **2011**, *52*, 1391–1401. [CrossRef]

51. Nagiah, N.; Madhavi, L.; Anitha, R.; Anandan, C.; Srinivasan, N.T.; Sivagnanam, U.T. Development and characterization of coaxially electrospun gelatin coated poly (3-hydroxybutyric acid) thin films as potential scaffolds for skin regeneration. *Mater. Sci. Eng. C* **2013**, *33*, 4444–4452. [CrossRef] [PubMed]

52. Puppi, D.; Zhang, X.; Yang, L.; Chiellini, F.; Sun, X.; Chiellini, E. Nano/microfibrous polymeric constructs loaded with bioactive agents and designed for tissue engineering applications: A review. *J. Biomed. Mater. Res. B Appl. Biomater.* **2014**, *102*, 1562–1579. [CrossRef] [PubMed]

53. Puppi, D.; Chiellini, F. 12-drug release kinetics of electrospun fibrous systems. In *Core-Shell Nanostructures for Drug Delivery and Theranostics*; Focarete, M.L., Tampieri, A., Eds.; Woodhead Publishing: Sawston, UK; Cambridge, UK, 2018; pp. 349–374.

54. Sombatmankhong, K.; Suwantong, O.; Waleetorncheepsawat, S.; Supaphol, P. Electrospun fiber mats of poly(3-hydroxybutyrate), poly(3-hydroxybutyrate-co-3-hydroxyvalerate), and their blends. *J. Polym. Sci. Part B Polym. Phys.* **2006**, *44*, 2923–2933. [CrossRef]

55. Sombatmankhong, K.; Sanchavanakit, N.; Pavasant, P.; Supaphol, P. Bone scaffolds from electrospun fiber mats of poly(3-hydroxybutyrate), poly(3-hydroxybutyrate-co-3-hydroxyvalerate) and their blend. *Polymer* **2007**, *48*, 1419–1427. [CrossRef]

56. Ramier, J.; Grande, D.; Bouderlique, T.; Stoilova, O.; Manolova, N.; Rashkov, I.; Langlois, V.; Albanese, P.; Renard, E. From design of bio-based biocomposite electrospun scaffolds to osteogenic differentiation of human mesenchymal stromal cells. *J. Mater. Sci. Mater. Med.* **2014**, *25*, 1563–1575. [CrossRef]

57. Grande, D.; Ramier, J.; Versace, D.L.; Renard, E.; Langlois, V. Design of functionalized biodegradable PHA-based electrospun scaffolds meant for tissue engineering applications. *New Biotechnol.* **2017**, *37*, 129–137. [CrossRef]

58. Zhu, S.; Yu, H.; Chen, Y.; Zhu, M. Study on the morphologies and formational mechanism of poly(hydroxybutyrate-co-hydroxyvalerate) ultrafine fibers by dry-jet-wet-electrospinning. *J. Nanomater.* **2012**, *2012*, 525419. [CrossRef]

59. Puppi, D.; Piras, A.M.; Detta, N.; Dinucci, D.; Chiellini, F. Poly(lactic-co-glycolic acid) electrospun fibrous meshes for the controlled release of retinoic acid. *Acta Biomater.* **2010**, *6*, 1258–1268. [CrossRef] [PubMed]

60. Daranarong, D.; Chan, R.T.H.; Wanandy, N.S.; Molloy, R.; Punyodom, W.; Foster, L.J.R. Electrospun polyhydroxybutyrate and poly(L-lactide-*co-ε*-caprolactone) composites as nanofibrous scaffolds. *Biomed. Res. Int.* **2014**, *2014*, 741408. [CrossRef] [PubMed]

61. El-Hadi, A.M. Improvement of the miscibility by combination of poly(3-hydroxy butyrate) phb and poly(propylene carbonate) ppc with additives. *J. Polym. Environ.* **2017**, *25*, 728–738. [CrossRef]

62. Nagiah, N.; Madhavi, L.; Anitha, R.; Srinivasan, N.T.; Sivagnanam, U.T. Electrospinning of poly (3-hydroxybutyric acid) and gelatin blended thin films: Fabrication, characterization, and application in skin regeneration. *Polym. Bull.* **2013**, *70*, 2337–2358. [CrossRef]

63. Zhijiang, C.; Yi, X.; Haizheng, Y.; Jia, J.; Liu, Y. Poly(hydroxybutyrate)/cellulose acetate blend nanofiber scaffolds: Preparation, characterization and cytocompatibility. *Mater. Sci. Eng. C* **2016**, *58*, 757–767. [CrossRef]

64. Sadeghi, D.; Karbasi, S.; Razavi, S.; Mohammadi, S.; Shokrgozar, M.A.; Bonakdar, S. Electrospun poly(hydroxybutyrate)/chitosan blend fibrous scaffolds for cartilage tissue engineering. *J. Appl. Polym. Sci.* **2016**, *133*. [CrossRef]

65. Ramier, J.; Boubaker, M.B.; Guerrouache, M.; Langlois, V.; Grande, D.; Renard, E. Novel routes to epoxy functionalization of PHA-based electrospun scaffolds as ways to improve cell adhesion. *J. Polym. Sci. Part A Polym. Chem.* **2014**, *52*, 816–824. [CrossRef]

66. Lemechko, P.; Ramier, J.; Versace, D.L.; Guezennec, J.; Simon-Colin, C.; Albanese, P.; Renard, E.; Langlois, V. Designing exopolysaccharide-graft-poly(3-hydroxyalkanoate) copolymers for electrospun scaffolds. *React. Funct. Polym.* **2013**, *73*, 237–243. [CrossRef]

67. Versace, D.-L.; Ramier, J.; Babinot, J.; Lemechko, P.; Soppera, O.; Lalevee, J.; Albanese, P.; Renard, E.; Langlois, V. Photoinduced modification of the natural biopolymer poly(3-hydroxybutyrate-co-3-hydroxyvalerate) microfibrous surface with anthraquinone-derived dextran for biological applications. *J. Mater. Chem. B* **2013**, *1*, 4834–4844. [CrossRef]

68. Versace, D.-L.; Ramier, J.; Grande, D.; Andaloussi, S.A.; Dubot, P.; Hobeika, N.; Malval, J.-P.; Lalevee, J.; Renard, E.; Langlois, V. Versatile photochemical surface modification of biopolyester microfibrous scaffolds with photogenerated silver nanoparticles for antibacterial activity. *Adv. Healthc. Mater.* **2013**, *2*, 1008–1018. [CrossRef]

69. Rodríguez-Tobías, H.; Morales, G.; Ledezma, A.; Romero, J.; Saldívar, R.; Langlois, V.; Renard, E.; Grande, D. Electrospinning and electrospraying techniques for designing novel antibacterial poly(3-hydroxybutyrate)/zinc oxide nanofibrous composites. *J. Mater. Sci.* **2016**, *51*, 8593–8609. [CrossRef]

70. Ramier, J.; Bouderlique, T.; Stoilova, O.; Manolova, N.; Rashkov, I.; Langlois, V.; Renard, E.; Albanese, P.; Grande, D. Biocomposite scaffolds based on electrospun poly(3-hydroxybutyrate) nanofibers and electrosprayed hydroxyapatite nanoparticles for bone tissue engineering applications. *Mater. Sci. Eng. C* **2014**, *38*, 161–169. [CrossRef] [PubMed]

71. Mangeon, C.; Mahouche-Chergui, S.; Versace, D.L.; Guerrouache, M.; Carbonnier, B.; Langlois, V.; Renard, E. Poly(3-hydroxyalkanoate)-grafted carbon nanotube nanofillers as reinforcing agent for PHAs-based electrospun mats. *React. Funct. Polym.* **2015**, *89*, 18–23. [CrossRef]

72. Yu, B.-Y.; Chen, P.-Y.; Sun, Y.-M.; Lee, Y.-T.; Young, T.-H. Response of human mesenchymal stem cells (hMSCs) to the topographic variation of poly(3-hydroxybutyrate-co-3-hydroxyhexanoate) (PHBHHx) films. *J. Biomater. Sci. Polym. Ed.* **2012**, *23*, 1–26. [CrossRef]

73. Tong, H.-W.; Wang, M. Electrospinning of aligned biodegradable polymer fibers and composite fibers for tissue engineering applications. *J. Nanosci. Nanotechnol.* **2007**, *7*, 3834–3840. [CrossRef]

74. Tong, H.-W.; Wang, M.; Lu, W.W. Electrospun poly(hydroxybutyrate-co-hydroxyvalerate) fibrous membranes consisting of parallel-aligned fibers or cross-aligned fibers: Characterization and biological evaluation. *J. Biomater. Sci. Polym. Ed.* **2011**, *22*, 2475–2497. [CrossRef]

75. Wang, Y.; Gao, R.; Wang, P.-P.; Jian, J.; Jiang, X.-L.; Yan, C.; Lin, X.; Wu, L.; Chen, G.-Q.; Wu, Q. The differential effects of aligned electrospun phbhhx fibers on adipogenic and osteogenic potential of mscs through the regulation of ppary signaling. *Biomaterials* **2012**, *33*, 485–493. [CrossRef]

76. Volova, T.; Goncharov, D.; Sukovatyi, A.; Shabanov, A.; Nikolaeva, E.; Shishatskaya, E. Electrospinning of polyhydroxyalkanoate fibrous scaffolds: Effects on electrospinning parameters on structure and properties. *J. Biomater. Sci. Polym. Ed.* **2014**, *25*, 370–393. [CrossRef]

77. Arslantunali, D.; Dursun, T.; Yucel, D.; Hasirci, N.; Hasirci, V. Peripheral nerve conduits: Technology update. *Med. Devices (Auckl.)* **2014**, *7*, 405–424.

78. Awad, N.K.; Niu, H.; Ali, U.; Morsi, Y.S.; Lin, T. Electrospun fibrous scaffolds for small-diameter blood vessels: A review. *Membranes* **2018**, *8*, 15. [CrossRef]

79. Standard F2792-12a. *Standard terminology for additive manufacturing technologies (Withdrawn 2015)*; ASTM International: West Conshohocken, PA, USA, 2012.

80. Mota, C.; Puppi, D.; Chiellini, F.; Chiellini, E. Additive manufacturing techniques for the production of tissue engineering constructs. *J. Tissue Eng. Regen. Med.* **2015**, *9*, 174–190. [CrossRef] [PubMed]

81. Chartrain, N.A.; Williams, C.B.; Whittington, A.R. A review on fabricating tissue scaffolds using vat photopolymerization. *Acta Biomater.* **2018**, *74*, 90–111. [CrossRef] [PubMed]

82. Williams, J.M.; Adewunmi, A.; Schek, R.M.; Flanagan, C.L.; Krebsbach, P.H.; Feinberg, S.E.; Hollister, S.J.; Das, S. Bone tissue engineering using polycaprolactone scaffolds fabricated via selective laser sintering. *Biomaterials* **2005**, *26*, 4817–4827. [CrossRef] [PubMed]

83. Butscher, A.; Bohner, M.; Hofmann, S.; Gauckler, L.; Müller, R. Structural and material approaches to bone tissue engineering in powder-based three-dimensional printing. *Acta Biomater.* **2011**, *7*, 907–920. [CrossRef] [PubMed]

84. Turner, B.N.; Strong, R.; Gold, S.A. A review of melt extrusion additive manufacturing processes: I. Process design and modeling. *Rapid Prototyp. J.* **2014**, *20*, 192–204. [CrossRef]

85. Puppi, D.; Chiellini, F. Wet-spinning of biomedical polymers: From single-fibre production to additive manufacturing of three-dimensional scaffolds. *Polym. Int.* **2017**, *66*, 1690–1696. [CrossRef]

86. Puppi, D.; Morelli, A.; Bello, F.; Valentini, S.; Chiellini, F. Additive manufacturing of poly(methyl methacrylate) biomedical implants with dual-scale porosity. *Macromol. Mater. Eng.* **2018**, *303*, 1800247. [CrossRef]

87. Oliveira, M.F.; Maia, I.A.; Noritomi, P.Y.; Nargi, G.C.; Silva, J.V.L.; Ferreira, B.M.P.; Duek, E.A.R. Construção de scaffolds para engenharia tecidual utilizando prototipagem rápida. *Matéria* **2007**, *12*, 373–382. [CrossRef]

88. Pereira, T.F.; Silva, M.A.C.; Oliveira, M.F.; Maia, I.A.; Silva, J.V.L.; Costa, M.F.; Thiré, R.M.S.M. Effect of process parameters on the properties of selective laser sintered poly(3-hydroxybutyrate) scaffolds for bone tissue engineering. *Virtual Phys. Prototyp.* **2012**, *7*, 275–285. [CrossRef]

89. Pereira, T.F.; Oliveira, M.F.; Maia, I.A.; Silva, J.V.L.; Costa, M.F.; Thiré, R.M.S.M. 3d printing of poly(3-hydroxybutyrate) porous structures using selective laser sintering. *Macromol. Symp.* **2012**, *319*, 64–73. [CrossRef]

90. Duan, B.; Wang, M.; Zhou, W.Y.; Cheung, W.L.; Li, Z.Y.; Lu, W.W. Three-dimensional nanocomposite scaffolds fabricated via selective laser sintering for bone tissue engineering. *Acta Biomater.* **2010**, *6*, 4495–4505. [CrossRef] [PubMed]

91. Duan, B.; Wang, M. Customized Ca–P/PHBV nanocomposite scaffolds for bone tissue engineering: Design, fabrication, surface modification and sustained release of growth factor. *J. R. Soc. Interface* **2010**, *7*, S615–S629. [CrossRef] [PubMed]

92. Duan, B.; Cheung, W.L.; Wang, M. Optimized fabrication of Ca–P/PHBV nanocomposite scaffolds via selective laser sintering for bone tissue engineering. *Biofabrication* **2011**, *3*, 015001. [CrossRef] [PubMed]

93. Kosorn, W.; Sakulsumbat, M.; Uppanan, P.; Kaewkong, P.; Chantaweroad, S.; Jitsaard, J.; Sitthiseripratip, K.; Janvikul, W. PCL/PHBV blended three dimensional scaffolds fabricated by fused deposition modeling and responses of chondrocytes to the scaffolds. *J. Biomed. Mater. Res. B Appl. Biomater.* **2017**, *105B*, 1141–1150. [CrossRef] [PubMed]

94. Mota, C.; Wang, S.Y.; Puppi, D.; Gazzarri, M.; Migone, C.; Chiellini, F.; Chen, G.Q.; Chiellini, E. Additive manufacturing of poly[(r)-3-hydroxybutyrate-co-(r)-3-hydroxyhexanoate] scaffolds for engineered bone development. *J. Tissue Eng. Regen. Med.* **2017**, *11*, 175–186. [CrossRef]

95. Menčík, P.; Přikryl, R.; Stehnová, I.; Melčová, V.; Kontárová, S.; Figalla, S.; Alexy, P.; Bočkaj, J. Effect of selected commercial plasticizers on mechanical, thermal, and morphological properties of poly(3-hydroxybutyrate)/poly(lactic acid)/plasticizer biodegradable blends for three-dimensional (3d) print. *Materials* **2018**, *11*, 1893. [CrossRef]

96. Wu, C.-S.; Liao, H.-T.; Cai, Y.-X. Characterisation, biodegradability and application of palm fibre-reinforced polyhydroxyalkanoate composites. *Polym. Degrad. Stab.* **2017**, *140*, 55–63. [CrossRef]

97. Wu, C.-S.; Liao, H.-T. Fabrication, characterization, and application of polyester/wood flour composites. *J. Polym. Eng.* **2017**, *37*, 689–698. [CrossRef]

98. Wu, C.S.; Liao, H.T. Interface design of environmentally friendly carbon nanotube-filled polyester composites: Fabrication, characterisation, functionality and application. *Express Polym. Lett.* **2017**, *11*, 187–198. [CrossRef]

99. Puppi, D.; Mota, C.; Gazzarri, M.; Dinucci, D.; Gloria, A.; Myrzabekova, M.; Ambrosio, L.; Chiellini, F. Additive manufacturing of wet-spun polymeric scaffolds for bone tissue engineering. *Biomed. Microdevices* **2012**, *14*, 1115–1127. [CrossRef]

100. Puppi, D.; Pirosa, A.; Morelli, A.; Chiellini, F. Design, fabrication and characterization of tailored poly[(r)-3-hydroxybutyrate-*co*-(r)-3-hydroxyexanoate] scaffolds by computer-aided wet-spinning. *Rapid Prototyp. J.* **2018**, *24*, 1–8. [CrossRef]

101. Puppi, D.; Morelli, A.; Chiellini, F. Additive manufacturing of poly(3-hydroxybutyrate-co-3-hydroxyhexanoate) /poly(ε-caprolactone) blend scaffolds for tissue engineering. *Bioengineering* **2017**, *4*, 49. [CrossRef] [PubMed]

102. Puppi, D.; Pirosa, A.; Lupi, G.; Erba, P.A.; Giachi, G.; Chiellini, F. Design and fabrication of novel polymeric biodegradable stents for small caliber blood vessels by computer-aided wet-spinning. *Biomed. Mater.* **2017**, *12*, 035011. [CrossRef] [PubMed]

103. Koller, M.; Maršálek, L.; de Sousa Dias, M.M.; Braunegg, G. Producing microbial polyhydroxyalkanoate (PHA) biopolyesters in a sustainable manner. *New Biotechnol.* **2017**, *37*, 24–38. [CrossRef]

9

Additive Manufacturing of Poly(3-hydroxybutyrate-co-3-hydroxyhexanoate)/ poly(ε-caprolactone) Blend Scaffolds for Tissue Engineering

Dario Puppi, Andrea Morelli and Federica Chiellini *

BIOLab Research Group, Department of Chemistry and Industrial Chemistry, University of Pisa,
UdR INSTM Pisa, via Moruzzi 13, 56124 Pisa, Italy; d.puppi@dcci.unipi.it (D.P.); a.morelli@dcci.unipi.it (A.M.)
* Correspondence: federica.chiellini@unipi.it

Academic Editor: Martin Koller

Abstract: Additive manufacturing of scaffolds made of a polyhydroxyalkanoate blended with another biocompatible polymer represents a cost-effective strategy for combining the advantages of the two blend components in order to develop tailored tissue engineering approaches. The aim of this study was the development of novel poly(3-hydroxybutyrate-*co*-3-hydroxyhexanoate)/ poly(ε-caprolactone) (PHBHHx/PCL) blend scaffolds for tissue engineering by means of computer-aided wet-spinning, a hybrid additive manufacturing technique suitable for processing polyhydroxyalkanoates dissolved in organic solvents. The experimental conditions for processing tetrahydrofuran solutions containing the two polymers at different concentrations (PHBHHx/PCL weight ratio of 3:1, 2:1 or 1:1) were optimized in order to manufacture scaffolds with predefined geometry and internal porous architecture. PHBHHx/PCL scaffolds with a 3D interconnected network of macropores and a local microporosity of the polymeric matrix, as a consequence of the phase inversion process governing material solidification, were successfully fabricated. As shown by scanning electron microscopy, thermogravimetric, differential scanning calorimetric and uniaxial compressive analyses, blend composition significantly influenced the scaffold morphological, thermal and mechanical properties. In vitro biological characterization showed that the developed scaffolds were able to sustain the adhesion and proliferation of MC3T3-E1 murine preosteoblast cells. The additive manufacturing approach developed in this study, based on a polymeric solution processing method avoiding possible material degradation related to thermal treatments, could represent a powerful tool for the development of customized PHBHHx-based blend scaffolds for tissue engineering.

Keywords: polyhydroxyalkanoates; poly(3-hydroxybutyrate-co-3-hydroxyhexanoate); poly(ε-caprolactone); polymers blend; tissue engineering; scaffolds; additive manufacturing; computer-aided wet-spinning

1. Introduction

Tissue engineering is a growing research area, with a few successful clinical results, aimed at developing reliable alternatives to conventional surgical strategies (e.g., auto- and allogenic tissue transplantation or artificial prosthesis implantation) for the treatment of human tissue and organ failure caused by defects, injuries or other types of damage [1]. Tissue engineering relies on the combination of cells, biomaterials and bioactive molecules to generate replacement biological tissues and organs for a wide range of medical conditions. The most common approach involves the employment of a highly porous biodegradable support, commonly referred to as the scaffold, which acts as a temporary template providing a cell adhesion substrate and mechanical support, and guiding the regeneration processes [2]. In the last two decades, a great variety of biodegradable materials and processing techniques have been investigated for the development of scaffolds with proper physico-chemical properties as well as macro-, micro- and nano-architecture features suitable for tissue growth in three dimensions [3].

Polyhydroxyalkanoates (PHAs) are microbial aliphatic polyesters widely investigated for biomedical applications due to their biodegradability and biocompatibility, as well as the wide range of mechanical and processing properties of the numerous homopolymers and copolymers belonging to this class of renewable polymers [4]. Different articles have reported on poly[(R)-3-hydroybutyrate] (PHB) and poly[(R)-3-hydroxybutyrate-co-(R)-3-hydroxyvalerate] (PHBV) investigations, both in vitro and in vivo, for bone tissue regeneration approaches [5–7]. Due to the relatively long alkyl side chain, poly[(R)-3-hydroxybutyrate-co-(R)-3-hydroxyhexanoate) (PHBHHx) exhibits lower crystallinity, a broader processing window and higher elasticity compared with PHB and PHBV [8]. Among the different investigated biomedical applications, PHBHHx has been proposed as scaffolding material for bone regeneration thanks to its piezoelectric behavior and cytocompatibility when cultured with osteoblasts and bone marrow cells [9–14]. In addition, recent articles showed that PHBHHx in the form of microgrooved membrane [15], aligned nanofibers [16] or carbon nanotubes-loaded composite materials [17] well supports the osteogenesis of human mesenchymal stem cells.

As defined by the American Society for Testing and Materials (ASTM), Additive Manufacturing (AM) refers to the process of joining materials to make objects from three-dimensional (3D) model data, usually layer upon layer, as opposed to subtractive manufacturing methodologies [18]. The introduction of a number of AM techniques, such as stereolithography and fused deposition modeling, into the tissue engineering field has allowed the enhancement of control over scaffold structure at different size scales (from macro- to micrometric scale) in terms of external shape and porous structure [19]. They involve a computer-controlled layered manufacturing process based on a sequential delivery of energy and/or materials starting from a 3D digital model to build up 3D polymeric scaffolds with a predefined geometry and internal porosity. Advanced computer-aided design and manufacturing approaches enable a high degree of automation, good accuracy and reproducibility for the fabrication of clinically-sized, anatomically-shaped scaffolds with a tailored porous structure characterized by a fully interconnected network of pores with customized size and shape. However, despite the promising results and widespread research on PHAs for tissue engineering applications, their narrow melt processing temperature window [20,21] has hindered the application of AM techniques for their processing into 3D porous scaffolds. Computer-aided wet-spinning (CAWS), a hybrid AM technique based on the computer-controlled deposition of a solidifying polymeric fiber extruded directly into a coagulation bath, was recently applied to process PHBHHx into 3D scaffolds with tailored geometry and networks of macropores as well as a homogenous microporous matrix [22,23].

The aim of this study was to investigate the suitability of the CAWS technique for the fabrication of scaffolds made of PHBHHx blended with poly(ε-caprolactone) (PCL). PCL is an aliphatic polyester that has been widely investigated for biomedical applications receiving FDA approval and CE Mark registration for a number of drug delivery and medical device applications [24]. Thanks to its good processing properties, tunable mechanical properties and slow biodegradation, PCL is seen as one of the most versatile scaffolding materials for the development of long-term biodegradable bone implants. Recent studies have investigated the blending of PHAs with PCL and other synthetic polyesters as a cost-effective strategy for combining the advantages of the two polymers and achieving additional desirable properties [21,25,26]. As an example, a research activity on PCL/PHBHHx blend membranes by solvent processing showed that by optimizing the weight ratio between the two components it was possible to enhance the resulting mechanical properties in comparison with PCL and PHBHHx alone [27]. Although the great versatility of the CAWS technique in customizing PHBHHx scaffold's shape and internal architecture, in the case of inter-fiber deposition distances larger than 200 μm (i.e., 500 and 1000 μm), a well-defined porosity along the Z axis was not achieved due to the slow solidification of the coagulating fiber. On the other hand, the optimization of PCL processing by CAWS has enabled the employment of large inter-fiber deposition distances (i.e., 500 and 1000 μm) for the fabrication of 3D scaffolds with a homogeneous porosity in the cross-section characterized by a Z axis pore size in the range of hundreds of micrometers [28–30]. Since a pore size larger than 100 μm is recommended to achieve enhanced bone tissue regeneration and vascularization [31], blending PHBHHx with a polymer showing better processing properties was investigated during the present study as an effective strategy to develop scaffolds meeting the aforementioned structural parameters requirement. For this purpose, the CAWS conditions for the fabrication of PHBHHx/PCL scaffolds with different ratios between the two blend components were investigated. Optimized PHBHHx/PCL scaffold prototypes were characterized in comparison with PHBHHx scaffolds for their morphology by means of scanning electron microscopy (SEM) under backscattered electron imaging, thermal properties by means of thermogravimetric analysis (TGA) and differential scanning calorimetry (DSC), and mechanical properties under compression using a uniaxial testing machine. The scaffold's biocompatibility was evaluated in vitro by employing the MC3T3-E1 murine preosteoblast cell line. Cell response, in terms of viability, proliferation and morphology was investigated by tetrazolium salts (WST-1) and confocal laser scanning microscopy (CLSM).

2. Materials and Methods

2.1. Materials

Poly(ε-caprolactone) (PCL, CAPA 6800, Mw = 80,000 g·mol^{-1}) was supplied by Perstorp UK Ltd (Warrington, Cheshire, UK) and used as received. Poly(3-hydroxybutyrate-co-3-hydroxyhexanoate) (PHBHHx, 12% mol HHx, Mw = 300,000 g·mol^{-1}) was kindly supplied by Tsinghua University (Beijing, China). PHBHHx was purified before use according to the following procedure: (i) the polymer was dissolved in 1,4 dioxane (5% w/v) under stirring at room temperature for 1 h; (ii) the solution was filtered under vacuum using filter paper; (iii) the filtrate was slowly dropped into 10-fold volume water to precipitate PHBHHx; (iv) after precipitation the polymer was collected by filtering; (v) the polymer was washed with distilled water and then ethanol, and vacuum dried and stored in a desiccator. All the solvents and chemical reagents were purchased from Sigma-Aldrich (Italy) and used as received without further purification.

2.2. Scaffolds Fabrication

PHBHHx solutions were prepared by dissolving the polymer in tetrahydrofuran (THF) at 32 °C under stirring for 2 h at a concentration of 25% w/v. For the preparation of PHBHHx/PCL solutions, PCL was dissolved in THF at 32 °C under stirring for 2 h and then the desired amount of PHBHHx was added to the polymer solution. The mixture was left under stirring for 2 h at 32 °C until a homogenous solution was obtained. Solutions with different PHBHHx/PCL weight ratios (3:1, 2:1 and 1:1) and a total concentration of the polymeric phase of 12% w/v were prepared.

Scaffolds were fabricated by means of a subtractive rapid prototyping system (MDX 40A, Roland MID EUROPE, Acquaviva Picena, Italy) modified in-house by replacing the milling head unit with a programmable syringe pump system (NE-1000; New Era Pump Systems Inc., Wantagh, NY, USA) to enable the deposition of polymeric solutions with a controlled 3D pattern (Figure 1) [29]. The 3D geometrical scaffold parameters were designed using an algorithm developed in Matlab software (The Mathworks, Inc., Natick, MA, USA). The desired polymeric solution was placed into a glass syringe fitted with a metallic needle (Gauge 23) and injected at a controlled feeding rate directly into an ethanol coagulation bath by using the syringe pump. Scaffold fabrication was carried out by employing a deposition trajectory aimed at the production of scaffolds with a 0–90° lay-down pattern, distance between fiber axis of 500 μm and layer thickness of 100 μm. The optimized initial distance between the tip of the needle and the bottom of the beaker (Z$_0$) was 1.5 mm. The effect of different processing parameters, such as the deposition velocity (V$_{dep}$) and the solution feed rate (F), on fiber collection and morphology was evaluated to produce blend scaffolds with a different PHBHHx/PCL ratio (Table 1). By employing the optimized fabrication parameters, cylindrical samples with a designed diameter of

15 mm and height of 5 mm were fabricated. The samples were removed from the coagulation bath, left under a fume hood for 24 h, placed in a vacuum chamber at about 0.5 mbar for 48 h and then stored in a desiccator for at least 72 h before characterization.

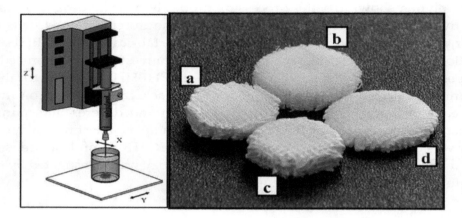

Figure 1. Schematics of the computer-aided wet-spinning (CAWS) process (left); representative image of the developed scaffolds (right): (**a**) PHBHHx; (**b**) PHBHHx/PCL 3:1; (**c**) PHBHHx/PCL 2:1; (**d**) PHBHHx/PCL 1:1.

Table 1. Optimized processing parameters, and scaffold structural parameters obtained from scanning electron microscopy (SEM) analysis.

Sample	F (mL·h^{-1})	Vdep (mm·min^{-1})	Fiber Diameter (μm)	Pore Size (μm)
PHBHHx	0.5	300	88 ± 12	485 ± 40
PHBHHx/PCL 3:1	1	560	116 ± 12	493 ± 29
PHBHHx/PCL 2:1	1	560	114 ± 15	470 ± 46
PHBHHx/PCL 1:1	1	560	120 ± 11	484 ± 18

Morphological parameters expressed as average ± standard deviation.

2.3. Morphological Characterization

The top-view and cross section (obtained by fracture in liquid nitrogen) of the scaffolds were analyzed by means of scanning electron microscopy (SEM, JEOL JSM 300, Tokyo, Japan) under backscattered electron imaging. The average fiber diameter and pore size, defined as inter-fiber distance, were measured by means of ImageJ 1.43u software on top-view micrographs with a 50X magnification. Data were calculated over 20 measurements per scaffold.

2.4. Thermal Analysis

Thermal properties of the scaffolds were evaluated by means of thermogravimetric analysis (TGA) and differential scanning calorimetry (DSC). TGA was performed using TGA Q500 instruments (TA Instruments, Milano, Italy) in the temperature range 30–600 °C, at a heating rate of 10 °C/min and under a nitrogen flow of 60 mL·min^{-1}. The scaffold's thermal decomposition were evaluated by analyzing weight and derivative weight profiles as functions of temperature. DSC analysis was performed using a Mettler DSC-822 instrument (Mettler Toledo, Novate Milanese (MI), Italy) in the range −100–200 °C, at a heating rate of 10 °C/min and a cooling rate of −20 °C/min, and under a nitrogen flow of 80 mL·min^{-1}. By considering the first and second heating cycle in the thermograms, glass transition temperature (T_g) was evaluated by analyzing the inflection point, while melting temperature (T_m) and enthalpy (ΔH) was evaluated by analyzing the endothermic peaks.

2.5. Mechanical Testing

The scaffold's mechanical properties were analyzed under compression using an Instron 5564 uniaxial testing machine (Instron Corporation, Norwood, MA, USA) equipped with a 2 kN load cell. After the treatment to remove residual solvents, as previously described, the samples were preconditioned at 25 °C and 50% of humidity for 48 h and then characterized at room temperature. The test was carried out on cylindrical samples with actual diameters of around 15 mm and actual heights of around 4 mm (50 layers). Six samples of each kind of scaffold were tested at a constant crosshead displacement of 0.4 mm·min^{-1} between two parallel steel plates up to 85% strain [32]. The stress was defined as the measured force divided by the total area of the apparent cross section of the scaffold, whilst the strain was evaluated as the ratio between the height variation and the initial height. Stress-strain curves were obtained from the software recording the data (Merlin, Series IX, Instron Corporation, Norwood, MA, USA). The compressive modulus was calculated as the slope of the initial linear region in the stress-strain curve, avoiding the toe region. Compressive yield strength and strain were considered at the yield point, and compressive strength was considered as the stress corresponding to 85% strain.

2.6. In Vitro Biological Evaluation

2.6.1. Cell Culture

Mouse calvaria-derived pre-osteoblast cell line MC3T3-E1 subclone 4 was obtained from the American Type Culture Collection (ATCC CRL-2593, Manassas, VA, USA) and cultured in Alpha Minimum Essential Medium (α—MEM, Sigma, Milan, Italy) supplemented with 2 mM·L-glutamine, 10% fetal bovine serum, 100 U/mL:100 µg/mL penicillin:streptomycin solution (GIBCO, Invitrogen Corporation, Milan, Italy) and antimycotic. Before experiments, cells were trypsinized with 0.25% trypsin-EDTA (GIBCO, Gaithersburg, MD, USA) solution and resuspended in complete α-MEM at a concentration of 3×10^4/mL. Scaffolds were seeded with 100 µL of cell suspension and the final volume was adjusted to 1 mL with complete medium. The specimens were then placed in an incubator with humidified atmosphere at 37 °C in 5% CO_2. Osteogenic differentiation was induced 24 h after seeding by culturing cells in osteogenic medium prepared with α–MEM supplemented with ascorbic acid (0.3 mM) and β—glycerolphosphate (10 mM). The culture medium was replaced every 48 h and biological characterizations were carried out weekly at days 7, 14, 21 and 28. Cells grown onto tissue culture polystyrene plates were used as control.

2.6.2. Cell Viability and Proliferation

Cell viability and proliferation were measured by using the (4-[3-(4-iodophenyl)-2-(4 nitrophenyl) -2H-5-tetrazolium]-1,3-benzene disulfonate) (WST-1) assay (Roche Molecular Biochemicals, Monza, Italy), which is based on the mitochondrial conversion of the tetrazolium salt WST-1 into soluble formazan in viable cells. WST-1 reagent diluted 1:10 was added to the culture and incubated for 4 h at 37 °C. Measurements of formazan dye absorbance were carried out with a Biorad microplate reader at 450 nm, with the reference wavelength at 655 nm. The in vitro biological test was performed on triplicate samples for each material.

2.6.3. Morphologic Characterizations by Confocal Laser Scanning Microscopy (CLSM)

The morphology of the cells grown on the prepared meshes was investigated by means of CLSM. Cells were fixed with 3.8% paraformaldehyde in PBS 0.01 M pH 7.4 (PBS 1X), permeabilized with a PBS 1X/Triton X-100 solution (0.2%) for 15 min and incubated with a solution of 4′-6-diamidino-2-phenylindole (DAPI; Invitrogen) and phalloidin-AlexaFluor488 (Invitrogen) in PBS 1X for 60 min at room temperature in the dark. After dye incubation, samples were washed with PBS 1X before being mounted on a glass slide and sealed with resin for microscopic observation. A Nikon Eclipse TE2000 inverted microscope equipped with an EZC1 confocal laser and Differential Interference Contrast

(DIC) apparatus was used to analyze the samples (Nikon, Tokyo, Japan). A 405 nm laser diode (405 nm emission) and an argon ion laser (488 nm emission) were used to excite DAPI and Alexa fluorophores, respectively. Images were captured with Nikon EZ-C1 software with identical settings for each sample. Images were further processed with GIMP (GNU Free Software Foundation) Image Manipulation Software and merged with Nikon ACT-2U software.

2.7. Statistical Analysis

The data are represented as mean ± standard deviation. Statistical differences were analyzed using one-way analysis of variance (ANOVA), and a Tukey test was used for post hoc analysis. A p-value < 0.05 was considered statistically significant.

3. Results and Discussion

3.1. Additive Manufacturing of Scaffolds

The fabrication process involved the deposition with a predefined pattern of an extruded polymeric solution into a coagulation bath to make a 3D scaffold using a layer-by-layer process. By optimizing the most influential manufacturing parameters, in terms of solution flow rate (F) and deposition velocity (V_{dep}) (Table 1), 3D cylindrical scaffolds were developed by processing solutions with different PHBHHx/PCL weight ratios (3:1, 2:1 or 1:1) (Figure 1).

Wet-spinning (WS) is a non-solvent-induced phase inversion technique suitable for the industrial production of continuous polymeric fibers through an immersion-precipitation process. Briefly, a polymeric solution is injected into a coagulation bath containing a non-solvent of the polymer, and the solution filament solidifies because of polymer desolvation caused by solvent/non-solvent exchange [33]. A number of studies have shown that 3D macroporous scaffolds made of synthetic or natural polymers can be obtained through physical bonding of fibers prefabricated by means of WS, using a glue and/or a thermomechanical treatment, or in a single-step fabrication process involving the continuous, randomly-oriented deposition of the solidifying fiber by means of a manually controlled motion of the coagulation bath [33–38]. Although 3D structures with high and interconnected porosity suitable for tissue regeneration processes have been developed, an accurate control over the scaffold macro- and microstructure has not been achieved by employing these methods. The CAWS technique has been proposed as a suitable AM approach to upgrade the fabrication process in terms of reproducibility, resolution, design freedom and automation degree [39]. Advanced scaffold structural features at different scale levels, in terms of external shape and internal porosity, have been developed by applying the CAWS technique to different biocompatible polymers, including PCL, three-arm star PCL, PHBHHx, chitosan/poly(γ-glutamate) polyelectrolyte complexes and a poly(ethylene oxide terephthalate)/poly(butylene terephthalate) block copolymer [22,23,28,29,40–44]. As previously discussed, the application of AM to PHAs is very limited due to the narrow thermal processing window of this class of polymers. The few exceptions of PHAs scaffolds manufactured by AM are represented by a set of nanocomposite PHA/tricalcium phosphate composite scaffolds fabricated by means of selective laser sintering [45–47] and PCL/PHBV blend scaffolds fabricated by fused deposition modeling [21]. The research activity reported in this study has led to the development of a novel AM process for the fabrication of PHA-based blend scaffolds by processing solutions containing PHBHHx and PCL in different ratios. This approach does not require thermal treatments that could cause polymer degradation as well as denaturation of bioactive agents possibly loaded into the scaffold [48]. The solvents employed are allowed by the European Medicine Agency as residues in medical products below recommended safety levels, and classified as solvents with low toxicity (i.e., ethanol) or to be limited (i.e., THF). Considering the small volume of THF and the high ethanol/THF volume ratio involved (>20), the high volatility of the two solvents, as well as the absence of thermal events in TGA and DSC curves that could be related to the evaporation of residual solvents in scaffolds, as discussed in one of the following sections, the process can be considered as meeting the basic requirements of good manufacturing practices.

3.2. Morphological Characterization

The morphology of the developed scaffolds was investigated by means of SEM analysis using backscattering electron imaging. Analysis of samples both in top view and cross section highlighted that the fabricated scaffolds were composed by a 3D layered structure of aligned fibers forming a fully interconnected network of macropores (Figure 2). Comparative analysis of SEM micrographs showed that the scaffold's fiber morphology and alignment were influenced by the composition of the wet-spinning solution. In addition, as observed in high magnification micrographs, the fibers had a microporous morphology both in the outer surface and in the cross-section due to the phase inversion process governing polymer solidification, as discussed elsewhere [28].

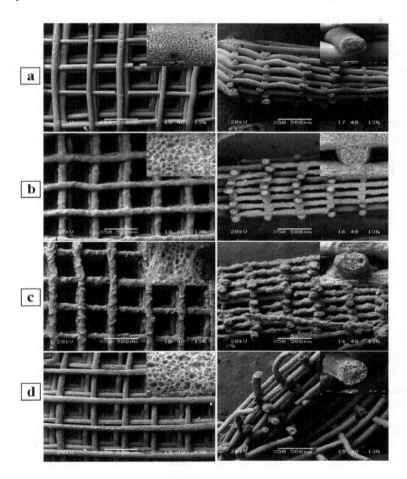

Figure 2. Representative top view (left) and cross-section (right) SEM micrographs of (**a**) PHBHHx; (**b**) PHBHHx/PCL 3:1, (**c**) PHBHHx/PCL 2:1, (**d**) PHBHHx/PCL 1:1. Inset high magnification micrographs show porosity of outer surface (left) and cross section (right) of single fibers.

PHBHHx/PCL blend scaffolds showed significantly larger fiber diameters (average value in the range of 100 to 135 μm) in comparison with plain PHBHHx scaffolds (average value of 88 μm) (Table 1). Differences in fiber diameter among the different blend scaffolds were not statistically significant. The pore size was in the range 400–500 μm, with no statistically significant differences among the different scaffolds.

Together with the possibility of easily loading a scaffold with a drug by simply adding it to the polymeric solution before processing [41], the main advantage of CAWS is represented by the obtainment of multi-scale scaffold porosities. In fact, porous structures fabricated by means of this technique are generally characterized by a fully interconnected network of macropores, with a size that can be tuned in the range of tens to hundreds of micrometers by varying the fiber lay-down pattern, and a local micro/nanosized porous morphology of the polymeric matrix that can be tailored by acting on different parameters of the phase separation process determining polymer solidification [39]. This multi-scale morphology control represents a powerful tool to tune key scaffold properties strictly related to porosity and surface roughness, such as biodegradation rate, mechanical behavior, and cell interactions.

3.3. Thermal Characterization

The thermal properties of the developed PHBHHx/PCL blend scaffolds were investigated by TGA and DSC in comparison with PHBHHx scaffolds and PCL raw polymer. TGA evaluation showed that the weight and derivative weight curves of the blend scaffolds were characterized by two main thermal decomposition events: the first one centered at around 290 °C ascribable to PHBHHx decomposition, and the other one centered at around 405 °C related to PCL decomposition (Figure 3). Thermal events relevant to evaporation of residual THF (boiling point of 66 °C) or ethanol (boiling point of 78 °C) were not detected. The thermograms of the PHBHHx scaffolds are characterized by a relatively high residue at 600 °C, that did not compromise scaffold's cytocompatibility as shown also by a previous study [22]. Nevertheless, future studies should investigate the reason and composition of this ash content and whether it may have an impact on use, particularly in extended biological testing.

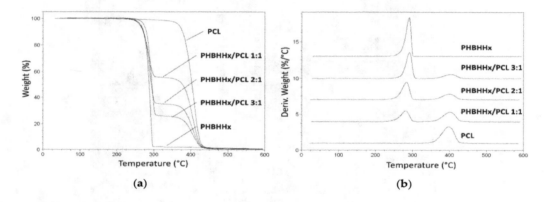

Figure 3. Thermogravimetric analysis (TGA) characterization: weight (**a**) and derivative weight (**b**) profiles vs temperature of the developed scaffolds.

The area under the first peak in derivative weight curves decreased on increasing PCL percentage, while that under the second peak increased by increasing PCL content. The resulting percentage weight losses during the first and second thermal events were close to the percentage weight in the starting polymeric solution of PHBHHx and PCL, respectively (Table 2).

Table 2. Data relevant to thermal decomposition obtained from TGA analysis.

Sample	1st Decomposition Step		2nd Decomposition STEP	
	Peak (°C)	Weight Loss (%)	Peak (°C)	Weight Loss (%)
PHBHHx	290.7 ± 1.8	97.6 ± 0.9	-	-
PHBHHx/PCL 3:1	291.3 ± 2.2	74.6 ± 1.2	405.9 ± 0.5	25.1 ± 0.4
PHBHHx/PCL 2:1	289.2 ± 1.4	65.4 ± 1.4	405.9 ± 0.7	34.2 ± 0.5
PHBHHx/PCL 1:1	285.2 ± 1.6	44.9 ± 0.8	405.3 ± 0.8	54.6 ± 0.8
PCL raw	-	-	406.6 ± 0.4	99.1 ± 0.4

Data expressed as average ± standard deviation ($n = 3$).

Representative DSC thermograms of the characterized samples are reported in Figure 4. The first heating cycle analysis was carried out to assess the thermal properties of the scaffolds in comparison to what was observed in the second heating cycle after blend melting and solidification to erase the prior thermal history. The glass transition and melting temperature of PCL (Tg_1 and Tm_1) and PHBHHx (Tg_2 and Tm_2), as well as their respective melting enthalpies (ΔH_1 and ΔH_2), were analyzed. An endothermic peak centered at around 60 °C ascribable to the melting of PCL crystalline domains is evident in both the first and second heating cycle thermograms of blend scaffolds. The endothermic peak related to the melting of PHBHHx crystalline domains is evident only in the first heating cycle thermograms, while in the second heating cycle thermograms, a pronounced glass transition of PHBHHx only is detectable. These results corroborate what was reported in previous articles about the appreciable crystalline degree of PHBHHx scaffolds by CAWS due to the relatively slow crystallization mechanism [22,23]. Polymer crystallinity influences different properties of scaffolds made of aliphatic polyesters, such as their mechanical behavior and biodegradation rate. Indeed, as widely reported in literature [3,49], crystalline and amorphous domains show different water diffusivity as well as different macromolecular deformation and rearrangement when subjected to mechanical solicitations. The quite broad endothermic peak in the first heating scan of PHBHHx scaffolds curve can be explained with the melting of different lamellar crystalline domains formed during polymer solidification, as suggested by previous articles on thermal characterization of PHBHHx films [50,51]. Endothermic peaks related to evaporation of residual solvents were not detectable in the first nor in the second heating scan.

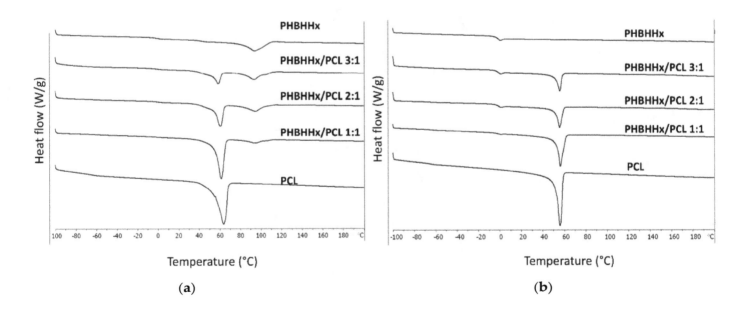

Figure 4. Representative differential scanning calorimetry (DSC) thermograms of the analyzed samples relevant to the first heating (**a**) and second heating (**b**) cycles.

By comparing data from either the first or the second DSC scan (Table 3), the effect of blend composition on endothermic peaks area was quantitatively confirmed through analysis of differences in enthalpies ΔH_1 and ΔH_2. In addition, Tg_1 and Tm_1 in the first heating cycle were significantly affected by PHBHHx/PCL ratio, in agreement with a previous article showing that by increasing PHBHHx content in the blend, the melting of the PCL component was shifted to lower temperatures [27], possibly due to a slight plasticization effect of PHBHHx on the PCL phase. Besides the previously mentioned effect on ΔH_1, differences in data from the second scan were not statistically significant [26].

Table 3. Data relevant to thermal characterization by DSC analysis.

Sample	1st Heating						2nd Heating			
	T_{g1} (°C)	T_{m1} (°C)	ΔH_1 (J/g)	T_{g2} (°C)	T_{m2} (°C)	ΔH_2 (J/g)	T_{g1} (°C)	T_{m1} (°C)	ΔH_1 (J/g)	T_{g2} (°C)
PHBHHx	—	—	—	-0.6 ± 0.2	93.7 ± 1.2	41.4 ± 1.8	—	—	—	-1.1 ± 0.6
PHBHHx/PCL 3:1	-70.7 ± 1.6	58.5 ± 0.9	18.6 ± 1.2	-0.2 ± 0.6	93.5 ± 1.6	18.4 ± 0.9	-66.2 ± 1.2	56.0 ± 1.2	15.4 ± 0.8	-1.1 ± 0.8
PHBHHx/PCL 2:1	-65.2 ± 1.4	60.7 ± 1.2	30.2 ± 1.8	-0.1 ± 0.3	94.3 ± 2.1	14.9 ± 1.1	-64.9 ± 1.3	55.8 ± 0.4	21.1 ± 1.4	-0.9 ± 0.4
PHBHHx/PCL 1:1	-60.2 ± 1.4	61.3 ± 0.8	53.4 ± 2.4	-0.6 ± 0.2	94.2 ± 1.9	7.5 ± 0.4	-63.4 ± 1.4	56.3 ± 1.6	36.0 ± 1.4	-1.4 ± 0.5
PCL raw	-61.4 ± 1.1	63.4 ± 0.5	96.5 ± 2.1	—	—	—	-64.7 ± 1.6	55.9 ± 1.5	83.0 ± 2.6	—

Data expressed as average ± standard deviation ($n = 3$).

3.4. Mechanical Characterization

The scaffold's mechanical properties were evaluated under compression using a uniaxial testing machine at a constant strain rate. Representative stress-strain compressive curves of PHBHHx and PHBHHx/PCL blend scaffolds are reported in Figure 5. They are characterized by three distinct regions: a roughly linear region, followed by a small plateau at fairly constant stress, and a final region of steeply rising stress. As suggested by previous papers reporting on mechanical characterization of polymeric scaffolds manufactured by CAWS [22,29,41], this three-region behavior can be explained with the sample response to the applied deformation at different structural scale levels. Indeed, the linear region is likely due to the initial response of the fiber–fiber contact points, the subsequent plateau region to the collapse of the pores network, and the final stress increase region to a further densification of the scaffold structure that behaves like a dense matrix.

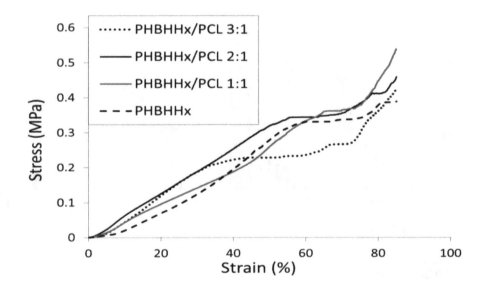

Figure 5. Representative stress-strain curve under compression (0.4 mm/min) of PHBHHx-based scaffolds.

The characterized samples showed compressive mechanical parameters (Table 4) of the same order of magnitude of PHBHHx scaffolds previously developed by CAWS [22,23] or solvent casting combined with salt-leaching techniques [52]. PHBHHx/PCL 3:1 scaffolds showed a comparable modulus but a marked drop in yield strain, yield stress and maximum stress in comparison to PHBHHx scaffolds. However, by increasing PCL content, the compressive modulus and the other mechanical parameters increased significantly. In fact, PHBHHx/PCL 2:1 and PHBHHx/PCL 1:1 scaffolds showed compressive modulus values (0.39 ± 0.14 and 0.37 ± 0.07, respectively) between that of PHBHHx scaffolds (0.16 ± 0.12 MPa) and that of PCL scaffolds with the same designed architecture (0.60 ± 0.20 MPa) [28]. In addition, PHBHHx/PCL 2:1 scaffolds showed a yield stress (0.36 ± 0.05 MPa) comparable to that of PHBHHx scaffolds and significantly higher than the other blend scaffolds. Overall, the developed scaffolds showed lower compressive strength in comparison to PHBV/PCL blend scaffolds by fused deposition modeling [21]. This difference should be mainly related to the multi-scale porous structure of scaffolds prepared using CAWS that, although the aforementioned advantages in providing a versatile tool for tuning scaffold properties and favor cells adhesion and interactions, is characterized by a higher void volume percentage in comparison to macroporous structures with a dense polymer matrix fabricated by means of melt processing.

Table 4. Compressive mechanical parameters of the developed scaffolds.

Scaffolds	Compressive Modulus (MPa)	Yield Strain (%)	Yield Stress (MPa)	Stress at 85% Strain (MPa)
PHBHHx	0.16 ± 0.12	56.8 ± 9.5	0.32 ± 0.02	0.47 ± 0.09
PHBHHx/PCL 3:1	0.17 ± 0.89	35.0 ± 9.4	0.18 ± 0.03	0.41 ± 0.09
PHBHHx/PCL 2:1	0.39 ± 0.14	57.9 ± 5.5	0.36 ± 0.05	0.48 ± 0.05
PHBHHx/PCL 1:1	0.37 ± 0.07	66.8 ± 5.5	0.36 ± 0.02	0.51 ± 0.07

Data expressed as average ± standard deviation ($n = 3$).

3.5. Biological Characterization

Investigations of viability and proliferation of cells seeded onto the developed scaffolds performed using the WST-1 assay showed an increase in the number of viable cells in all the tested samples during the 28 days of culture (Figure 6).

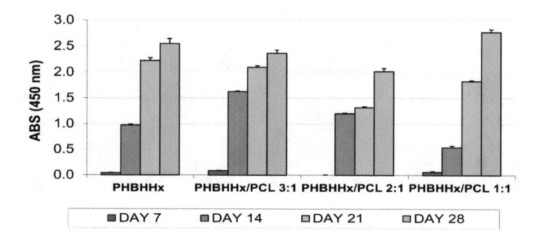

Figure 6. MC3T3-E1 cell proliferation on PHBHHx and PHBHHx/PCL based scaffolds.

The time-dependent changes in proliferation and differentiation can be divided into distinct stages of pre-osteoblast development [53]. The initial phase of osteoblast development is characterized by active replication of undifferentiated cells. At day 7, the observed limited cell proliferation was probably due to the large inter-fiber distance of the tested scaffolds that did not retain cells during the seeding procedure. During the second week, the cultures display a rapid increase in cell proliferation (Figure 6). The obtained data confirm a crucial role of cell–material interaction and cell density on cell adhesion, which influence cell proliferation during the initial stage of culturing [54].

Confocal Laser Scanning Microscopy (CLSM) characterization was employed to observe cell morphology on the scaffolds at different time points. Actin filament and nuclei were stained by Phalloidin-AlexaFluor488 and DAPI, and visualized as green and blue fluorescence respectively. Figure 7 shows the cell morphology and distribution of cells on the investigated samples after 7, 14, 21, and 28 days of culture. A good surface colonization of the scaffolds by MC3T3-E1 cells, with a variable shape and spreading can be observed, especially starting from the second week of culture. F-actin organization was consistent with early stages of cell adaptation to the material [55], exhibiting great stress fibers stretched along the cytoplasm, and a low cell number coherent with the quantitative proliferation data (Figure 6). Similar results for number and morphology of cells were detected by comparing different types of samples at early stages. After 14 days of culture the cell density appears to be increased, and at day 28 cells completely spread on the polymeric structure, with large cell cluster formations with inter-cellular connections (Figure 7).

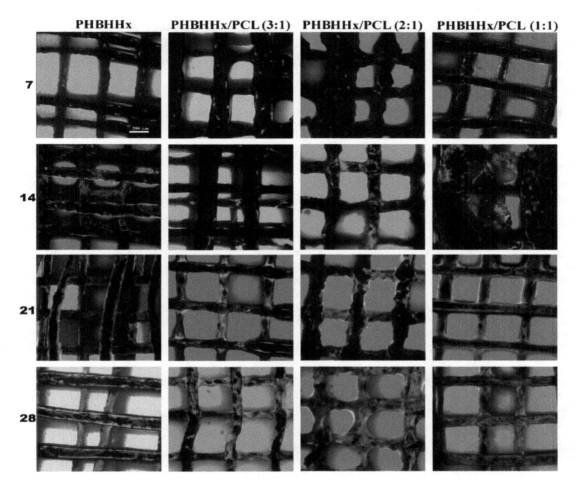

Figure 7. Confocal Laser Scanning Microscopy (CLSM) microphotographs showing MC3T3-E1 cell cultured on PHBHHx and PHBHHx/PCL based scaffolds, at different end-points.

These observations were in accordance with the differentiation pathway proposed for the preosteoblasts in vitro, where after an early growing latency, morpho-functional cellular aggregates are developed and single cell morphology is not distinguishable [56]. At the final phase of culturing, samples exhibited a nearly full cellular colonization of the available fiber surface by a wide continuous cell culture net.

4. Conclusions

The main result attained during the reported research activity is the development of an AM process based on the processing of polymeric solutions for the fabrication of PHBHHx-based blend scaffolds. This represents a novel approach to combining the advantages of PHA blending with other biocompatible polymers and the versatility of AM in supplying advanced fabrication tools for the development of scaffolds with customized macro- and microstructure. The developed manufacturing process meets both the product specification and good manufacturing practice requirements. In fact, it allows a good control of scaffold composition, external shape and internal porosity, it does not require thermal treatments that could cause material degradation, and involves the use of solvents allowed in medical device manufacturing that are completely removed from the scaffold during the fabrication and post-processing treatment.

The characterization analyses highlighted the versatility of the developed manufacturing process by demonstrating how PHBHHx/PCL blend scaffold composition, morphological features, thermal properties and mechanical parameters could be tuned in certain ranges by varying the ratio between the two blend components in the starting solution. In addition, the results obtained from the performed preliminary biological evaluations indicated that the developed scaffolds are able to sustain a good cell adhesion and proliferation, and after 28 days of culture, scaffolds were fully colonized by MC3T3-E1 preosteoblast cells.

As shown by recent studies, the CAWS technique is well suited for the development of PHBHHx scaffolds with a complex shape resembling that of an anatomical part and a tailored porous structure with advanced architectural features at different scale levels (e.g., longitudinal macrochannel, local micro/nanoporosity) designed to enhance tissue regeneration processes [22,23]. The developed PHBHHx/PCL scaffolds can therefore represent advanced prototypes for the development of sophisticated PHAs-based blend constructs with tailored composition, anatomical shape, macroporosity and nanoporous morphology.

Acknowledgments: The financial support of the University of Pisa PRA-2016-50 project entitled "Functional Materials" and the Tuscany Region (Italy) funded Project "Nuovi Supporti Bioattivi a Matrice Polimerica per la Rigenerazione Ossea in Applicazioni Odontoiatriche (R.E.O.S.S.)" as part of the program POR CReO FESR 2007–2013—Le ali alle tue idee are gratefully acknowledged. PHBHHx was kindly supplied by Guo-Qiang Chen of Tsinghua University (Beijing, China) within the framework of the EC-Funded project Hyanji Scaffold in the People Program of the 7FP (2010–2013). Mairam Myrzabekova is acknowledged for her contribution during scaffolds preparation. Matteo Gazzarri and Cristina Bartoli are acknowledged for their contribution during biological characterization. Paolo Narducci is acknowledged for his support in recording SEM images.

Author Contributions: Dario Puppi and Federica Chiellini conceived and designed the experiments; Dario Puppi and Andrea Morelli performed the experiments; Dario Puppi, Andrea Morelli and Federica Chiellini analyzed the data; Dario Puppi and Federica Chiellini wrote the paper.

References

1. Puppi, D.; Chiellini, F.; Dash, M.; Chiellini, E. Biodegradable Polymers for Biomedical Applications. In *Biodegradable Polymers: Processing, Degradation and Applications*; Felton, G.P., Ed.; Nova Science Publishers, Inc.: Hauppauge, NY, USA, 2011; pp. 545–604.

2. Woodruff, M.A.; Lange, C.; Reichert, J.; Berner, A.; Chen, F.; Fratzl, P.; Schantz, J.-T.; Hutmacher, D.W. Bone tissue engineering: From bench to bedside. *Mater. Today* **2012**, *15*, 430–435. [CrossRef]

3. Puppi, D.; Chiellini, F.; Piras, A.M.; Chiellini, E. Polymeric materials for bone and cartilage repair. *Prog. Polym. Sci.* **2010**, *35*, 403–440. [CrossRef]

4. Morelli, A.; Puppi, D.; Chiellini, F. Polymers from Renewable Resources. *J. Renew. Mater.* **2013**, *1*, 83–112. [CrossRef]

5. Doyle, C.; Tanner, E.T.; Bonfield, W. In vitro and in vivo evaluation of polyhydroxybutyrate and of polyhydroxybutyrate reinforced with hydroxyapatite. *Biomaterials* **1991**, *12*, 841–847. [CrossRef]

6. Chen, G.Q.; Wu, Q. The application of polyhydroxyalkanoates as tissue engineering materials. *Biomaterials* **2005**, *26*, 6565–6578. [CrossRef] [PubMed]

7. Jack, K.S.; Velayudhan, S.; Luckman, P.; Trau, M.; Grøndahl, L.; Cooper-White, J. The fabrication and characterization of biodegradable HA/PHBV nanoparticle-polymer composite scaffolds. *Acta Biomater.* **2009**, *5*, 2657–2667. [CrossRef] [PubMed]

8. Gao, Y.; Kong, L.; Zhang, L.; Gong, Y.; Chen, G.; Zhao, N.; Zhang, X. Improvement of mechanical properties of poly(dl-lactide) films by blending of poly(3-hydroxybutyrate-co-3-hydroxyhexanoate). *Eur. Polym. J.* **2006**, *42*, 764–775. [CrossRef]

9. Wang, Y.W.; Wu, Q.; Chen, G.Q. Attachment, proliferation and differentiation of osteoblasts on random biopolyester poly(3-hydroxybutyrate-co-3-hydroxyhexanoate) scaffolds. *Biomaterials* **2004**, *25*, 669–675. [CrossRef]

10. Wang, Y.W.; Yang, F.; Wu, Q.; Cheng, Y.C.; Yu, P.H.; Chen, J.; Chen, G.Q. Effect of composition of poly(3-hydroxybutyrate-co-3-hydroxyhexanoate) on growth of fibroblast and osteoblast. *Biomaterials* **2005**, *26*, 755–761. [CrossRef] [PubMed]

11. Yang, M.; Zhu, S.; Chen, Y.; Chang, Z.; Chen, G.; Gong, Y.; Zhao, N.; Zhang, X. Studies on bone marrow stromal cells affinity of poly (3-hydroxybutyrate-co-3-hydroxyhexanoate). *Biomaterials* **2004**, *25*, 1365–1373. [CrossRef] [PubMed]

12. Jing, X.; Ling, Z.; Zhenhu An, Z.; Guoqiang, C.; Yandao, G.; Nanming, Z.; Xiufang, Z. Preparation and evaluation of porous poly(3-hydroxybutyrate-co-3-hydroxyhexanoate) hydroxyapatite composite scaffolds. *J. Biomater. Appl.* **2008**, *22*, 293–307. [CrossRef] [PubMed]

13. Garcia-Garcia, J.M.; Garrido, L.; Quijada-Garrido, I.; Kaschta, J.; Schubert, D.W.; Boccaccini, A.R. Novel poly(hydroxyalkanoates)-based composites containing Bioglass (R) and calcium sulfate for bone tissue engineering. *Biomed. Mater.* **2012**, *7*, 054105. [CrossRef] [PubMed]

14. Ke, S.; Yang, Y.; Ren, L.; Wang, Y.; Li, Y.; Huang, H. Dielectric behaviors of PHBHHx–BaTiO$_3$ multifunctional composite films. *Compos. Sci. Technol.* **2012**, *72*, 370–375. [CrossRef]

15. Wang, Y.; Jiang, X.-L.; Yang, S.-C.; Lin, X.; He, Y.; Yan, C.; Wu, L.; Chen, G.-Q.; Wang, Z.-Y.; Wu, Q. MicroRNAs in the regulation of interfacial behaviors of MSCs cultured on microgrooved surface pattern. *Biomaterials* **2011**, *32*, 9207–9217. [CrossRef] [PubMed]

16. Wang, Y.; Gao, R.; Wang, P.-P.; Jian, J.; Jiang, X.-L.; Yan, C.; Lin, X.; Wu, L.; Chen, G.-Q.; Wu, Q. The differential effects of aligned electrospun PHBHHx fibers on adipogenic and osteogenic potential of MSCs through the regulation of PPARγ signaling. *Biomaterials* **2012**, *33*, 485–493. [CrossRef] [PubMed]

17. Wu, L.-P.; You, M.; Wang, D.; Peng, G.; Wang, Z.; Chen, G.-Q. Fabrication of carbon nanotube (CNT)/poly(3-hydroxybutyrate-co-3-hydroxyhexanoate) (PHBHHx) nanocomposite films for human mesenchymal stem cell (hMSC) differentiation. *Polym. Chem.* **2013**, *4*, 4490–4498. [CrossRef]

18. ASTM. *International F2792—12a Standard Terminology for Additive Manufacturing Technologies*; ASTM: West Conshohocken, PA, USA, 2012.

19. Mota, C.; Puppi, D.; Chiellini, F.; Chiellini, E. Additive manufacturing techniques for the production of tissue engineering constructs. *J. Tissue Eng. Regen. Med.* **2015**, *9*, 174–190. [CrossRef] [PubMed]

20. Leroy, E.; Petit, I.; Audic, J.L.; Colomines, G.; Deterre, R. Rheological characterization of a thermally unstable bioplastic in injection molding conditions. *Polym. Degrad. Stab.* **2012**, *97*, 1915–1921. [CrossRef]

21. Kosorn, W.; Sakulsumbat, M.; Uppanan, P.; Kaewkong, P.; Chantaweroad, S.; Jitsaard, J.; Sitthiseripratip, K.; Janvikul, W. PCL/PHBV blended three dimensional scaffolds fabricated by fused deposition modeling and responses of chondrocytes to the scaffolds. *J. Biomed. Mater. Res. B* **2016**. [CrossRef]

22. Mota, C.; Wang, S.Y.; Puppi, D.; Gazzarri, M.; Migone, C.; Chiellini, F.; Chen, G.Q.; Chiellini, E. Additive manufacturing of poly[(R)-3-hydroxybutyrate-co-(R)-3-hydroxyhexanoate] scaffolds for engineered bone development. *J. Tissue Eng. Regen. Med.* **2017**, *11*, 175–186. [CrossRef] [PubMed]

23. Puppi, D.; Pirosa, A.; Morelli, A.; Chiellini, F. Design, fabrication and characterization of tailored poly[(R)-3-hydroxybutyrate-*co*-(R)-3-hydroxyexanoate] scaffolds by Computer-aided Wet-spinning. *Rapid Prototyp. J.* **2018**, *24*. unpublished.

24. Woodruff, M.A.; Hutmacher, D.W. The return of a forgotten polymer-Polycaprolactone in the 21st century. *Prog. Polym. Sci.* **2010**, *35*, 1217–1256. [CrossRef]

25. Zhao, Q.; Wang, S.; Kong, M.; Geng, W.; Li, R.K.Y.; Song, C.; Kong, D. Phase morphology, physical properties, and biodegradation behavior of novel PLA/PHBHHx blends. *J. Biomed. Mater. Res. B* **2011**, *100B*, 23–31. [CrossRef] [PubMed]

26. Chiono, V.; Ciardelli, G.; Vozzi, G.; Sotgiu, M.G.; Vinci, B.; Domenici, C.; Giusti, P. Poly(3-hydroxybutyrate-co-3-hydroxyvalerate)/poly(ε-caprolactone) blends for tissue engineering applications in the form of hollow fibers. *J. Biomed. Mater. Res. A* **2008**, *85A*, 938–953. [CrossRef] [PubMed]

27. Lim, J.; Chong, M.S.K.; Teo, E.Y.; Chen, G.-Q.; Chan, J.K.Y.; Teoh, S.-H. Biocompatibility studies and characterization of poly(3-hydroxybutyrate-co-3-hydroxyhexanoate)/polycaprolactone blends. *J. Biomed. Mater. Res. B* **2013**, *101B*, 752–761. [CrossRef] [PubMed]

28. Puppi, D.; Mota, C.; Gazzarri, M.; Dinucci, D.; Gloria, A.; Myrzabekova, M.; Ambrosio, L.; Chiellini, F. Additive manufacturing of wet-spun polymeric scaffolds for bone tissue engineering. *Biomed. Microdevices* **2012**, *14*, 1115–1127. [CrossRef] [PubMed]

29. Mota, C.; Puppi, D.; Dinucci, D.; Gazzarri, M.; Chiellini, F. Additive manufacturing of star poly(ε-caprolactone) wet-spun scaffolds for bone tissue engineering applications. *J. Bioact. Compat. Polym.* **2013**, *28*, 320–340. [CrossRef]

30. Dini, F.; Barsotti, G.; Puppi, D.; Coli, A.; Briganti, A.; Giannessi, E.; Miragliotta, V.; Mota, C.; Pirosa, A.; Stornelli, M.R.; et al. Tailored star poly (ε-caprolactone) wet-spun scaffolds for in vivo regeneration of long bone critical size defects. *J. Bioact. Compat. Polym.* **2016**, *31*, 15–30. [CrossRef]

31. Karageorgiou, V.; Kaplan, D. Porosity of 3D biomaterial scaffolds and osteogenesis. *Biomaterials* **2005**, *26*, 5474–5491. [CrossRef] [PubMed]

32. ASTM. *D1621—10 "Standard Test Method for Compressive Properties of Rigid Cellular Plastics"*; ASTM: West Conshohocken, PA, USA, 2010.

33. Puppi, D.; Piras, A.M.; Chiellini, F.; Chiellini, E.; Martins, A.; Leonor, I.B.; Neves, N.; Reis, R. Optimized electro- and wet-spinning techniques for the production of polymeric fibrous scaffolds loaded with bisphosphonate and hydroxyapatite. *J. Tissue Eng. Regen. Med.* **2011**, *5*, 253–263. [CrossRef] [PubMed]

34. Tuzlakoglu, K.; Alves, C.M.; Mano, J.F.; Reis, R.L. Production and characterization of chitosan fibers and 3-D fiber mesh scaffolds for tissue engineering applications. *Macromol. Biosci.* **2004**, *4*, 811–819. [CrossRef] [PubMed]

35. Gomes, M.E.; Holtorf, H.L.; Reis, R.L.; Mikos, A.G. Influence of the porosity of starch-based fiber mesh scaffolds on the proliferation and osteogenic differentiation of bone marrow stromal cells cultured in a flow perfusion bioreactor. *Tissue Eng.* **2006**, *12*, 801–809. [CrossRef] [PubMed]

36. Malheiro, V.N.; Caridade, S.G.; Alves, N.M.; Mano, J.F. New poly(ε-caprolactone)/chitosan blend fibers for tissue engineering applications. *Acta Biomater.* **2010**, *6*, 418–428. [CrossRef] [PubMed]

37. Neves, S.C.; Moreira Teixeira, L.S.; Moroni, L.; Reis, R.L.; Van Blitterswijk, C.A.; Alves, N.M.; Karperien, M.; Mano, J.F. Chitosan/Poly(ε-caprolactone) blend scaffolds for cartilage repair. *Biomaterials* **2011**, *32*, 1068–1079. [CrossRef] [PubMed]

38. Puppi, D.; Dinucci, D.; Bartoli, C.; Mota, C.; Migone, C.; Dini, F.; Barsotti, G.; Carlucci, F.; Chiellini, F. Development of 3D wet-spun polymeric scaffolds loaded with antimicrobial agents for bone engineering. *J. Bioact. Compat. Polym.* **2011**, *26*, 478–492. [CrossRef]

39. Puppi, D.; Chiellini, F. Wet-spinning of Biomedical Polymers: From Single Fibers Production to Additive Manufacturing of 3D Scaffolds. *Polym. Int.* **2017**. [CrossRef]

40. Puppi, D.; Migone, C.; Grassi, L.; Pirosa, A.; Maisetta, G.; Batoni, G.; Chiellini, F. Integrated three-dimensional fiber/hydrogel biphasic scaffolds for periodontal bone tissue engineering. *Polym. Int.* **2016**, *65*, 631–640. [CrossRef]

41. Puppi, D.; Piras, A.M.; Pirosa, A.; Sandreschi, S.; Chiellini, F. Levofloxacin-loaded star poly(ε-caprolactone) scaffolds by additive manufacturing. *J. Mater. Sci. Mater. Med.* **2016**, *27*, 1–11. [CrossRef] [PubMed]

42. Puppi, D.; Migone, C.; Morelli, A.; Bartoli, C.; Gazzarri, M.; Pasini, D.; Chiellini, F. Microstructured chitosan/poly(γ-glutamic acid) polyelectrolyte complex hydrogels by computer-aided wet-spinning for biomedical three-dimensional scaffolds. *J. Bioact. Compat. Polym.* **2016**, *31*, 531–549. [CrossRef]

43. Chiellini, F.; Puppi, D.; Piras, A.M.; Morelli, A.; Bartoli, C.; Migone, C. Modelling of pancreatic ductal adenocarcinoma in vitro with three-dimensional microstructured hydrogels. *RSC Adv.* **2016**, *6*, 54226–54235. [CrossRef]

44. Neves, S.C.; Mota, C.; Longoni, A.; Barrias, C.C.; Granja, P.L.; Moroni, L. Additive manufactured polymeric 3D scaffolds with tailored surface topography influence mesenchymal stromal cells activity. *Biofabrication* **2016**, *8*, 025012. [CrossRef] [PubMed]

45. Duan, B.; Wang, M.; Zhou, W.Y.; Cheung, W.L.; Li, Z.Y.; Lu, W.W. Three-dimensional nanocomposite scaffolds fabricated via selective laser sintering for bone tissue engineering. *Acta Biomater.* **2010**, *6*, 4495–4505. [CrossRef] [PubMed]

46. Duan, B.; Wang, M. Customized Ca–P/PHBV nanocomposite scaffolds for bone tissue engineering: Design, fabrication, surface modification and sustained release of growth factor. *J. R. Soc. Interface* **2010**, *7*, S615–S629. [CrossRef] [PubMed]

47. Bin, D.; Wai Lam, C.; Min, W. Optimized fabrication of Ca–P/PHBV nanocomposite scaffolds via selective laser sintering for bone tissue engineering. *Biofabrication* **2011**, *3*, 015001.

48. Puppi, D.; Zhang, X.; Yang, L.; Chiellini, F.; Sun, X.; Chiellini, E. Nano/microfibrous polymeric constructs loaded with bioactive agents and designed for tissue engineering applications: A review. *J. Biomed. Mater. Res. B* **2014**, *102*, 1562–1579. [CrossRef] [PubMed]

49. Middleton, J.C.; Tipton, A.J. Synthetic biodegradable polymers as orthopedic devices. *Biomaterials* **2000**, *21*, 2335–2346. [CrossRef]

50. Yang, H.-X.; Sun, M.; Zhou, P. Investigation of water diffusion in poly(3-hydroxybutyrate-co-3-hydroxyhexanoate) by generalized two-dimensional correlation ATR–FTIR spectroscopy. *Polymer* **2009**, *50*, 1533–1540. [CrossRef]

51. Ding, C.; Cheng, B.; Wu, Q. DSC analysis of isothermally melt-crystallized bacterial poly(3-hydroxybutyrate-co-3-hydroxyhexanoate) films. *J. Therm. Anal. Calorim.* **2011**, *103*, 1001–1006. [CrossRef]

52. Wang, Y.W.; Wu, Q.; Chen, J.; Chen, G.Q. Evaluation of three-dimensional scaffolds made of blends of hydroxyapatite and poly(3-hydroxybutyrate-co-3-hydroxyhexanoate) for bone reconstruction. *Biomaterials* **2005**, *26*, 899–904. [CrossRef] [PubMed]

53. Wutticharoenmongkol, P.; Pavasant, P.; Supaphol, P. Osteoblastic Phenotype Expression of MC3T3-E1 Cultured on Electrospun Polycaprolactone Fiber Mats Filled with Hydroxyapatite Nanoparticles. *Biomacromolecules* **2007**, *8*, 2602–2610. [CrossRef] [PubMed]

54. Kommareddy, K.P.; Lange, C.; Rumpler, M.; Dunlop, J.W.C.; Manjubala, I.; Cui, J.; Kratz, K.; Lendlein, A.; Fratzl, P. Two stages in three-dimensional in vitro growth of tissue generated by osteoblastlike cells. *Biointerphases* **2010**, *5*, 45–52. [CrossRef] [PubMed]

55. Hutmacher, D.W.; Schantz, T.; Zein, I.; Ng, K.W.; Teoh, S.H.; Tan, K.C. Mechanical properties and cell cultural response of polycaprolactone scaffolds designed and fabricated via fused deposition modeling. *J. Biomed. Mater. Res.* **2001**, *55*, 203–216. [CrossRef]

56. Quarles, L.D.; Yohay, D.A.; Lever, L.W.; Caton, R.; Wenstrup, R.J. Distinct proliferative and differentiated stages of murine MC3T3-E1 cells in culture: An in vitro model of osteoblast development. *J. Bone Min. Res.* **1992**, *7*, 683–692. [CrossRef] [PubMed]

10

Biotechnological Production of Poly(3-Hydroxybutyrate-*co*-4-Hydroxybutyrate-*co*-3-Hydroxyvalerate) Terpolymer by *Cupriavidus* sp. DSM 19379

Dan Kucera [1,2]**, Ivana Novackova** [1]**, Iva Pernicova** [1,2]**, Petr Sedlacek** [2] **and Stanislav Obruca** [1,2,*]

[1] Faculty of Chemistry, Brno University of Technology, Purkynova 118, 612 00 Brno, Czech Republic
[2] Material Research Centre, Faculty of Chemistry, Brno University of Technology, Purkynova 118, 612 00 Brno, Czech Republic
* Correspondence: obruca@fch.vut.cz

Abstract: The terpolymer of 3-hydroxybutyrate (3HB), 3-hydroxyvalerate (3HV), and 4-hydroxybutyrate (4HB) was produced employing *Cupriavidus* sp. DSM 19379. Growth in the presence of γ-butyrolactone, ε-caprolactone, 1,4-butanediol, and 1,6-hexanediol resulted in the synthesis of a polymer consisting of 3HB and 4HB monomers. Single and two-stage terpolymer production strategies were utilized to incorporate the 3HV subunit into the polymer structure. At the single-stage cultivation mode, γ-butyrolactone or 1,4-butanediol served as the primary substrate and propionic and valeric acid as the precursor of 3HV. In the two-stage production, glycerol was used in the growth phase, and precursors for the formation of the terpolymer in combination with the nitrogen limitation in the medium were used in the second phase. The aim of this work was to maximize the Polyhydroxyalkanoates (PHA) yields with a high proportion of 3HV and 4HB using different culture strategies. The obtained polymers contained 0–29 mol% of 3HV and 16–32 mol% of 4HB. Selected polymers were subjected to a material properties analysis such as differential scanning calorimetry (DSC), thermogravimetry, and size exclusion chromatography coupled with multi angle light scattering (SEC-MALS) for determination of the molecular weight. The number of polymers in the biomass, as well as the monomer composition of the polymer were determined by gas chromatography.

Keywords: polyhydroxyalkanoates; terpolymer; P(3HB-*co*-3HV-*co*-4HB); *Cupriavidus malaysiensis*

1. Introduction

Polyhydroxyalkanoates (PHA) represent a very attractive family of materials which are considered as an alternative to petrochemical polymers in applications which may benefit from their fully biodegradable and biocompatible nature. PHA are produced via fermentation since they are biosynthesized by numerous prokaryotes in the form of intracellular granules primarily as storage of carbon and energy [1]. Nevertheless, according to recent findings, PHA also plays a crucial role in the stress robustness and resistance of bacterial cells against various stress factors [2,3].

PHA are disadvantaged in competition with petrochemical polymers by their high-production cost. Since a substantial amount of the final cost is attributed to the cost of the carbon substrate, there are many attempts to produce PHA from inexpensive or even waste products in the food industry [4] such as waste lipids [5,6], crude glycerol formed as a side product of biodiesel production [7,8], various lignocellulose materials [9], or even carbon dioxide [10,11].

In general, the material properties of PHA strongly depend upon monomer composition. The homopolymer of 3-hydroxybutyrate (3HB), poly(3-hydroxybutyrate) (P3HB) is the most studied member of the PHA family, as it possesses numerous desirable properties. It is very interesting that the material in the native intracellular granules is completely amorphous and demonstrates extraordinary properties resembling super-cooled liquid [12]; nevertheless, when extracted from bacterial biomass, it quickly crystalizes. Therefore, its application potential is limited mainly by its high crystallinity, which reduces flexibility and elongation of the material. Nevertheless, the properties of the materials could be tuned when other monomer structures are incorporated into the polymer chain by feeding microbial culture with a suitable precursor(s). Therefore, copolymers containing, aside from 3HB, 3-hydroxyvalarate (3HV) subunits could be gained when microbial culture is cultivated in the presence of a suitable precursor with an odd number of carbon atoms such as propanol, propionate, pentanol, valerate, etc. The resulting copolymer poly(3-hydroxybutyrate-co-3-hydroxyvalerate) (P[3HB-co-3HV]) reveals substantially improved material properties and decreased crystallinity [13]. Similarly, some bacterial strains exposed to 1,4-butanediol or γ-butyrolactone (GBL) are able to biosynthesize copolymers containing 3HB and 4-hydroxybutyrate (4HB) monomer units. The copolymer poly(3-hydroxybutyrate-co-4-hydroxybutyrate) (P[3HB-co-4HB]) reveals mechanical properties, which resemble thermoplastic elastomers [14]. Moreover, PHA possessing 4HB subunits demonstrate increased biodegradability because lipases, which with PHA depolymerases, also have the ability to degrade P(3HB-co-4HB) [15], show higher activity at a higher fraction of 4HB [16]. Therefore, they find numerous high-value applications in the medical field [17]. Of course, terpolymer P(3HB-co-3HV-co-4HB) containing all of the above-mentioned monomer subunits demonstrate even superior properties and could be used in numerous fields and applications [18].

There are several reports dealing with the production of P(3HB-co-3HV-co-4HB) terpolymers employing various microorganisms. *Cupriavidus necator* (formerly *Alcaligenes eutrophus*, *Ralstonia eutropha* and *Wautersia eutropha*) was capable of desirable terpolymer production when cultivated on GBL and propionate; it was observed that propionate served not only as a 3HV precursor but it also increased the efficiency of 4HB incorporation into the terpolymer chain [19]. Similarly, Cavalheiro et al. produced P(3HB-co-3HV-co-4HB) by *Cupriavidus necator* using crude glycerol as the main carbon source, GBL as the 4HB precursor, and the 3HV-related precursor compound propionic acid [20]. Also, *Haloferax mediterranei* could be employed for the production of the terpolyester poly(3HB-co-3HV-co-4HB) without the need for a specific 3HV precursor which is based on the extraordinary metabolism of this microorganism, since it is capable of 3HV production from structurally unrelated carbon sources such as sugars or glycerol [21]. Finally, Ramachandran et al. used *Cupriavidus* sp. USMAA2-4 (now designated as *Cupriavidus malaysiensis* DSM 19379) for the terpolymer production from oleic acid and various 4HB and 3HV precursors [22].

In this work, we attempted to develop an efficient process of P(3HB-co-3HV-co-4HB) production employing *Cupriavidus malaysiensis* DSM 19379. We aimed at the maximization of both PHA yields, as well as 3HV and 4HB monomer fractions in the polymer to achieve desired material properties of the produced materials. Various culture strategies were used for this purpose.

2. Materials and Methods

2.1. Microorganisms and Cultivation

Cupriavidus malaysiensis USMAA2-4 (DSM 19379) was purchased from Leibnitz Institute DSMZ-German Collection of Microorganism and Cell Cultures, Braunschweig, Germany. The nutrient broth (Himedi—10 g/L Peptone, 10 g/L Beef Extract, 5 g/L NaCl) (NB) medium was used for the inoculum development. The mineral salt medium (MSM) for cultivation was composed of 3 g/L $(NH_4)_2SO_4$, 1.02 g/L KH_2PO_4, 11.1 g/L $Na_2HPO_4 \cdot 12 H_2O$, 0.2 g/L $MgSO_4 \cdot 7 H_2O$, and 1 mL/L of microelement solution, the composition of which was as follows: 9.7 g/L $FeCl_3 \cdot 6 H_2O$, 7.8 g/L $CaCl_2 \cdot 2 H_2O$, 0.156 g/L $CuSO_4 \cdot 5 H_2O$, 0.119 g/L $CoCl_2 \cdot 2 H_2O$, 0.118 g/L $NiCl_2 \cdot 4 H_2O$, and 1 L

0.1 M HCl. The following carbon sources were used to prepare the production media: GBL (8 g/L) (Sigma Aldrich, Steinheim, Germany); ε-caprolactone (8 g/L) (Sigma Aldrich, Steinheim, Germany); 1,4-butanediol (8 g/L) (Sigma Aldrich, Schnelldorf, Germany); 1,6-hexanediol (8 g/L) (Sigma Aldrich, Schnelldorf, Germany); fructose (20 g/L); glucose (20 g/L); sunflower oil (20 g/L); glycerol p.a. (20 g/L) (Lach-ner, Neratovice, Czechia). Carbon sources, salt solutions, and microelement solutions were autoclaved separately (121 °C, 20 min) and then aseptically reconstituted at room temperature prior to the inoculation (inoculum ratio was 10 vol%). The cultivations were performed in Erlenmeyer flasks (volume 250 mL) containing 100 mL of MSM. The temperature was set at 30 °C, the agitation at 180 rpm. The cells were harvested after 72 h of cultivation as described in Section 2.2. For a successful centrifugation process, the medium was heated to 70 °C for 15 min.

2.1.1. Single-Stage Cultivation Mode

GBL or 1,4-butanediol were used to prepare the production media in the same way as described in Section 2.1. Propionic acid (Sigma Aldrich, Schnelldorf, Germany) and valeric acid (Sigma Aldrich, Schnelldorf, Germany) as 3HV precursors were added at a concentration 1 g/L to media after 24 h of cultivation to minimize their toxic effect on growth of the microbial culture. After another 48 h of cultivation, the cells were harvested. The total length of the cultivation was 72 h. As a control, we chose to cultivate without adding any of the precursors of 3HV.

2.1.2. Two-Stage Cultivation Mode

Glycerol (20 g/L) or combination of glycerol and 1,4-butanediol (12 and 8 g/L, respectively) were used to prepare the production media based on MSM. After 48 h of cultivation (30 °C, 180 rpm), biomass was separated by centrifugation (6000 rpm, 4 °C) and aseptically transferred to fresh MSM with 0.1 g/L $(NH_4)_2SO_4$, 8 g/L 1,4-butanediol and 1 g/L valeric acid. Cultivation without valeric acid served as a control. The cells were harvested after another 48 h of cultivation (30 °C, 180 rpm).

2.2. Determination of the CDM and PHA Content

To determine the biomass concentration and PHA content in cells, samples (10 mL) were centrifuged (6000 rpm) and then the cells were washed with distilled water. The biomass concentration expressed as cell dry mass (CDM) was analyzed as reported previously [23]. The PHA content of dried cells was analyzed by gas chromatography (GC) (Trace GC Ultra, Thermo Scientific, Waltham, MA, USA) as reported by Brandl et al. [24]. Commercially available P(3HB-co-3HV) (Sigma Aldrich, Schnelldorf, Germany) composed of 88 mol% 3HB and 12 mol% 3HV were used as a standard and benzoic acid (LachNer, Neratovice, Czechia) was used as an internal standard. In addition to the quantification of total PHA in biomass, GC was also used to determine the monomeric composition and to determine the molar content of individual monomers in the obtained polymers.

2.3. Polymer Characterization

Following four polymers obtained by *Cupriavidus malaysiensis*, USMAA2-4 (DSM 19379) using different substrates and cultivation strategies were selected due to various 4HB content for polymer characterization: Sample 1—single-stage, fructose (20 g/L); Sample 2—single-stage, 1,4-butanediol + valeric acid; Sample 3—single-stage, 1,4-butanediol + propionic acid; Sample 4—two-stage, glycerol (20 g/L), and then 1,4-butanediol + valeric acid.

To determine the molecular weight of PHA, approximately 20 mg CDM was washed in 5 mL chloroform at 70 °C for 24 h under continuous stirring. Solid residues were separated by filtration and, finally, the solvent was removed by evaporation at 70 °C to a constant weight. The obtained polymer was also used for DSC analysis. After that, 5 mg of the polymer was solubilized in 1 mL of HPLC-grade chloroform and passed through syringe filters (nylon membrane, pore size 0.45 μm). Samples were analyzed by gel Size Exclusion Chromatography (Agilent, Infinity 1260 system containing PLgel MIXED-C column) coupled with Multiangle Light Scattering (Wyatt Technology, Dawn Heleos II, Goleta, CA, USA) and Differential Refractive Index (Wyatt Technology, Optilab T-rEX, Goleta, CA, USA) detection [24]. The weight-average molecular weight (Mw) and polydispersity index (Đ) were determined using the ASTRA software (Wyatt Technology, Goleta, CA, USA) based on Zimm´s equations.

Melting behavior of the isolated PHA polymers was analyzed by means of a differential scanning calorimeter (DSC) Q2000 (TA Instruments, New Castle, DE, USA) equipped with an RCS90 cooling accessory as previously described by Kucera et al. [25]. Phase transitions of mercury and indium were used for the calibration in the applied temperature range. Approximately 5 mg of sample was placed in hermetically sealed Tzero aluminum pans, and the measurement was carried out under a dynamic nitrogen atmosphere. To ensure the same thermal history of all samples prior to the evaluation of their melting behavior, each sample was first heated at 10 °C/min to 190 °C and subsequently cooled down to −30 °C at the same cooling rate. Then the sample was heated again (10 °C/min to 200 °C) and the thermogram, recorded in this second heating step, was further evaluated.

Thermogravimetric analysis of the isolated polymers was performed on Q5000 TGA analyzer (TA Instruments, New Castle, DE, USA). During the analysis, a known weight of a sample (ca 5 um) was heated at 10 °C/min to 800 °C under oxidative atmosphere (air). The major decomposition step, characterized by a rapid fall in the sample weight in the temperature range 250 °C to 350 °C, was further processed using TGA data evaluation software Universal Analysis 2000 (TA Instruments, New Castle, DE, USA). The automated evaluation of the weight change provided two characteristic temperatures of the degradation step: onset temperature of the thermal decomposition (Td_{onset}) and temperature corresponding to the maximal rate of the weight change (Td_{max}).

3. Results and Discussion

3.1. Biosynthesis of P(3HB-co-4HB) Copolymer

Cupriavidus malaysiensis DSM 19379 was employed to produce polyhydroxyalkanoates (PHA) using different carbon sources. This bacterium was isolated from water samples collected from Sg. Pinang river, Penang, Malaysia based on its ability to produce various types of PHA, including copolymers containing 4HB [26]. According to our results shown in Table 1, P3HB or P(3HB-*co*-4HB) were produced according to the type of the substrate. The bacterial strain was capable to produce copolymer P(3HB-*co*-4HB) only in the presence of precursors structurally related to 4HB such as GBL, 1,4-butandiol, ε-caprolactone, or 1,6-hexanediol e.g., diols and carboxylic acids possessing hydroxy group at last carbon atom which is agreement with results of Rahayu et al. [27] The highest PHA titers were achieved when four-carbon precursors of 4HB such as 1,4-butanediol or GBL were used. When such a structural motif was lacking, the strain accumulated homopolymer consisting exclusively of 3HB subunits. In the results, the strain appears to be unable to utilize oil because the CDM yield was low, and GC did not reveal PHA in the cell structure. There is a significant difference between utilization of fructose and glucose. While the yield of CDM and PHA with fructose was 10.78 g/L and 7.54 g/L, respectively, the yield with glucose was only 2.29 g/L CDM and 0.23 g/L PHA. This is not a very surprising result, also the closely related wild-type strain *Cupriavidus necator* H16 is not able to efficiently utilize glucose because it does not possess the activity of 6-phosphofructokinase [28]. *Cupriavidus malaysiensis* USMAA2-4 was also able to utilize glycerol reaching relatively high biomass titers; nevertheless, PHA production was the lowest among the substrates used which enabled PHA biosynthesis.

Table 1. Substrates for P3HB and P(3HB-*co*-4HB) production by *Cupriavidus* sp. DSM 19379.

Substrate	CDM (g/L)	PHA (wt%)	PHA (g/L)	4HB (mol%)	3HB (mol%)
fructose	10.78 ± 0.06	69.95 ± 0.42	7.54 ± 0.10	0	100
glucose	2.29 ± 0.06	10.03 ± 0.06	0.23 ± 0.02	0	100
glycerol	4.60 ± 0.04	5.30 ± 0.05	0.24 ± 0.02	0	100
sunflower oil	1.33 ± 0.05	0	0	0	0
GBL	4.50 ± 0.02	35.84 ± 0.92	1.61 ± 0.12	22.18 ± 1.06	77.82 ± 1.06
1,4-butanediol	4.01 ± 0.02	11.67 ± 0.06	0.47 ± 0.03	23.12 ± 1.61	76.88 ± 1.61
ε-caprolactone	0.22 ± 0.04	42.80 ± 0.61	0.10 ± 0.04	68.89 ± 1.12	31.11 ± 1.12
1,6-hexanediol	2.64 ± 0.01	39.83 ± 0.95	1.05 ± 0.07	34.35 ± 0.96	65.65 ± 0.96

3.2. Biosynthesis of P(3HB-co-3HV-co-4HB) Terpolymer through Single-Stage Cultivation

The following experiments were focused on the production of the terpolymer P(3HB-*co*-3HV-*co*-4HB). To obtain the desired material, 1,4-butanediol and GBL have been selected as carbon sources since the bacteria can utilize these substances for growth but also incorporate them into the copolymer P(3HB-*co*-4HB). Sodium propionate and valeric acid were tested in this experiment as odd carbon atom precursors for the synthesis of 3HV monomer incorporated into the terpolymer chain. Results of the single-stage terpolymer production including yields of CDM and PHA are shown in Table 2.

Table 2. Single stage terpolymer production (72 h cultivation, application of 3HV precursor at the 24 h of cultivation 1 g/L).

Primary Substrate	3HV Precursor	CDM (g/L)	PHA (g/L)	PHA (wt%)	3HB (mol%)	4HB (mol%)	3HV (mol%)
GBL	none	3.64 ± 0.03	0.81 ± 0.05	22.14 ± 0.01	68.40 ± 0.23	31.60 ± 0.23	0
	propionic acid	5.06 ± 0.37	0.62 ± 0.06	12.16 ± 0.00	69.18 ± 0.22	23.41 ± 0.05	7.41 ± 0.16
	valeric acid	7.97 ± 1.85	0.82 ± 0.09	10.41 ± 0.01	56.22 ± 0.32	25.85 ± 0.40	17.92 ± 0.07
1,4-butanediol	none	7.41 ± 0.51	1.05 ± 0.19	14.44 ± 0.02	68.97 ± 2.26	31.03 ± 2.26	0
	propionic acid	8.19 ± 0.35	1.65 ± 0.43	20.01 ± 0.04	63.81 ± 1.71	27.87 ± 0.10	8.32 ± 1.80
	valeric acid	8.68 ± 0.14	1.79 ± 0.88	20.52 ± 0.10	60.63 ± 2.90	24.72 ± 7.42	14.65 ± 4.53

In the resulting Table 2, CDM column shows that it generally achieved better growth using 1,4-butanediol as carbon sources than with GBL. Surprisingly, with the addition of the precursors for terpolymer synthesis, the CDM gain was higher. Valeric acid appears to be superior in the production of the P(3HB-*co*-4HB-*co*-3HV) terpolymer. With the addition of this precursor, significant growth was achieved with both GBL and 1,4-butanediol. The highest biomass concentration was obtained using 1,4-butanediol in combination with valeric acid, with a biomass yield of 8.68 g/L.

The highest PHA production was achieved in combination with valeric acid. The PHA yields were 0.82 g/L and 1.79 g/L for GBL and 1,4-butanediol, respectively. Thus, the combination of 1,4-butanediol with valeric acid again appears to be the best for production terpolymer in the single-stage strategy. Regarding the composition of the polymers obtained in this experiment, the terpolymer was synthesized using both precursors of 3HV. However, a higher 3HV fraction was obtained using valeric acid. In the case of propionate, generation of 3HV requires activity of 3-ketothiolase coupling propionyl-CoA and acetyl-CoA such as BktB in *C. necator* H16 [29]. On the contrary, conversion of valerate into 3HV could be relatively simply performed within the first "turn" of β-oxidation. It is likely that *Cupriavidus* DSM 19379 reveals relatively lower 3-ketohiolase activity as compared to the activity of the β-oxidation pathway, and therefore, valerate seems to be superior to the 3HV precursor for terpolymer synthesis as compared with propionate. In the case of terpolymer composition, the highest 3HV content was achieved using GBL together with valeric acid. The monomeric composition of the P(3HB-*co*-3HV-*co*-4HB) terpolymer was 56.22, 17.92, and 25.85 mol%, respectively. The polymer produced by *Cupriavidus malaysiensis* DSM 19379 using 1,4-butanediol in combination with valeric acid had almost the same composition. Nevertheless, it should be pointed out that the overall PHA productivity gained in the single-stage process was relatively low. The PHA content was about 20 weight percent of CDM and gained PHA titers were, therefore, also low. Hence, we attempted to improve the productivity of the culture by employing the two-stage cultivation.

3.3. Biosynthesis of the P(3HB-co-3HV-co-4HB)Tterpolymer through the Two-Stage Cultivation

To enhance PHA productivity, we performed an additional experiment in which cultivation was performed in two steps. In the first step, we aimed at a cultivation of maximal biomass using glycerol (20 g/L) as a cheap carbon source. According to our results, glycerol stimulates growth of the bacterium, but it is not converted into P(3HB) which could be taken as an advantage since the production of a desirable terpolymer with low 3HB fraction could be achieved in the second step. In addition, glycerol (12 g/L) was also mixed with 1,4-butanediol (8 g/L) in a parallel series of cultivations. The second stage was performed in the cultivation media with nitrogen limitation and 1,4-butanediol, and most importantly, 1,4-butanediol and valeric acids were used as 4HB and 3HV precursors, respectively. Valeric acid was chosen as the precursor of the 3HV since it was identified as the superior 3HV precursor for the investigated culture. The first phase of cultivation served to obtain a high amount of PHA-poor biomass. PHA production was then achieved by nitrogen limitation in the second phase. All results are shown in Table 3.

Table 3. Two-stage terpolymer production (48 h at glycerol or glycerol + 1,4-butanediol, after that transfer to nitrogen-limited medium with precursor of 3HV.

Primary Substrate	Secondary Precursor	CDM (g/L)	PHA (g/L)	PHA (wt%)	3HB (mol%)	4HB (mol%)	3HV (mol%)
Glycerol	1,4-butanediol	1.60 ± 0.03	0.84 ± 0.02	52.25 ± 0.12	80.85 ± 0.68	18.09 ± 0.26	1.06 ± 0.43
	1,4-butanediol + valeric acid	2.73 ± 0.58	1.42 ± 0.25	52.12 ± 1.76	53.78 ± 0.61	16.76 ± 0.87	29.46 ± 0.26
Glycerol + 1,4-butanediol	1,4-butanediol	3.26 ± 0.11	2.09 ± 0.01	64.14 ± 2.38	77.89 ± 0.53	21.60 ± 0.54	0.51 ± 0.01
	1,4-butanediol + valeric acid	5.94 ± 0.14	4.14 ± 0.05	69.64 ± 0.73	65.68 ± 1.02	16.46 ± 1.28	17.86 ± 0.26

From the results of this experiment shown above, it could be seen that the bacterial strain grew best when, in the first step, glycerol was used in combination with 1,4-butanediol and in the second one, 1,4-butanediol with valeric acid. The biomass gain was 5.94 g/L. Conversely, the smallest growth was achieved by cultivation using glycerol followed by 1,4-butanediol, where only 1.60 g/L CDM was obtained. CDM and PHA analysis was also performed on cultures after the first stage of the two-stage production. The assumption that at this stage a biomass with a low PHA content would be obtained, has been fulfilled. The glycerol-based medium reached 3.1 g/L CDM containing 5.3% PHB. Using a substrate containing 14-BD, we obtained 3.8 g/L CDM containing 24.0% P(3HB-co-4HB).

From the results, *Cupriavidus malaysiensis* DSM 19379 can efficiently synthesize the desired terpolymer P(3HB-co-3HV-co-4HB), PHA contents in bacterial cells are substantially higher when two-stage cultivation strategy was adopted. The highest weight fraction, 69.64 wt%, as well as the highest PHA gain, 4.14 g/L, was achieved when glycerol was used together with 1,4-butanediol in the first step and 1,4-butanediol with valeric acid in the second. Regarding polymer composition, good results were achieved when 1,4-butanediol was used in combination with valeric acid in the second step. When only glycerol was used in the first step, we obtained a terpolymer composed of 53.78 mol% 3HB, 16.76 mol% 4HB, and 29.46 mol% 3HV. Using glycerol together with 1,4-butanediol in the first step, a terpolymer composed of 3HB 65.68 mol%, 4HB 16.46 mol%, 3HV 17.86 mol% was subsequently obtained. It seems that a combination of glycerol and 1,4-butanediol in the first step of cultivation and 1,4-butanediol and valeric acid in the second stage of the cultivation is a very promising strategy which results in very high PHA titers and high PHA content in the cells and also high portions of 4HB and 3HV in the terpolymer structure.

3.4. Characteristics of Isolated Polymers

Differential scanning calorimetry and thermogravimetry were chosen to study the thermal properties of the polymers; size exclusion chromatography was used to determine molecular weight and polydispersity index of the polymers. The following samples of isolated polymers were selected for analysis. Sample No. 1 is a control polymer containing almost exclusively 3HB monomer units. Sample No. 2 was collected by cultivation using a combination of 1,4-butanediol and valeric acid; the proportion of 3-hydroxyvalerate in this sample is 14.65 mol%. Sample No. 3 was obtained from cultivation using 1,4-butanediol and sodium propionate, and the concentration of 3HV was 8.32 mol%. The last sample was isolated from a cell suspension-cultured to produce a terpolymer, using glycerol followed by 1,4-butanediol together with valeric acid. In this sample, the 3HV molar ratio was highest at all, namely 29.46 mol%. The results are placed in Table 4. From the thermograms recorded by differential scanning calorimetry, we determined glass transition temperature (Tg) and melting point (Tm). The total heat of fusion ΔH, was also determined via integration of the melting endotherm. Using thermogravimetry, the degradation onset temperature (Td_{onset}) and the temperature that corresponds to the maximal rate of sample decomposition (Td_{max}) were determined.

Table 4. Properties of the selected materials.

Sample	3HB (mol%)	4HB (mol%)	3HV (mol%)	Mw (kDa)	Đ (-)	Tg (°C)	Tm (°C)	ΔH (J/g)	Td_onset (°C)	Td_max (°C)
1	99.33	0.67	0	155.97	1.04	-	155.79 / 168.70	4.70 / 64.89	271.88	287.94
2	60.63	24.72	14.65	258.66	1.02	24.78	161.34	2.80	271.48	293.49
3	63.81	27.87	8.32	314.60	1.01	26.19	161.67	3.04	275.24	300.83
4	53.78	16.76	29.46	137.89	1.17	29.00	164.63	12.69	271.36	295.53

Comparison of DSC thermograms of the four isolated polymers is shown in Figure 1. In Sample No. 1, there is a sharp melting endotherm which appears at about 170 °C, which is typical of polyhydroxybutyrate. The peak area corresponds to the heat released in this process. The large area of the melting endotherm indicates a high tendency of the polymer to crystallize spontaneously which in turn causes no significant signs of glass transition and cold crystallization are found in its thermogram as compared to the other three analyzed samples. Further, the sample is characterized by a double peak at the melting point, indicating that the polymer crystallites are present in two forms with distinct thermal stability. On the other hand, for all the terpolymer samples (Samples 2–4), it can be seen at first sight that much less intensive melting peak is shown on the curves. Furthermore, apparent glass transition and cold crystallization of the polymer chains altogether indicates significantly reduced the tendency for spontaneous crystallization. In other words, involvement of and additional monomer to the copolymer structure resulted in a more amorphous structure. Incorporating 3HV into the polymer structure also caused a decrease in melting point to about 161 °C. Fahima Azira et al. [30] produced terpolymer P(3HB-co-3HV-co-4HB) and the melting points ranged from 160 to 164 °C.

The SEC-MALS technique was used to measure weight average molecular weight (Mw) of obtained polymers and values ranged from 137 to 314 kDa. The highest value was measured for the sample with the highest molar ratio of 4HB and the lowest for the sample with the highest molar ratio of 3HV. The Mw values measured are typical for the bacterial strain used and consistent with other studies [22].

Thermogravimetric analysis was performed in order to compare the thermal stability of the produced polymers. In a respective thermogram, decomposition of a polymer is represented by the onset temperature of the decomposition (the temperature at which the polymer starts to decompose, Td_{onset}) and by the temperature which corresponds to the maximal rate of the decomposition (Td_{max}). Among the isolated P(3HB-co-3HV-co-4HB) terpolymer samples, the highest degradation temperature (i.e., the highest thermal stability of the polymer) was measured for Sample No. 3 composed of

63.81 mol% 3HB, 27.87 mol% 4HB, and 8.32 mol% 3HV. This is the sample with the lowest 3HV but the highest 4HB. This suggests that a higher proportion of 4HB in the terpolymer leads to the higher thermal stability of the polymer. Thus, sample 3 has the most promising properties from a technological point of view because its melting point was set at near lowest, 161.67 °C, and the degradation temperature to highest 300.83 °C. The wide temperature window between melting temperature and degradation temperature is important for polymer processing. When working with the melt, it is important that it does not decompose.

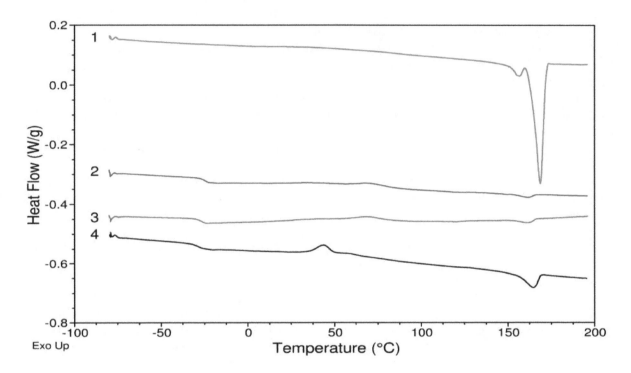

Figure 1. Results of DSC analysis of isolated polymers.

4. Conclusions

To sum-up, in flask experiments, we have developed a two-stage cultivation strategy which is based on the application of glycerol and 1,4-butanediol as the carbon substrates in the first stage of cultivation, after that the cells are transferred into nitrogen-limited cultivation media with 1,4-butanediol and valeric acids. This cultivation strategy provides high PHA yields and PHA content in bacterial cells. Moreover, the P(3HB-co-3HV-co-4HB) terpolymer with low 3HB fraction and high 3HV and 4HB contents is obtained. The material properties of obtained polymers were consistent with materials produced in previous studies aimed at the production of P(3HB-co-3HV-co-4HB) terpolymers. In our future experiments, we will transfer the process into laboratory bioreactors to evaluate its suitability for industrial production of PHA.

Author Contributions: Conceptualization—D.K. and S.O.; Data curation—D.K. and I.P.; Formal analysis—D.K. and P.S.; Funding acquisition—S.O.; Investigation—S.O.; Methodology—S.O.; Project administration—S.O.; Resources—S.O.; Supervision—S.O.; Validation—I.N. and I.P.; Visualization—D.K.; Writing—original draft—D.K.; Writing—review & editing—P.S. and S.O.

Acknowledgments: Authors kindly thank to Leona Kubikova for all the help with TGA and DSC measurement and to Michal Kalina for determination of molecular weight analysis by SEC-MALS.

References

1. Kourmentza, C.; Plácido, J.; Venetsaneas, N.; Burniol-Figols, A.; Varrone, C.; Gavala, H.N.; Reis, M.A. Recent advances and challenges towards sustainable polyhydroxyalkanoate (PHA) production. *Bioengineering* **2017**, *4*, 55. [CrossRef]

2. Obruca, S.; Sedlacek, P.; Koller, M.; Kucera, D.; Pernicova, I. Involvement of polyhydroxyalkanoates in stress resistance of microbial cells: Biotechnological consequences and applications. *Biotechnol. Adv.* **2018**, *36*, 856–870. [CrossRef] [PubMed]

3. Slaninova, E.; Sedlacek, P.; Mravec, F.; Mullerova, L.; Samek, O.; Koller, M.; Hesko, O.; Kucera, D.; Marova, I.; Obruca, S. Light scattering on PHA granules protects bacterial cells against the harmful effects of UV radiation. *Appl. Microbiol. Biotechnol.* **2018**, *102*, 1923–1931. [CrossRef]

4. Haas, C.; Steinwandter, V.; De Apodaca, E.D.; Madurga, B.M.; Smerilli, M.; Dietrich, T.; Neureiter, M. Production of PHB from chicory roots - Comparison of three *Cupriavidus necator* strains. *Chem. Biochem. Eng. Q.* **2015**, *29*, 99–112. [CrossRef]

5. Verlinden, R.A.J.; Hill, D.J.; Kenward, M.A.; Williams, C.D.; Piotrowska-Seget, Z.; Radecka, I.K. Production of polyhydroxyalkanoates from waste frying oil by *Cupriavidus necator*. *Amb Express* **2011**, *1*, 1–8. [CrossRef] [PubMed]

6. Ciesielski, S.; Mozejko, J.; Pisutpaisal, N. Plant oils as promising substrates for polyhydroxyalkanoates production. *J. Clean. Prod.* **2015**, *106*, 408–421. [CrossRef]

7. Jiang, G.; Hill, D.J.; Kowalczuk, M.; Johnston, B.; Adamus, G.; Irorere, V.; Radecka, I. Carbon sources for polyhydroxyalkanoates and an integrated biorefinery. *Int. J. Mol. Sci.* **2016**, *17*, 1157. [CrossRef] [PubMed]

8. Moita, R.; Freches, A.; Lemos, P.C. Crude glycerol as feedstock for polyhydroxyalkanoates production by mixed microbial cultures. *Water Res.* **2014**, *58*, 9–20. [CrossRef]

9. Obruca, S.; Benesova, P.; Marsalek, L.; Marova, I. Use of lignocellulosic materials for PHA production. *Chem. Biochem. Eng. Q.* **2015**, *29*, 135–144. [CrossRef]

10. Meixner, K.; Kovalcik, A.; Sykacek, E.; Gruber-Brunhumer, M.; Zeilinger, W.; Markl, K.; Haas, C.; Fritz, I.; Mundigler, N.; Stelzer, F.; et al. Cyanobacteria Biorefinery—Production of poly(3-hydroxybutyrate) with *Synechocystis salina* and utilisation of residual biomass. *J. Biotechnol.* **2018**, *265*, 46–53. [CrossRef]

11. Troschl, C.; Meixner, K.; Drosg, B. Cyanobacterial PHA Production—Review of Recent Advances and a Summary of Three Years' Working Experience Running a Pilot Plant. *Bioengineering* **2017**, *4*, 26. [CrossRef]

12. Sedlacek, P.; Slaninova, E.; Enev, V.; Koller, M.; Nebesarova, J.; Marova, I.; Hrubanova, K.; Krzyzanek, V.; Samek, O.; Obruca, S. What keeps polyhydroxyalkanoates in bacterial cells amorphous? A derivation from stress exposure experiments. *Appl. Microbiol. Biotechnol.* **2019**, *103*, 1905–1917. [CrossRef]

13. Koller, M. Chemical and biochemical engineering approaches in manufacturing polyhydroxyalkanoate (PHA) biopolyesters of tailored structure with focus on the diversity of building blocks. *Chem. Biochem. Eng. Q.* **2018**, *32*, 413–438. [CrossRef]

14. Lee, W.H.; Azizan, M.N.M.; Sudesh, K. Effects of culture conditions on the composition of poly(3-hydroxybutyrate-*co*-4-hydroxybutyrate) synthesized by *Comamonas acidovorans*. *Polym Degrad Stab.* **2004**, *84*, 129–134. [CrossRef]

15. Rodríguez-Contreras, A.; Calafell-Monfort, M.; Marqués-Calvo, M.S. Enzymatic degradation of poly(3-hydroxybutyrate-*co*-4-hydroxybutyrate) by commercial lipases. *Polym. Degrad. Stabil.* **2012**, *97*, 597–604. [CrossRef]

16. Saito, Y.; Nakamura, S.; Hiramitsu, M.; Doi, Y. Microbial synthesis and properties of poly(3-hydroxybutyrate-*co*-4-hydroxybutyrate). *Polym. Int.* **1996**, *39*, 167–174. [CrossRef]

17. Singh, A.K.; Srivastava, J.K.; Chandel, A.K.; Sharma, L.; Mallick, N.; Singh, S.P. Biomedical applications of microbially engineered polyhydroxyalkanoates: an insight into recent advances, bottlenecks, and solutions. *Appl. Microbiol. Biotechnol.* **2019**, *103*, 2007–2032. [CrossRef]

18. Chanprateep, S.; Kulpreecha, S. Production and characterization of biodegradable terpolymer poly(3-hydroxybutyrate-*co*-3-hydroxyvalerate-*co*-4-hydroxybutyrate) by *Alcaligenes* sp. A-04. *J. Biosci. Bioeng.* **2006**, *101*, 51–56. [CrossRef]

19. Lee, Y.H.; Kang, M.S.; Jung, Y.M. Regulating the molar fraction of 4-hydroxybutyrate in poly(3-hydroxybutyrate-4-hydroxybutyrate) biosynthesis by *Ralstonia eutropha* using propionate as a stimulator. *J. Biosci. Bioeng.* **2000**, *89*, 380. [CrossRef]

20. Cavalheiro, J.M.; Raposo, R.S.; de Almeida, M.C.M.; Cesário, M.T.; Sevrin, C.; Grandfils, C.; Da Fonseca, M.M.R. Effect of cultivation parameters on the production of poly(3-hydroxybutyrate-*co*-4-hydroxybutyrate) and poly(3-hydroxybutyrate-4-hydroxybutyrate-3-hydroxyvalerate) by *Cupriavidus necator* using waste glycerol. *Biores. Technol.* **2012**, *111*, 391. [CrossRef]

21. Hermann-Krauss, C.; Koller, M.; Muhr, A.; Fasl, H.; Stelzer, F.; Braunegg, G. Archaeal production of polyhydroxyalkanoate (PHA) co-and terpolyesters from biodiesel industry-derived by-products. *Archaea* **2013**, *2013*, 129268. [CrossRef]

22. Ramachandran, H.; Iqbal, N.M.; Sipaut, C.S.; Abdullah, A.A.A. Biosynthesis and characterization of poly (3-hydroxybutyrate-*co*-3-hydroxyvalerate-*co*-4-hydroxybutyrate). Terpolymer with various monomer compositions by *Cupriavidus* sp. USMAA2-4. *Appl. Biochem. Biotechnol.* **2011**, *164*, 867–877. [CrossRef]

23. Obruca, S.; Marova, I.; Melusova, S.; Mravcova, L. Production of polyhydroxyalkanoates from cheese whey employing *Bacillus megaterium* CCM 2037. *Ann. Microbiol.* **2011**, *61*, 947–953. [CrossRef]

24. Brandl, H.; Gross, R.A.; Lenz, R.W.; Fuller, R.C. *Pseudomonas oleovorans* as a source of poly(beta-hydroxyalkanoates) for potential application as a biodegradable polyester. *Appl. Environ. Microb.* **1988**, *54*, 1977–1982.

25. Kucera, D.; Pernicová, I.; Kovalcik, A.; Koller, M.; Mullerova, L.; Sedlacek, P.; Mravec, F.; Nebesarova, J.; Kalina, M.; Marova, I.; et al. Characterization of the promising poly(3-hydroxybutyrate) producing halophilic bacterium *Halomonas halophila*. *Biores. Technol.* **2018**, *256*, 552–556. [CrossRef]

26. Amirul, A.A.; Yahya, A.R.M.; Sudesh, K.; Azizan, M.N.M.; Majid, M.I.A. Biosynthesis of poly(3-hydroxybutyrate-*co*-4-hydroxybutyrate) copolymer by *Cupriavidus* sp. USMAA1020 isolated from Lake Kulim, Malaysia. *Biores. Technol.* **2008**, *99*, 4903–4909.

27. Rahayu, A.; Zaleha, Z.; Yahya, A.R.M.; Majid, M.I.A.; Amirul, A. A Production of copolymer poly(3-hydroxybutyrate-co-4-hydroxybutyrate) through a one-step cultivation process. *World J. Microbiol. Biotechnol.* **2008**, *24*, 2403–2409. [CrossRef]

28. Lopar, M.; Špoljarić, I.V.; Cepanec, N.; Koller, M.; Braunegg, G.; Horvat, P. Study of metabolic network of *Cupriavidus necator* DSM 545 growing on glycerol by applying elementary flux modes and yield space analysis. *J. Ind. Microbiol. Biotechnol.* **2014**, *41*, 913–930. [CrossRef]

29. Lindenkamp, N.; Peplinski, K.; Volodina, E.; Ehrenreich, A.; Steinbuchel, A. Impact of multiple beta-ketothiolase deletion mutations in *Ralstonia eutropha* H16 on the composition of 3-mercaptopropionic acid-containing copolymers. *Appl. Environ. Microbiol.* **2010**, *76*, 5373–5382. [CrossRef]

30. Fahima Azira, T.M.; Nursolehah, A.A.; Norhayati, Y.; Majid, M.I.A.; Amirul, A.A. Biosynthesis of Poly(3-hydroxybutyrate-co-3-hydroxyvalerate-co-4-hydroxybutyrate) terpolymer by *Cupriavidus* sp. USMAA2-4 through two-step cultivation process. *World J. Microbiol. Biotechnol.* **2011**, *27*, 2287–2295. [CrossRef]

11

Molecular Diagnostic for Prospecting Polyhydroxyalkanoate-Producing Bacteria

Eduarda Morgana da Silva Montenegro [1], Gabriela Scholante Delabary [1],
Marcus Adonai Castro da Silva [1], Fernando Dini Andreote [2] and André Oliveira de Souza Lima [1,*]

[1] Centro de Ciências Tecnológicas da Terra e do Mar, Universidade do Vale do Itajaí, R. Uruguai 458, 88302-202 Itajaí-SC, Brazil; dudamorgana@gmail.com (E.M.S.M.); gabidelabary@hotmail.com (G.S.D.); marcus.silva@univali.br (M.A.C.S.)

[2] Department of Soil Science, "Luiz de Queiroz" College of Agriculture, University of São Paulo, Piracicaba-SP 13418-260, Brazil; fdandreo@gmail.com

[*] Correspondence: andreolima@gmail.com

Academic Editor: Martin Koller

Abstract: The use of molecular diagnostic techniques for bioprospecting and microbial diversity study purposes has gained more attention thanks to their functionality, low cost and quick results. In this context, ten degenerate primers were designed for the amplification of polyhydroxyalkanoate synthase (*phaC*) gene, which is involved in the production of polyhydroxyalkanoate (PHA)—a biodegradable, renewable biopolymer. Primers were designed based on multiple alignments of *phaC* gene sequences from 218 species that have their genomes already analyzed and deposited at Biocyc databank. The combination of oligos *phaCF3/phaCR1* allowed the amplification of the expected product (PHA synthases families types I and IV) from reference organisms used as positive control (PHA producer). The method was also tested in a multiplex system with two combinations of initiators, using 16 colonies of marine bacteria (pre-characterized for PHA production) as a DNA template. All amplicon positive organisms ($n = 9$) were also PHA producers, thus no false positives were observed. Amplified DNA was sequenced ($n = 4$), allowing for the confirmation of the *phaC* gene identity as well its diversity among marine bacteria. Primers were also tested for screening purposes using 37 colonies from six different environments. Almost 30% of the organisms presented the target amplicon. Thus, the proposed primers are an efficient tool for screening bacteria with potential for the production of PHA as well to study PHA genetic diversity.

Keywords: bioprospecting; biopolymer; environmental diversity

1. Introduction

Bioplastics polymers have emerged as an alternative to the excessive use of polymers from petrochemical origin, which represent a problem in terms of waste management and environmental impact [1,2]. Polyhydroxyalkanoate (PHA) is among these polymers and has attracted increasing attention due to its properties and suitability for biodegradation, as well as its biocompatibility and thermoplastic characteristics [3,4]. These biopolymers accumulate in the cytoplasm of cells in the form of granules due to nutritional limitations that restrict growth [5]. They are generally associated with carbon reserves or excess in the medium, as well as reduced energy equivalents [6]. Although the procedure for the formation and accumulation of biopolymers is well-known, the main impediment to employing biopolymers is the large scale and the high cost of PHA production, which is nine times more expensive than the production of synthetic plastics [7,8]. In this sense, the bioprospection of bacteria capable of producing these biopolymers in greater quantities from the conversion of cheaper and renewable substrates is necessary, aiming a greater production and consequent reduction in cost [9–11].

The genes responsible for PHA synthesis can be classified into four different classes, according to the organization of gene locus and the structural and functional properties of enzymes PHA synthase [12,13]. Class I is represented by gene *phaC* of *Cupriavidus necator*, and class II by *Pseudomonas*, where PHA synthase is encoded by *phaC1* and *phaC2* [14]. Class III synthase is composed of the genes *phaC* and *phaE* and can be found in the model organism *Allochromatium vinosum* [15]. Class IV synthase is represented by *Bacillus megaterium*, in which the main genes are *phaC* and *phaR* [16]. Among the genes involved in PHA production, *phaC* is the most important since it encodes the key enzyme for PHA synthesis, thus justifying its choice as an indicative of possible producers of these biopolymers [11].

PHA-producing organisms can be identified and evaluated by different methods [17]. Among the traditional methods, the most frequently used are based on microscopy and specific dyes, such as the lipophilic dye Sudan Black B [18], the fluorescent dye from the Nile [19], and the Nile Red dye [20]. The traditional identification techniques require specific conditions for each bacterium and therefore become more laborious, and moreover offer no specificity and may indicate false positives [13,21,22]. In this context, molecular methods appear as an effective tool for the selection and diagnosis of PHA-producing bacteria, for agility of results, ease of handling, and low cost. Among these techniques, polymerase chain reaction (PCR) is simple and efficient for such a diagnosis [23], as it involves the use of specific primers for the locus of the gene responsible for PHA synthase, the biosynthesis of interest [13,21,22,24–27]. Thus, the present work aims to design primers capable of identifying bacteria that produce different classes of PHAs, as well as the prospection of environments for the pre-selection of the producing organisms and thus the analysis of the environmental diversity of such organisms.

2. Experimental Procedures

2.1. Bacteria Strains and Media

The reference bacteria strains *Bacillus pumilus* ATCC 14884, *Bacillus thurigiensis var. israelensis* 4Q2-72, *Bacillus megaterium* ATCC 14581, *Bacillus cereus* ATCC 14579, *Chromobacterium violaceum* (CV11), and *Cupriavidus necator* DSM 545, were used as positive controls for the presence of the gene *phaC* and a negative control was made by using DNA from *Escherichia coli* DH5α. Genomic DNA was isolated (DNeasy Blood & Tissue Kit, Qiagen, Hilden, Germany), quantified after agarose gel electrophoresis (Kodak 1D v.3.5.5b, Kodak, Rochester, NY, USA), and used as a DNA template for PCR (approximately 50 ng per reaction).

Genomic DNA from twelve marine bacteria, *Pseudomonas* sp. (LAMA 572), *Halomonas hydrothermalis* (LAMA 685), *Micrococcus luteus* (LAMA 702), *Brevibacterium* sp. (LAMA 758), *Halomonas* sp. (LAMA 761), LAMA 677, LAMA 726, LAMA 729, LAMA 737, LAMA 748, LAMA 760, LAMA 765, LAMA 790, and LAMA 896, previously recognized (tested by staining with Nile Red in seven culture media) as PHA producers, were used in the developed PHA PCR. Also, negative marine bacteria were applied; *Idiomarina loihiensis* and *Terribacillus saccharophilus*. To check the efficiency in pre-selection of PHA-producer bacteria isolated from the environment, 37 newly isolated bacteria from soils were tested. Genomic DNA of these bacteria was obtained as described, and similarly used for amplifications.

2.2. Design and Evaluation of Primers for the Amplification of the Gene phaC

The protocol for primer design was similar to that previously described Lima & Garcês [28]. A total of 218 sequences of the superfamily *phaC* gene were retrieved from the BioCyc Database Collection [29] and analyzed in Megan 4 program [30]. Sequences were aligned and phylogeny was inferred on the basis of neighbor-joining trees built from a similarity matrix determined by the Kimura-2 parameter. The sequences were also analyzed in the amino acid level, which was used to allocate them into the classes of PHA synthase. This was performed using the tool Conserved Domain search, available on NCBI [31]. The description of conserved regions was also evidenced by sequence alignment using the ClustalW algorithm [32] at Unipro UGENE 1.26.1 [33]. The best regions were selected for the primer design, using as parameters a high degree of identity, regions without gaps, and few degenerate bases.

The primer sequences were determined with the online program OligoAnalyzer version 3.1 (Integrate DNA Technologies, Coralville, IA, USA), in which important parameters for the efficiency of PCR reaction were defined, such as the melting temperature (Tm) and the percentage of C+G. The primers drawn were also evaluated at CLC Genomics Workbench 4.8 (CLC bio, Cambridge, MA, USA), enabling the visualization of which primer would produce more annealing results to all the gene sequences used during primer design. The parameter considered in this evaluation was the possibility of up to two degenerate bases for each primer.

2.3. Amplification of phaC Gene by PCR

The program used for the amplification of *phaC* gene fragments with all primer combinations was a cycle of 94 °C for 4 min, followed by 35 cycles of 94 °C for 45 s, 61 °C for 20 s, 72 °C for 10 s and a final extension of 72 °C for 2 min. As a template, extracted DNA or bacteria cells isolated from the environment were used for *phaC* amplification. Once the best set of primers was established, the program used for *phaC* gene amplification using reference strains was adjusted to one cycle of 94 °C for 4 min, followed by 10 cycles of 94 °C for 30 s, 68 °C for 20 s, as well as 25 cycles of 94 °C for 12 s, 65 °C for 10 s, 72 °C for 7 s and a final extension of 72 °C for 2 min. All reactions were held in the thermocycler Eppendorf Mastercycler Gradient, consisting of 20 μL containing 1× PCR amplification buffer (Invitrogen), 0.2 mM of each dNTP, 0.5 μM of each primer, 1U Taq DNA polymerase (Invitrogen), 2 mM MgCl$_2$, and the template DNA. PCR amplicons were observed by electrophoresis in 2% agarose gel further stained with ethidium bromide, and viewed under a UV transilluminator.

2.4. DNA Sequencing

The amplified fragments obtained from marine bacteria LAMA 677, LAMA 737, LAMA 748, LAMA 760, and the reference bacteria *C. violaceum* were purified (QIAquick PCR Purification Kit, Qiagen, Hilden, Germany) and sequenced in an ABI-Prism 3100 Genetic Analyzer at ACTGene (Alvorada, RS, Brazil). The identity of the sequences was evaluated through the Genomics Workbench 4.8 program accessing the tool of comparison BLASTX (Nacional Center for Biotechnology Information, Bethesda, MA, USA) [34]. The gene sequences retrieved by BLAST, in addition to the newly sequenced DNA, were pooled and analyzed for phylogenetic tree classification using multiple alignment calculated by the ClustalW algorithm [33] in Geneious v. 5.5.3 (Biomatters, Auckland, New Zealand).

3. Results

3.1. Design of Primers for Gene phaC Amplification

The phylogenetic classification of the 218 sequences of the gene *phaC* (1 sequence = 1 specie) used for primer design indicated a high percentage of organisms belonging to the phylum Proteobacteria (alpha, beta, and gamma) (Figure 1). The phyla Firmicutes and Spirochaetales were also presented, as well as organisms of the orders Chroococcales, Chloroflexales, and Actinomycelates. The domain Archea appears uniquely represented by organisms belonging to the order Halobacteriales (Figure 1). The analysis at the amino acid level showed that the regions of conserved domains were characteristic for three classes of PHA synthases; the classes I, II and III.

The application of the designed workflow resulted in the generation of 10 primers, as well as their determined characteristics (sequences, annealing temperatures, relative location to the consensus sequence) and compatibility for annealing with the 218 used sequences (Table 1). Among these, some sets were first selected to amplify the target gene *phaC*. For instance, primers *phaCF3* and *phaCR1* were selected due to their capacity to anneal with a large number of sequences. Also, primer *phaCF1* was picked due to its relative position to *phaCR1* and its ability to be used in multiplex PCR (Figure 2). These two combinations also resulted in the generation of small fragments; 304 bp for primers *phaCF1/phaCR1* and 239 bp for *phaCF3/phaCR*, which is desired when one is looking for fast detection and maximum amplification efficiency. For a shorter PCR period, amplicons are less likely to vary in size among distinct template sequences. Even so, it is important to consider that amplicon size may vary among different organisms, due to modifications that occurr during evolution. However, variation can be observed, for example, for amplifications with primers *phaCF1/phaCR1*, which resulted in fragments varying from 242 to 316 bp.

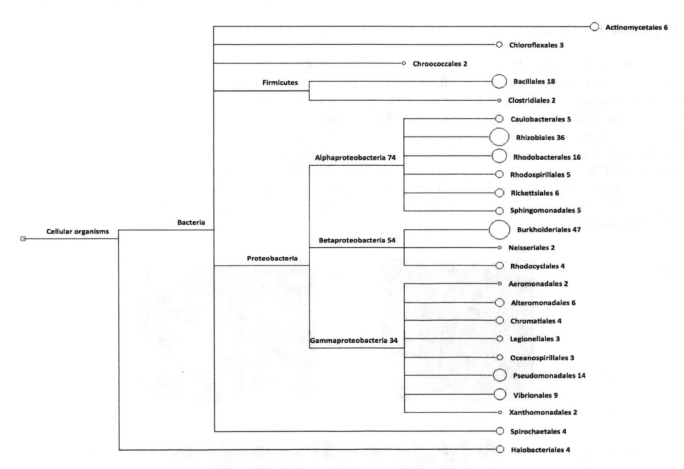

Figure 1. Analysis of the gene sequences used to design the primers, and their phylogenetic classification by the Megan 4 program.

Table 1. Primers designed showing: nomenclature, nucleotide sequences, melting temperatures, initial and final positions correspondent to the consensus sequence and the number of aligned sequences according to the database of organisms used in the study.

Primer ID	Sequence	Tm	Consensus Position	Aligned Sequences *
phaCF1	5'TGATSSAGCTGATCCAGTAC3'	53.9°	489–508	18
phaCF2	5'CCGCTGCTGATCGTBCCGCC3'	65.5°	539–558	41
phaCF3	5'CCGCCSTGGATCAACAAGT3'	58.0°	554–572	61
phaCF4	5'CTACATCCTCGACCTGMAGCCGGA3'	63.1°	574–597	24
phaCF5	5'GGCTACTGCRTCGGCGGCAC3'	65.1°	773–792	47
phaCF6	5'TGGAACDSCGACDCCACCAAC3'	61.6°	1078–1098	0
phaCF7	5'CGACRCCACCAACMTGCCGGG3'	65.8°	1086–1106	4
phaCR1	5'GTGCCGCCGAYGCAGTAGCC3'	65.1°	773–792	47
phaCR2	5'CCCGGCAKGTTGGTGGYGTCG3'	65.8°	1086–1106	4
phaCR3	5'CAGTSCGGCCACCAGSWGCC3'	66.3°	1432–1451	0

* Considering a maximum of two mismatches.

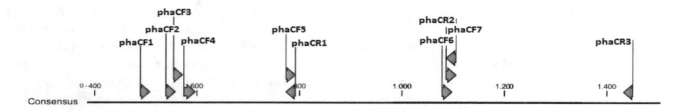

Figure 2. Positioning of the forward and reverse primers accordingly to the consensus sequence.

3.2. Partial Amplification of phaC Gene

Although a particular set of primers (*phaCF1, phaCF3,* and *phaCR1*) were revealed to be attractive for *phaC* amplification during in silico analysis, their efficiency and specificity has to be determined in vitro. Therefore, a total of seven pairs of primers were tested using LAMA 677 (PHA producer). Positive results were observed for three of these combinations (Figure 3A), with a remarkable match for functioning of sets previously elected by bioinformatics tools. This pair (*phaCF3/phaCR1*) was further tested using genomic DNA from already known *phaC* carrier species: the model organisms *B. pumilus* (ATCC 14884), *B. thuringiensis* var. israelensis (4Q2-72), *B. megaterium* (ATCC 14581), *B. cereus* (ATCC 14579), and *C. necator* (DSM 545). The expected fragments of 239bp were generated for all organisms tested (Figure 3B) and no amplicon was obtained from the negative control *E. coli* DH5α (data not shown).

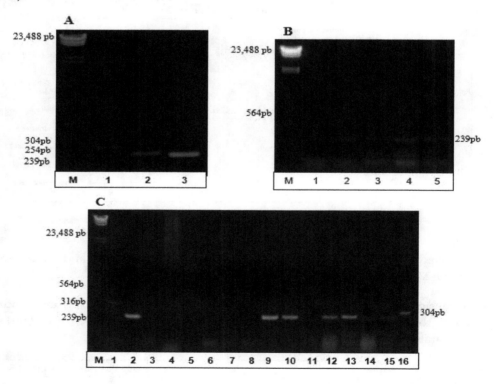

Figure 3. *phaC* gene amplification. Line M: ladder *λ*-hind III. (**A**) *phaC* amplicons for combinations of primers with positive results (organism LAMA 677): 1. *phaCF1/phaCR1*; 2. *phaCF2/phaCR1*; 3. *phaCF3/phaCR1*. (**B**) *phaC* amplicons in: 1. *Bacillus pumilus*; 2. *B. thurigiensis*; 3. *B. megaterium*; 4. *B. cereus* ATCC; 5. *C. necator.* (**C**) *phaC* amplicons in: 1. *Pseudomonas* sp. (LAMA 572); 2. LAMA 677; 3. *Halomonas hydrothermalis* (LAMA 685); 4. LAMA 691; 5. LAMA 694; 6. *Micrococcus luteus* (LAMA 702); 7. LAMA 726; 8. LAMA 729; 9. LAMA 737; 10. LAMA 748; 11. *Brevibacterium* sp. (LAMA 758); 12. LAMA 760; 13. LAMA 790; 14. *Halomonas* sp. (LAMA 761); 15. LAMA 765; 16. LAMA 896.

Once it was verified that primers were efficient in recognizing PHA-producing bacteria, they were also tested for the detection of potential new polymer producers isolated from environmental samples. For this purpose, two sets of primers (*phaCF1/phaCR1* and *phaCF3/phaCR1*) were used. The use of these primers in multiplex reactions allows for the increase of coverage for the detection of PHA producers. In this context, 16 marine organisms (14 positive and two negative PHA producers) and 37 environmental isolates (unknown PHA production) were screened. The proposed PCR protocol was able to detect *phaC* in nine marine isolates. No false positives were identified, highlighting the specificity of the primers designed. The positive control *C. violaceum* was amplified efficiently. When applying the *phaC* multiplex-PCR with the 37 environmental isolates, *phaC* amplicons were observed in approximately 30% of them (data not shown), revealing the great potential of this method for the screening of PHA producers.

3.3. DNA Sequencing and phaC *Gene Identification*

Amplicons from different reactions were sequenced and compared to the Genbank database. The sequences identities were compatible with *phaC* genes/proteins previously described. This indicates the specificity and efficiency of the proposed method. The originated sequences also allowed the taxonomic classification of organisms harboring the *phaC* gene (Figure 4). The differential allocation of positive isolates supports the inference that the developed tool is able to detect most of the *phaC* gene diversity that resides in bacterial cells belonging to distinct taxa.

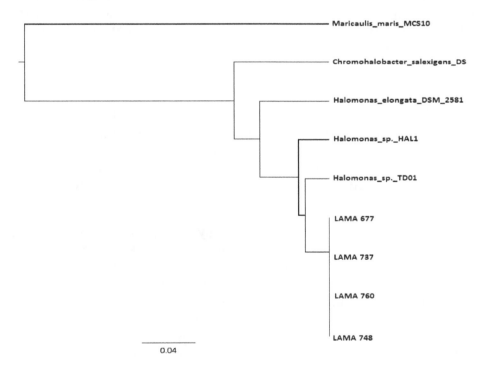

Figure 4. Sub-tree alignment of sequenced amplicons (LAMA 677, LAMA 737, LAMA 748, and LAMA 760), showing organisms with more genetic similarity. The analysis were through the software Geneious V.5.5.3.

4. Discussion

The use of molecular tools for the detection of organisms with particular features can aid in the field of biotechnology. These methodologies have been used for the determination of the microbial potential to produce PHA at the genetic level as well as to determine how much PHA can be produced by a given organism [17]. Here, we use the same approach to describe newly designed primers for the assessment of bacteria able to produce PHAs. Our approach is based on the growth of records for *phaC* gene sequences, which subsidizes our primer design [17,35]. An innovative aspect is the use of PCR to detect all distinct classes of PHA synthases; Shamala et al. [21] drew primers based only one gene sequence of *B. megaterium*, not obtaining a large breadth of results, and this approach was also employed by Solaiman & Ashby [13], as a result of the simplicity of the method used for primer design. Sheu et al. [24] used the *phaC* gene sequences from 13 gram-positive bacteria to design primers, and presented a greater breadth of results, which were capable of detecting organisms belonging PHA synthase class I and II. The present study used a wide variety of gene sequences belonging to different organisms, resulting in the generation of a precise tool for the detection of organisms with the potential for the production of PHAs from classes I and IV. This method was tested in reference organisms, and was also employed for the screening of new isolates, working in both systems for the detection of the targeted gene.

Colony PCR proved efficient, as this method used amplified regions of interest without necessary DNA extraction methods, which has also been suggested by Sheu et al. [24], making faster work of the screening of environmental organisms. Lane & Benton [23] obtained good results using the same method to determine if six cyanobacteria contained the *phaC* gene. Sasidharam et al. [22] identified the potential of *Vibrio azureus* BTKB33 isolated from marine sediments through PCR confirmation of PHA synthase class I. The use of the PCR technique considerably reduced the number of isolates and thus optimized the process. In addition to the traditional PCR, a multiplex PCR was performed. This methodology used more than one combination of primers to obtain a wider range of results and did not generate false positives, indicating that the use of specific primers for the samples and the chosen conditions were appropriate for the technique [17]. Castroverde et al. [36] showed the efficiency of identifying pathogens using three primers combined in a single PCR. The combination of primers used in the pre-selection of soil organisms in different environments was efficient, and the fragments showed the expected size in approximately 30% of the isolates. These results show the efficiency of using primers designed in the pre-selection of bacteria with the potential for PHA production in samples isolated from the environment. Tzu & Semblante [35] proved the efficiency of multiplex PCR by demonstrating that the primer set was more efficient than the primers tested individually, increasing the detection sensitivity of PHA synthases of classes I and II up to almost 90%. Class I and II PHA synthases were detected from alphaproteobacteria, betaproteobacteria, and gammaproteobacteria, indicating the wide diversity of PHA-accumulating bacteria in wastewater treatment from activated sludge.

Molecular detection of genes involved in PHA synthesis also allows for the prospection of PHA-producing organisms, as well as furthers the understanding and study of gene diversity and evolution [37,38]. Discrepancies in the phylogenetic trees for *phaA*, *phaB*, and *phaC* genes of the PHA biosynthesis have led to the suggestion that horizontal gene transfer may be a major contributor for their evolution [39]. In this way, the use of degenerate primers to study the genetic diversity of genes of biotechnological interest has been gaining prominence, as it aims to define the knowledge of conserved and variable regions of the gene, as well as the structural and functional organization of the enzyme. In the work described by Cheng et al. [40], degenerate primers are used to study the diversity of the subtilase gene with metagenomic DNA samples. This also indicates the potential use of primers in the study of environmental samples taken directly, through DNA metagenomics, which allows to access much of the genetic diversity present in the sample, since organisms that would not be cultivated in the laboratory can be studied directly by the DNA present in the sample. Tai et al. [41] successfully used a culture-independent approach for the detection of the presence of *phaC* genes in limestone soil using primers targeting the class I and II PHA synthases, reassuring the relevance of the approach used in our study.

The related sequences found in studies of diversity still have the potential to be used in genetic improvement programs by site-directed mutations, such as the DNA shuffling technique. Wang et al. [42] pointed out the efficiency of the variant technique called DNA family shuffling for metagenomic studies of homologous genes with specific primers, showing yet another possible application for the primers designed in the present study.

5. Conclusions

This study presents a powerful molecular tool for the identification and bioprospecting of bacteria that have the potential to produce PHAs. The tool also shows high potential for the identification of marine bacteria and pre-screening of environmental bacteria that have *phaC* gene, as well as for use in analyses of environmental diversity.

Acknowledgments: ICGEB/CNPq (Brazil, Process 577915/2008-8) and CNPq/INCT-Mar COI (Brazil, Process 565062/2010-7) supported this work. We also thank CNPq for the scholarship provided to A.O.S.L (Process 311010/2015-6) and G.S.D. (Process 400551/2014-4), as well as Santa Catarina State Govern for the E.M.S. scholarship.

Author Contributions: Eduarda Morgana da Silva Montenegro and Andre Oliveira de Souza Lima conceived and designed the study and experiments. Eduarda Morgana da Silva Montenegro performed experiments with Andre Oliveira de Souza Lima supervision. Marcus Adonai Castro da Silva and Fernando Dini Andreote contributed with analysis. Eduarda Morgana da Silva Montenegro, Andre Oliveira de Souza Lima and Gabriela Scholante Delabary wrote the paper with suggestions/corrections of Fernando Dini Andreote and Marcus Adonai Castro da Silva.

References

1. Song, J.H.; Murphy, R.J.; Narayan, R.; Davies, G.B.H. Biodegradable and compostable alternatives to conventional plastics. *Philos. Trans. R. Soc. Lond. B Biol. Sci.* **2009**, *364*, 2127–2139. [CrossRef] [PubMed]
2. Barnes, D.K.A.; Galgani, F.; Thompson, R.C.; Barlaz, M. Accumulation and fragmentation of plastic debris in global environments. *Philos. Trans. R. Soc. Lond. B Biol. Sci.* **2009**, *364*, 1985–1998. [CrossRef] [PubMed]
3. Steinbüchel, A.; Füchtenbush, B. Bacterial and other biological systems for polyester production. *Trends Biotechnol.* **1998**, *16*, 419–427. [CrossRef]
4. Rehm, B.H. Biogenesis of microbial polyhydroxyalkanoate granules: A platform technology for the production of tailor-made bioparticles. *Curr. Issues Mol. Biol.* **2007**, *9*, 41–62. [PubMed]
5. Bugnicourt, E.; Cinelli, P.; Lazzeri, A.; Alvarez, V.A. Polyhydroxyalkanoate (PHA): Review of synthesis, characteristics, processing and potential applications in packaging. *Express Polym. Lett.* **2014**, *8*, 791–808. [CrossRef]
6. Godbole, S. Methods for identification, quantification and characterization of polyhydroxyalkanoates. *Int. J. Bioassays* **2016**, *5*, 4977–4983. [CrossRef]
7. Khardenavis, A.A.; Kumar, M.S.; Mudliar, S.N.; Chakrabarti, T. Biotechnological conversion of agro industrial wastewaters into biodegradable plastic, poly-β-hydroxybutyrate. *Biosour. Technol.* **2007**, *98*, 3579–3584. [CrossRef] [PubMed]
8. Tan, G.Y.A.; Chen, C.L.; Li, L.; Ge, L.; Wang, L.; Razaad, I.M.N.; Li, Y.; Zhao, L.; Mo, Y.; Wang, J.-Y. Start are search on biopolymer polyhydroxyalkanoate (PHA): A review. *Polymers* **2014**, *6*, 706–754. [CrossRef]
9. Lee, S.Y.; Choi, J.; Wong, H.H. Recent advances in polyhydroxyalkanoate production by bacterial fermentation: Mini-review. *Int. J. Biol. Macromol.* **1999**, *25*, 31–36. [CrossRef]
10. Khanna, S.; Srivastava, A.K. Recent advances in microbial polyhydroxyalkanoates. *Process Biochem.* **2005**, *40*, 607–619. [CrossRef]
11. Silva, A.L.; dos Santosa, E.C.; dos Santosa, Í.A.; Lópeza, A.M. Seleção polifásica de microrganismos produtores de polihidroxialcanoatos. *Quim. Nova.* **2016**, *39*, 782–788.
12. Rehm, B.H. Polyester synthases: Natural catalysts for plastics. *Biochem. J.* **2003**, *376*, 15–33. [CrossRef] [PubMed]
13. Solaiman, D.K.; Ashby, R.D. Rapid genetic characterization of poly(hydroxyalkanoate) synthase and its applications. *Biomacromolecules* **2005**, *6*, 532–537. [CrossRef] [PubMed]
14. Hein, S.; Paletta, J.R.; Steinbüchel, A. Cloning, characterization and comparison of the *Pseudomonas mendocina* polyhydroxyalkanoate synthases *Pha*C1 and *Pha*C2. *Appl. Microbiol. Biotechnol.* **2002**, *58*, 229–236. [PubMed]

15. Yuan, W.; Jia, Y.; Tian, J.; Snell, K.D.; Muh, U.; Sinskey, A.J.; Lambalot, R.H.; Walsh, C.T.; Stubbe, J. Class I and III polyhydroxyalkanoate synthases from *Ralstonia eutropha* and *Allochromatium vinosum*: Characterization and substrate specificity studies. *Arch. Biochem. Biophys.* **2001**, *394*, 87–98. [CrossRef] [PubMed]

16. McCooL, G.J.; Cannon, M.C. *PhaC* and *PhaR* are required for polyhydroxyalkanoic acid synthase activity in *Bacillus megaterium*. *J. Bacteriol.* **2001**, *183*, 4235–4243. [CrossRef] [PubMed]

17. Koller, M.; Rodríguez-Contreras, A. Techniques for tracing PHA-producing organisms and for qualitative and quantitative analysis of intra-and extracellular PHA. *Eng. Life Sci.* **2015**, *15*, 558–581. [CrossRef]

18. Murray, R.G.E.; Doetsch, R.N.; Robinow, C.F. Determinative and cytological light microscopy. *Am. Soc. Microbiol.* **1994**, *1*, 21–41.

19. Ostle, A.G.; Holt, J.G. Nile Blue A as a fluorescent stain for polybeta-hydroxybutyrate. *Appl. Environ. Microbiol.* **1982**, *44*, 238–241. [PubMed]

20. Spiekermann, P.; Rehm, B.H.; Kalscheuer, R.; Baumeister, D.; Steinbüchel, A. A sensitive, viable-colony staining method using Nile red for direct screening of bacteria that accumulate polyhydroxyalkanoic acids and other lipid storage compounds. *Arch. Microbiol.* **1999**, *171*, 73–80. [CrossRef] [PubMed]

21. Shamala, T.R.; Chandrashekar, A.; Vijayendra, S.V.; Kshama, L. Identification of polyhydroxyalkanoate (PHA)-producing Bacillus spp. using the polymerase chain reaction (PCR). *J. Appl. Microbiol.* **2003**, *94*, 369–374. [CrossRef] [PubMed]

22. Sasidharan, R.S.; Bhat, S.G.; Chandrasekaran, M. Amplification and sequence analysis of *phaC* gene of polyhydroxybutyrate producing *Vibrio azureus* BTKB33 isolated from marine sediments. *Ann. Microbiol.* **2016**, *66*, 299–306. [CrossRef]

23. Lane, C.E.; Benton, M.G. Detection of the enzymatically-active polyhydroxyalkanoate synthase subunit gene, *phaC*, in cyanobacteria via colony PCR. *Mol. Cell. Probes* **2015**, *29*, 454–460. [CrossRef] [PubMed]

24. Sheu, D.S.; Wang, Y.T.; Lee, C.Y. Rapid detection of polyhydroxyalkanoate accumulating bacteria isolated from the environment by colony PCR. *Microbiology* **2000**, *146*, 2019–2025. [CrossRef] [PubMed]

25. Solaiman, D.K.; Ashby, R.D.; Foglia, T.A. Rapid and specific identification of medium-chain-length polyhydroxyalkanoate synthase gene by polymerase chain reaction. *Appl. Microbiol. Biotechnol.* **2000**, *53*, 690–694. [CrossRef] [PubMed]

26. Kung, S.S.; Chuang, Y.C.; Chen, C.H.; Chien, C.C. Isolation of polyhydroxyalkanoates-producing bacteria using a combination of phenotypic and genotypic approach. *Lett. Appl. Microbiol.* **2007**, *44*, 364–371. [CrossRef] [PubMed]

27. Desetty, R.D.; Mahajan, V.S.; Khan, B.M.; Rawal, S.K. Isolation and heterologous expression of PHA synthesising genes from Bacillus thuringiensis R1. *World J. Microbiol. Biotechnol.* **2008**, *24*, 1769–1774. [CrossRef]

28. Lima, A.O.S.; Garcês, S.P.S. Intragenic Primer Design: Bringing Bioinformatics Tools to the Class. *Biochem. Mol. Biol. Educ.* **2006**, *34*, 332–337. [CrossRef] [PubMed]

29. Biocyc Database Collection. Available online: http://biocyc.org (accessed on 20 October 2016).

30. Huson, D.H.; Auch, A.; Qi, J.; Schuster, S.C. Megan Analysis of Metagenome Data. *Genome Res.* **2011**, *17*, 377–386. [CrossRef] [PubMed]

31. Nacional Center for Biotechnology Information. Available online: https://www.ncbi.nlm.nih.gov/Structure/cdd/wrpsb.cgi (accessed on 11 November 2016).

32. Larkin, M.A.; Blackshields, G.; Brown, N.P.; Chenna, R.; McGettigan, P.A.; McWilliam, H.; Valentin, F.; Wallace, I.M.; Wilm, A.; Lopez, R.; et al. ClustalW and ClustalX version 2. *Bioinformatics* **2007**, *23*, 2947–2948. [CrossRef] [PubMed]

33. Fursov, M.Y.; Oshchepkov, D.Y; Novikova, O.S. UGENE: Interactive computational schemes for genome analysis. In Proceedings of the Fifth Moscow International Congress on Biotechnology, Moscow, Russia, 16–20 March 2009; Volume 3, pp. 14–15.

34. Altschul, S.F.; Madden, T.L.; Schaffer, A.A.; Zhang, J.; Zhang, Z.; Miller, W.; Lipman, D.J. Gapped BLAST and PSI-BLAST: A new generation of protein database search programs. *Nucleic Acids Res.* **1997**, *25*, 3389–3402. [CrossRef] [PubMed]

35. Tzu, H.Y.; Semblante, G.U. Detection of polyhydroxyalkanoate-accumulating bacteria from domestic wastewater treatment plant using highly sensitive PCR primers. *J. Microbiol. Biotechnol.* **2012**, *22*, 1141–1147.

36. Castroverde, C.D.M.; San Luis, B.B.; Monsalud, R.G.; Hedreyda, C.T. Differential detection of vibrios pathogenic to shrimp by multiplex PCR. *J. Gen. Appl. Microbiol.* **2006**, *52*, 273–280. [CrossRef] [PubMed]

37. Sujatha, K.; Mahalakshmi, A.; Shenbagarathai, R. Molecular characterization of *Pseudomonas* sp. LDC-5 involved in accumulation of poly 3 hydroxybutyrate and medium-chain-length poly 3-hydroxyalkanoates. *Arch. Microbiol.* **2007**, *188*, 451–462. [CrossRef] [PubMed]

38. Aneja, K.K.; Ashby, R.D.; Solaiman, D.K.Y. Altered composition of *Ralstonia eutropha* poly (hydroxyalkanoate) through expression of PHA synthase from *Allochromatium vinosum* ATCC 35206. *Biotechnol. Lett.* **2009**, *31*, 1601–1612. [CrossRef] [PubMed]

39. Kalia, V.C.; Lal, S.; Cheema, S. Insight in to the phylogeny of polyhydroxyalkanoate biosynthesis: Horizontal gene transfer. *Gene* **2007**, *389*, 19–26. [CrossRef] [PubMed]

40. Cheng, X.; Gao, M.; Wang, M.; Liu, H.; Sun, J.; Gao, J. Subtilase genes diversity in the biogas digester microbiota. *Curr. Microbiol.* **2011**, *62*, 1542–1547. [CrossRef] [PubMed]

41. Tai, Y.T.; Foong, C.P.; Najimudin, N.; Sudesh, K. Discovery of a new polyhydroxyalkanoate synthase from limestone soil through metagenomic approach. *J. Biosci. Bioeng.* **2016**, *121*, 355–364. [CrossRef] [PubMed]

42. Wang, Q.; Wu, H.; Wang, A.; Du, P.; Pei, X.; Li, H.; Yin, X.; Huang, L.; Xiong, X. Prospecting Metagenomic Enzyme Subfamily Genes for DNA Family Shuffling by a Novel PCR-based Approach. *J. Biol. Chem.* **2010**, *285*, 41509–41516. [CrossRef] [PubMed]

12

Polyhydroxyalkanoate Biosynthesis at the Edge of Water Activity-Haloarchaea as Biopolyester Factories

Martin Koller [1,2]

[1] Office of Research Management and Service, c/o Institute of Chemistry, University of Graz, NAWI Graz, Heinrichstrasse 28/III, 8010 Graz, Austria; martin.koller@uni-graz.at

[2] ARENA—Association for Resource Efficient and Sustainable Technologies, Inffeldgasse 21b, 8010 Graz, Austria

Abstract: Haloarchaea, the extremely halophilic branch of the Archaea domain, encompass a steadily increasing number of genera and associated species which accumulate polyhydroxyalkanoate biopolyesters in their cytoplasm. Such ancient organisms, which thrive in highly challenging, often hostile habitats characterized by salinities between 100 and 300 g/L NaCl, have the potential to outperform established polyhydroxyalkanoate production strains. As detailed in the review, this optimization presents due to multifarious reasons, including: cultivation setups at extreme salinities can be performed at minimized sterility precautions by excluding the growth of microbial contaminants; the high inner-osmotic pressure in haloarchaea cells facilitates the recovery of intracellular biopolyester granules by cell disintegration in hypo-osmotic media; many haloarchaea utilize carbon-rich waste streams as main substrates for growth and polyhydroxyalkanoate biosynthesis, which allows coupling polyhydroxyalkanoate production with bio-economic waste management; finally, in many cases, haloarchaea are reported to produce copolyesters from structurally unrelated inexpensive substrates, and polyhydroxyalkanoate biosynthesis often occurs in parallel to the production of additional marketable bio-products like pigments or polysaccharides. This review summarizes the current knowledge about polyhydroxyalkanoate production by diverse haloarchaea; this covers the detection of new haloarchaea producing polyhydroxyalkanoates, understanding the genetic and enzymatic particularities of such organisms, kinetic aspects, material characterization, upscaling and techno-economic and life cycle assessment.

Keywords: Archaea; bioeconomy; biopolyester; downstream processing; extremophiles; haloarchaea; *Haloferax*; halophiles; polyhydroxyalkanoates; salinity

1. Introduction

The first description of a biological polymer with plastic-like properties was published in the 1920s, when Maurice Lemoigne detected light-refractive intracellular inclusion bodies [1], today referred to as "granules"—or, more recently "carbonosomes" [2]—n resting cultures of the Gram-positive bacterium *Bacillus megaterium*. Based on the acidic degradation product of these inclusions, 3-hydroxybutyrate (3HB), Lemoigne correctly assumed the microscopically observed intracellular product to be the polymer of 3HB, namely poly(3-hydroxybutyrate) (PHB). In the meantime, PHB and its related homo- and heteropolyesters, as a group labelled as polyhydroxyalkanoates (PHA), have attracted global attention as biological, bio-based, biocompatible and biodegradable alternatives to established plastics of petrochemical origin in many sectors of the rocketing plastic market [3,4]. PHA consist of a variety of diverse building blocks, which make their material properties highly versatile [5], and can be produced biotechnologically by different continuous or discontinuous fermentation approaches and feeding strategies [6]. In principle, short chain length PHA (*scl*-PHA) are distinguished from medium chain length PHA (*mcl*-PHA). While *scl*-PHA typically constitute thermoplastic materials, *mcl*-PHA are known as materials with elastomeric and latex-like properties and are often of a sticky nature. Among scl-PHA, the homopolyester PHB and the copolyester poly(3-hydroxybutyrate-co-3-hydroxyvalerate) (PHBHV) are best described; in this context, increasing 3-hydroxyvalerate (3HV) fractions in the copolyester decreases melting temperature and crystallinity, which makes such PHBHV copolyesters easier to process than the rather crystalline and brittle PHB, a material of restricted applicability [3,4].

Apart from wild-type and genetically engineered eubacteria and recombinant yeasts, plants, and microalgae, PHA biosynthesis takes place also in the cytoplasm of various extremely halophilic species from the Archaea domain, the so called "haloarchaea". Exclusively scl-PHA production is reported for haloarchaea, while for eubacteria, both scl- and mcl-PHA production is reported [3,4]. Extremely challenging habitats include environments where such highly adaptive survivalists are typically isolated; illustrative examples are the Great Salt Lake, the Dead Sea, hypersaline anoxic deep-sea basins, solar saltern crystallizers, hypersaline soil samples, salt mine boreholes, salt production pans, or even alpine dry salt rocks. The taxonomic classification of these extremely salt-demanding, typically aerobic organisms is by no means a trivial task and is based on steadily refined knowledge about the genomics, proteomics, metabolomics, and lipidomics of these organisms. Traditionally, haloarchaea are members of the family Halobacteriaceae, which belongs to the order Halobacteriales, which in turn is part of class III (Halobacteria) consisting of two major clades A and B, of the phylum and (sub)kingdom of Euryarchaeota, which belongs to the domain of Archaea (according to International Committee on Systematics of Prokaryotes, Subcommittee on the taxonomy of Halobacteriaceae; cited by [7]). Later, members of the class Halobacteria were re-grouped into three orders: a revised order Halobacteriales and two new orders, Haloferacales and Natrialbales, which encompass the novel families Haloferacaceae and Natrialbaceae [8]. More recently, based on phylogenetic analyses and conserved molecular characteristics, it was suggested to divide the order "Halobacteriales" into the families Halobacteriaceae, Haloarculaceae, and Halococcaceae, and the order "Haloferacales" into the families Haloferacaceae and Halorubraceae [9]. These are the currently valid designations of the families where haloarchaea demonstrated to produce PHA are grouped. Figure 1 provides a schematic overview about the phylogenetic classification of the haloarchaeal species discussed in the present review.

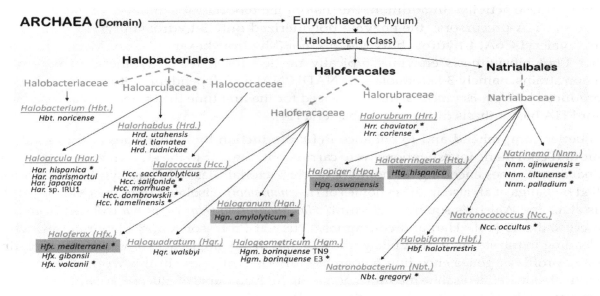

Figure 1. Extract from the phylogenetic tree of haloarchaea, selecting those species reported to accumulate PHA biopolyesters. Colored (orange) background highlights the limited number of species to date cultivated on bioreactor scale to study PHA production. The asterisks indicate 3-hydroxyvalerate production by the strain from structurally unrelated substrates. (Bold: orders; red: families; *italics and underlined*: genera; *italics*: species).

Talking about PHA biosynthesis by haloarchaea, it took the scientific community until 1972, when Kirk and Ginzburg carried out morphological characterizations of a Dead Sea isolate, which was labeled "*Halomonas* sp." by these authors. This organism was cultivated on a highly saline medium containing 200 g/L NaCl. By using freeze-fracture and freeze-etch techniques, the authors revealed plastic-like cytoplasmic inclusion bodies, which were extracted from microbial biomass and investigated by X-ray diffractometry. Grounded solely on these examinations, the authors correctly recognized this material as the biopolyester PHB, the material already known at the time as a carbon and energy storage product for many eubacteria, as reported by Lemoigne [1] and succeeding generations of scholars. In any case, this study by Kirk and Ginzburg was the very first unambiguous description of PHA production by an archaeon [10]. Regarding the production strain "*Halomonas* sp.", it took nearly three decades until this isolate was classified as *Haloarcula* (*Har.*) *marismortui*, its currently valid species name, in a report published by Nicolaus et al. [11].

2. Genetic and Enzymatic Particularities of Haloarchaeal PHA Biosynthesis

Generally, PHA synthases, the enzymes catalyzing the polymerization of PHA precursors (hydroxyacyl-CoAs like acetyl-CoA, propionyl-CoA, etc.) found in haloarchaea are grouped in the Class III of PHA synthases [12]. Class III PHA synthases were identified in several eubacteria such as *Allochromatium vinosum* (previously known as *Chromatium vinosum*) or *Thiocapsa pfennigii*; Class III PHA synthases polymerize short hydroxyacyl-CoAs, namely those not longer than 3-hydroxyvaleryl-CoA; moreover, such synthases are typically composed of two subunits: the catalytically active subunit PhaC (molar mass ranging from 40–53 kDa) and the structural subunit PhaE (molar mass 20–40 kDa), which is also indispensable for polymerization. Together, the two subunits form a biocatalytic cluster, the so called "PhaEC complex" [13]. Hezayen et al. were the first scientists who revealed the special features of PHA synthases in haloarchaea. When studying "strain 56" (today classified as *Halopiger* (*Hpq.*) *aswanensis*), a species isolated from hypersaline soil collected near Aswan, Egypt, which thrives best with 250 g/L NaCl, Hezayen et al. discovered a PHA synthase covalently bound to the PHA granules. This enzyme exposed particular features in comparison to PHA synthases in eubacteria described earlier; the new enzyme displayed high thermostability up to 60 °C, with strongly increasing activity at higher salinity; especially, Mg^{2+} ion concentration had a significant effect on synthase activity. In addition, this halophilic biocatalyst exhibited a remarkably narrow spectrum of PHA-precursors: the enzyme polymerized only 3-hydroxybutyryl-CoA, but neither 3-hydroxyvaleryl-CoA, 4-hydroxybutyryl-CoA, nor 3-hydroxyhexanoyl-CoA. Most extraordinarily, no other PHB biosynthesis enzymes typically needed for PHA biosynthesis in bacterial PHA production strains, namely 3-ketothiolase or NADH/NADPH-dependent acetoacetyl–CoA reductase, were produced by *Hpq. aswanensis*; this evidenced for the first time that haloarchaea use a metabolic route for PHA biosynthesis different to eubacteria [14].

To better comprehend and to enhance PHA production by haloarchaea, various subsequent genomic and enzymatic investigations were carried out. Similar to the studies with *Hpq. aswanensis*, the haloarchaeal genes encoding for homologues of bacterial Class III PhaC synthase enzymes were identified by Baliga et al. also in the genomes of *Har. marismortui* isolated from the Dead Sea [15], or by Bolhuis et al. in "Walsby's square bacterium" *Haloquadratum* (*Hqr.*) *walsbyi* isolated from different saltern crystallizers [16]. Han and colleagues, active at Professor Xiang's laboratories, which are world-leading in the study of PHA biosynthesis genes in haloarchaea, explored for the first time the expression profile of genes encoding haloarchaeal PHA synthases. In this work, *Har. marismortui*, when cultivated in defined saline medium containing high amounts of glucose as carbon source, is able to accumulate PHB fractions in cell dry mass (CDM) up to 21 wt.%. As a major result, the neighboring genes *phaE_{Hm}* and *phaC_{Hm}* were identified by molecular characterization of the *phaEC_{Hm}* operon; these two genes encode two Class III PHA synthase subunits, and are triggered by only one single promoter. It was shown that these genes are constitutively expressed, both under balanced and nutrient-limited cultivation conditions. Remarkably, in contrast to the non-granule associated gene *PhaE_{Hm}*, *PhaC_{Hm}* is strongly connected to the PHA granules. Inserting $phaE_{Hm}$ or $phaC_{Hm}$ genes into the closely related strain *Har. hispanica*, which contains highly homologue $phaEC_{Hh}$ genes, considerably increased PHB biosynthesis. Particularly, the co-expression of both genes resulted in the highest PHB productivity; in contrast, deleting $phaEC_{Hh}$ genes from the *Har. hispanica* genome ("knocking out") totally terminates PHA production. By transferring $phaEC_{Hm}$ genes into such knockout mutants fully restored the activity of PHA synthase and PHA accumulation. These studies validated for the first time the high significance of *phaEC* genes for PHA biosynthesis in haloarchaea [17].

Lu et al. carried out groundbreaking work with *Haloferax* (*Hfx.*) *mediterranei* to elucidate the genetic and enzymatic PHA biosynthesis background of this strain. Using thermal asymmetric interlaced PCR, these authors were able to clone the $phaEC_{Hme}$ gene cluster of strain *Hfx. mediterranei* CGMCC 1.2087. By Western blotting, it was shown that, analogous to the above described findings for *Har. marismortui*, both $phaE_{Hme}$ (about 21 kDa) and $phaC_{Hme}$ (about 53 kDa) genes were constitutively expressed, and both synthases were strongly connected to the PHA granules. Interestingly, the strain synthesized poly(3-hydroxybutyrate-*co*-3-hydroxyvalerate) (PHBHV) copolyesters in both nutrient-limited (supplemented with 1% starch, production of up to 24 wt.% PHBHV in CDM) and nutrient-rich (up to 18 wt.% PHBHV in CDM) media in shaking flask experiments. Knockout of $phaEC_{Hme}$ genes in this strain completely stopped PHBHV biosynthesis; PHBHV biosynthesis capability was re-established only after complementation with the complete $phaEC_{Hme}$ gene cluster, but not when transferring either $phaE_{Hme}$ or $phaC_{Hme}$ alone. It is worth noting that the described PhaC synthase subunits were considerably longer at their carbon-end than reported for bacterial PHA synthases; this C-terminal extension of $PhaC_{Hme}$ was shown to be indispensable for the enzymes´ in vivo activity at high salinity. Moreover, a 1:1 mixture of isolated $PhaE_{Hme}/PhaC_{Hme}$ enzymes displayed substantial PHA synthesis activity in vitro. These outcomes showed that also *Hfx. mediterranei* possesses the novel type of class III PHA synthases typical for haloarchaea, which are assembled by $PhaC_{Hme}$ and $PhaE_{Hme}$ subunits [18]; this corresponds to the above described discoveries for PHA synthases in *Har. hispanica* and *Har. marismortui* [17].

By further sequencing of the *Hfx. mediterranei* CGMCC 1.2087 genome, Han and colleagues identified three more "hidden" *phaC* genes (*phaC1*, *phaC2*, and *phaC3*), which encode possible PhaC synthases. The three "cryptic" genes were distributed all over the whole *Hfx. mediterranei* genome. Similar to $PhaC_{Hme}$ (molar mass 54.8 kDa), PhaC1 (49.7 kDa) and PhaC3 (62.5 kDa) exhibited conserved Class III PHA synthase motifs, which was not the case for PhaC2 (40.4 kDa). Moreover, the longer C-terminus of the other three PhaC enzymes was not found in PhaC2. It was revealed via reverse transcription PCR (RT-PCR) that among all four genes, only $phaC_{Hme}$ was transcribed in the wild-type strain under conditions supporting PHA biosynthesis. Astonishingly, heterologous co-expression of $phaE_{Hme}$ with each *phaC* gene in the PHA-negative mutant *Har. hispanica* PHB-1 revealed that all PhaCs, except PhaC2, effect PHBHV synthesis, though with different 3HV portions in the copolyesters. These products were characterized, revealing that thermal properties (melting point, crystallinity, glass transition temperature, etc.) and molecular mass strongly depend on the 3HV fraction in the copolyester. Briefly, the study defined three novel "hidden" *phaC* genes in *Hfx. mediterranei*, and suggested that genetic engineering of these "cryptic" *phaC* genes might have biotechnological applicability in terms of designing PHBHV copolyesters of tailored material properties based on the fine-tuning 3HV contents [19]. In 2012, Han et al. were able to report the complete genome sequence of *Hfx. mediterranei* CGMCC 1.2087, which has a size of 3,904,707 bp and consists of one chromosome and three mega-plasmids, by using a combination of 454 pyrosequencing and Sanger sequencing [20]. Shortly thereafter, Ding et al. deciphered the complete sequence of the *Har. hispanica* ATCC 43049 genome; unexpectedly, these authors noticed substantial differences when comparing this sequence with the gene sequence of *Har. hispanica* ATCC 33960 [21], the model organism used for molecular characterization studies by Han et al. described above [19]. In any case, the works presented by Han et al. [19] and Ding et al. [21] demonstrate that clustered *phaEC* genes encoding Class III PHA synthases are typical features of PHA-producing haloarchaea. This was substantiated by Han and colleagues, who screened PHA synthase genes in haloarchaeal PHA producers from 12 genera; the authors demonstrated the wide distribution of *phaEC* genes among haloarchaea. Compared to their bacterial counterparts, haloarchaeal PHA synthases differ significantly in both molecular weight and some conserved motifs. Therefore, Han and colleagues proposed to classify haloarchaeal PHA synthases as "subtype IIIA", while type III PHA synthases from bacteria were proposed as "subtype IIIB" [22].

Genome analysis of *Hfx. mediterranei* has also evidenced eight potential 3-ketothiolase genes in *H. mediterranei*, which might express enzymes responsible for the condensation of two acetyl-CoA molecules to acetoacetyl-CoA, or one acetyl-CoA and one propionyl-CoA to 3-ketovaleryl-CoA. It was shown that only the 3-ketothiolases encoded by *HFX_6004-HFX_6003* and *HFX_1023-HFX_1022* are involved in the biosynthesis of PHBHV. Knockout of *HFX_6004-6003* leads to the accumulation of PHB homopolyester without the 3HV building blocks, while simultaneous knockout of *HFX_6004-6003* and *HFX_1023-1022* stopped the strain´s ability to produce PHA. This was the first report on haloarchaeal 3-ketothiolases, which reveled considerable differences to their bacterial relatives in subunit composition and catalytic residue [23]. Finally, genes encoding for PHA-specific acetoacetyl-CoA reductases, catalyzing the reduction of ketoacyl-CoAs to hydroxyacyl-CoAs as the substrates of PHA synthases were discovered and characterized in *Har. hispanica* [24] and *Hfx. mediterranei* [25]. Further, these enzymes displayed considerable differences to their eubacterial counterparts encoded by *phaB* genes. Only recently, Xiang summarized the genomic and enzymatic particularities of PHA biosynthesis by *Hfx. mediterranei* in a comprehensive way [26]. Figure 2 provides a simplified schematic of the pathways leading to PHB and PHBHV by eubacteria and haloarchaea, respectively. For haloarchaea, especially the multiple propionyl-CoA supplying pathways are highlighted.

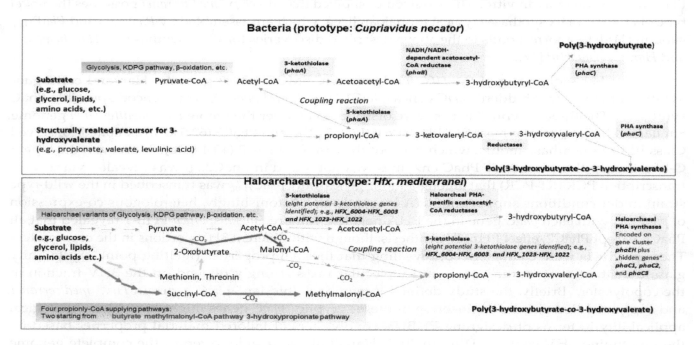

Figure 2. Simplified illustration of PHB and PHBHV biosynthesis by eubacteria (upper part, in grey; prototype organism: *C. necator*) and haloarchaea (lower part, in pink; prototype: *Hfx. mediterranei*). Enzymes and genes (in italics) involved in the PHA biosynthesis steps (starting from acetyl-CoA and propionyl-CoA) are in green text boxes. Special emphasis is dedicated to the propionyl-CoA supplying pathways in haloarchaea: Propionyl-CoA is generated (a) beginning with the coupling of pyruvate and acetyl-CoA, and the decarboxylation of 2-oxobutyrate (marked in brown), (b) starting from the conversion of the amino acids methionine or threonine to 2-oxobutyrate (marked in yellow), (c) starting from succinyl-CoA via methylmalonyl-CoA (marked in green), or (d) starting with carboxylation of acetyl-CoA to malonyl-CoA (marked in purple). Based on [26].

3. *Haloferax mediterranei*—The Prototype PHA Production Strain among Haloarchaea

Hfx. mediterranei was the first haloarchaeon for which PHA-accumulation kinetics were studied in detail. *Hfx. mediterranei* was among the 19 organisms first isolated in 1980 by Rodriguez-Valera et al. from samples collected from the evaporation ponds of solar salterns near Alicante, Spain, and designated as strain "Q4". In this publication, the authors reported on the isolation of "moderate and extremophile bacteria", without discriminating between halophilic eubacteria and haloarchaea. However, this study already proposed the possibility to enrich slowly growing extremophiles from mixed microbial cultures by carrying out chemostat continuous cultivations at a low dilution rate (D) and high substrate concentration. Moreover, the authors supposed that moderate halophiles by trend prefer a rather low temperature for growth, while extreme halophiles grow best at higher temperatures. Further, this study showed for the first time that the pigmentation of extremely halophilic organisms is more pronounced at elevated temperatures, and especially so at high salinity. In their study, Rodriguez-Valera et al. described already crucial characteristics of the most outstanding among the isolates, strain "Q4", namely negative Gram-staining, pinkish pigmentation, formation of pleomorphic rods, an optimum salinity of 250 g/L NaCl with salinity range of 100–300 g/L, and a maximum specific growth rate ($\mu_{max.}$) of 0.05 1/h. Yet, this study did not search for PHA accumulation by this strain [27]. In a subsequent study, these researchers mentioned that this new isolate exhibited substantial physiological and morphological differences to other "halobacteria" described, and recommended that "R-4" (previously strain "Q4") should be grouped into the new species *Halobacterium (Hbt.) mediterranei* [28].

In 1986, Fernandez-Castillo et al. recognized for the first time granular PHA inclusions in cells of this intriguing strain when exploring it in an experimental series with other extremely halophilic isolates (in this study termed "halobacteria"), viz. *Hbt. gibonsii, Hbt. halobium, Hbt. hispanicum*, and *Hbt. volcanii*, which were farmed in rather simple cultivation setups performed in aerated and magnetically stirred glass vessels. All these strains except *Hbt. halobium* showed PHA accumulation when growing on media containing 250 g/L salts, 10 g/L glucose, and 1 g/L yeast extract; however, *Hbt. mediterranei* by far outperformed the other strains in terms of PHA formation [29]. Today, these isolates are classified as the strains *Hfx. gibbonsii, Hfx. volcanii, Har. hispanica*, and *Har. marismortui*. These updated species names are based on a new numerical taxonomic classification based on the polar lipids of "halobacteria"; this classification was performed during the studies of Torreblanca et al.; as result, the original genus *Halobacterium* was divided into the three new genera *Haloarcula, Halobacterium*, and *Haloferax*; strain "*Halobacterium mediterranei*" (isolate R-4, originally "Q4") got the new species name "*Haloferax mediterranei*" [30]. For this strain, higher PHA contents in biomass of 17 wt.% were obtained using glucose than with other substrates (citrate, cellobiose, and glycerol). Further isolates used in the study (*Hfx. volcanii, Hfx. gibbonsii*, and *Har. hispanica*) exhibited only minor PHA fractions in CDM of 7 wt.%, 12 wt.%, and 24 wt.%, respectively, when thriving in a medium with 250 g/L NaCl, 10 g/L glucose, and 1 g/L yeast extract. Notably, in this study, authors reported that exclusively PHB homopolyester ("poly-β-butyric acid") was produced of by all of these strains, despite the fact that the products were subjected towards ^{13}C-NMR characterization, which revealed that PHA constituents other than 3HB were present in some of the isolated PHA samples [29]. According to today's knowledge, particularly *Hfx. mediterranei* synthetizes PHBHV copolyesters under the described cultivation conditions (sufficient supply of sugars or glycerol) [31]. Importantly, this study suggested for the first time the disruption of haloarchaeal cells by exposing them to hypotonic media (distilled water) for the facile recovery of PHA granules without the use of organic solvents, which substantially facilitates downstream processing in economic and environmental terms [29]. This study can be considered the ignition spark for setting *Hfx. mediaterranei* at the pole position of research activity with respect to PHA production by haloarchaea.

Data on the detailed exploration of PHA production by other haloarchaea are still rather scarce because, as reported by Lillo and Rodriguez-Valera [32] and Rodriguez-Valera and Lillo [31], *Hfx. mediterranei* displays higher specific growth and PHA production rates in comparison to other haloarchaea reported to accumulate PHA; this consequently is beneficial for the volumetric productivity of the bioprocess. Detailed insights into kinetics and optimized cultivation process parameters for PHA production with this strain were provided by a key publication by Lillo and Rodriguez-Valera. These authors studied continuous chemostat cultivations performed at a dilution rate D of 0.12 1/h at 38 °C. Using 20 g/L glucose and high salinity (250 g/L marine salts), 3.5 g/L PHA were produced. Replacing glucose with inexpensive starch resulted in an almost duplication of the PHA concentration (6.5 g/L), and also demonstrated the high α-amylase activity of the strain. These authors already determined that the temperature optima for growth and PHA biosynthesis by this strain are not identical [32]. Further, Antón et al. demonstrated experimentally that the organism requires highly saline nutrient media containing at least 200 g/L NaCl for optimum growth; such high salinity de facto excludes the risk of contamination with foreign germs, which is a significant gain when carrying out large-scale production setups under reduced sterility precautions [33]. This high robustness of *Hfx. mediterranei* cultivation setups against microbial rivals was later substantiated by Hermann-Krauss and colleagues, who carried out fed-batch cultivations with this strain without any sterilization provisions neither for the cultivation medium nor the bioreactor; even after several days, no infection by other microbes was detectable [34]. In contrast to high medium salinity, the cytoplasm of *Hfx. mediterranei* contains high quantities of KCl to generate high inner osmotic pressure, hence, to balance the outer osmotic pressure; this strategy of the strain to cope with such extremely high extracellular salinity [20], the so-called "salt-in" strategy, is a typical feature of haloarchaea. This requires an adaptation of the proteome, e.g., high surface charge of enzymes, to conserve the proper conformation and activity of enzymes at the edge of salt saturation. This approach drastically differs from strategies known of halophilic eubacteria, which accumulate soluble osmolytes such as ectoins as a reaction to excessive extracellular salinity [35].

PHA is not the only intriguing polymeric product produced by *Hfx. mediterranei*. In 1988, Antón et al. reported that the strain excretes also an extracellular polymer, which can be recovered from solution by precipitation with cold ethanol. This extracellular polysaccharide (EPS) causes the typical mucous appearance of *Hfx. mediterranei* colonies grown on solid nutrients [33]. This EPS is an anionic, sulfated polymer, consisting of a regular trisaccharide-repeating unit with one mannose and two 2-acetamido-2-deoxyglucuronic acid monomers and one sulfate ester bond per trisaccharide unit. Rheologically, the polymer displays xanthan-like characteristics, which has attracted interest in it as a thickening and gelling agent in food technology [36]. Later, the interrelation between parallel production and in vivo degradation of the two polymers (PHA and EPS) was investigated. It was revealed that intracellular PHA degradation is a rather slow process, even under carbon-limited conditions; technologically, this allows postponing cell inactivation and harvest after complete depletion of the carbon source without risking significant product degradation. In this study, it was also demonstrated that pronounced EPS production takes place when feeding the strain with defined carbon sources like carbohydrates, but not when supplying complex substrates like yeast extract; this trend is analogous to the strain's PHA accumulation profile: defined carbon sources result in high PHA biosynthesis, while complex substrates favor biomass growth [37]. More details were reported by Cui et al., who studied the salinity effect on PHA and/or EPS biosynthesis as a tool to direct the carbon flux towards one product or another. These cultivations were performed in 1.2 L airlift bioreactors. In a nutshell, high salinity inhibited EPS biosynthesis, but preferred PHA accumulation. Increasing NaCl concentration from 75 g/L to 250 g/L, EPS production slightly dropped from 37 wt.% to 32 wt.%. With 71 wt.%, the PHA fraction in biomass reached its highest value at a salinity of 250 g/L NaCl; this demonstrated that a high salinity boosts PHA production at the expense of EPS formation. Technologically, these results enable a regulation of the carbon flux in *Hfx. mediterranei* by adapting the salinity of the cultivation medium in order to enhance the biosynthesis of either PHA or EPS, both constituting industrially applicable products [38].

Figure 3a shows *Hfx. mediterranei* colonies grown on solid agar medium with yeast extract and enzymatically hydrolyzed whey permeate as substrates; the mucous and pinkish character of colonies due to the production of EPS and pigments (C50 carotenoids) is apparent. Figure 3b presents liquid samples taken from the beginning through to the end of a bioreactor cultivation of *Hfx. mediterranei* under controlled conditions in a 200 L (working volume; total volume: 300 L) pilot scale bioreactor (L 1523, Bioengineering, Wald, CH) using the same substrates: yeast extract (initial concentration 6.25 g/L) and enzymatically hydrolyzed whey permeate (initial concentration 50 g/L, corresponds to 10 g/L of an equimolar glucose/galactose mixture); a refeed of hydrolyzed whey permeate was done according to HPLC analysis of the cultivation broth after each sampling, while a re-feed of yeast extract solution was done drop-wise according to the reaction of the dissolved oxygen probe during the first phase of the cultivation (until t = 28 h) in order to provoke enhanced PHA biosynthesis. The pH-value and dissolved oxygen were permanently controlled and recorded online and PHA, EPS and protein concentrations were determined after each sampling. The samples show increasing coloration and viscosity until the end of the cultivation, when the mass fraction of PHA (a PHBHV copolyester containing 10 mol.% of 3HV amounted to 67 wt.%; final concentrations of 7.2 g/L and 1 g/L were obtained for PHBHV and EPS, respectively.

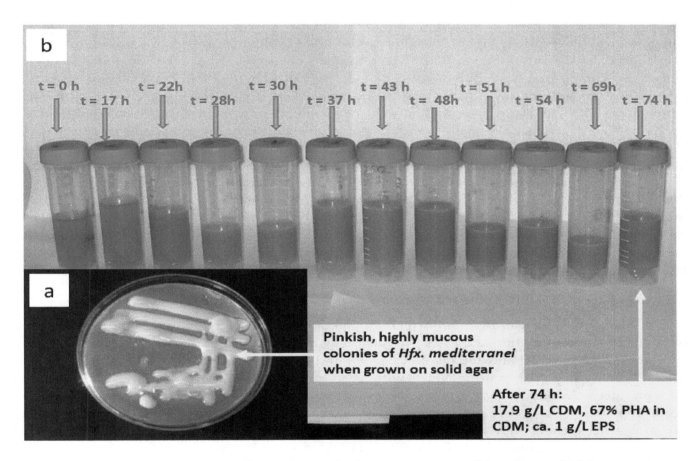

Figure 3. (**a**) Pinkish, mucous *Hfx. mediterranei* colonies grown on solid medium. (**b**) Macroscopic appearance of samples taken from a *Hfx. mediterranei* bioreactor cultivation from the beginning (t = 0 h) until the end (t = 74 h) of the process. Own pictures of the author M. Koller.

As stated above, when using simple carbon sources like carbohydrates, *Hfx. mediterranei* does not produce the homopolyester PHB as typical for the majority of wild type eubacteria, but a PHBHV copolyester. *Hfx. mediterranei* was the first strain at all, for which PHBHV copolyester production from structurally unrelated carbon sources was reported. For other strains, in vivo incorporation of 3-hydroxyvalerate (3HV) in growing PHA chains is dependent on the supply of precursors structurally related to 3HV, such as propionic acid, valeric acid, or levulinic acid. These precursors contribute considerably to the costs of PHBHV production [4]. Only decades later, the metabolic background of this particular feature was revealed by bioinformatic analysis of the *Hfx. mediterranei* genome sequence, when Han et al. proposed four active pathways in *Hfx. mediterranei*, which synthesize the 3HV-precursor propionyl-CoA. The first two pathways involve the conversion of 2-oxobutyrate (either from starting pyruvate and acetyl-CoA or starting from threonine and methionine) to propionyl-CoA. The third pathway, the so called methylmalonyl-CoA pathway, starts from the isomerization of succinyl-CoA to methylmalonyl-CoA, which gets decarboxylated to propionyl-CoA. Finally, the 3-hydroxypropionate pathway starts with carboxylation of acetyl-CoA forming malonyl-CoA, which gets reduced in a cascade of catalytic steps to propionyl-CoA. In this context, coupling of propionyl-CoA and acetyl-CoA generates 3-ketovaleryl-CoA, while condensation of two acetyl-CoA molecules, the universal, central metabolite, generates acetoacetyl-CoA. Both reactions are catalyzed by the enzyme 3-ketothioase (in older literature: β-ketothiolase). Subsequently, 3-ketovaleryl-CoA is reduced by reductases to 3HV-CoA, which acts as substrate of PHA synthases for 3HV polymerization in growing PHA chains, while acetoacetyl-CoA is transformed into 3-hydroxybuytryl-CoA, the active form of the PHB monomer 3HB [39]. Technologically important: PHBHV copolyesters, characterized by their lower crystallinity and higher difference between meting temperature and degradation temperature, are more easily processed by injection molding, melt extrusion, or other polymer processing techniques if compared with the typically highly crystalline and brittle PHB homopolyester. Moreover, due to their pronounced amorphous domains, PHBHV copolyesters are more prone to (bio)degradation in vivo and during composting if compared with PHB [40]. In particular, the PHBHV copolyester produced by *Hfx. mediterranei* typically exhibits material features desired for processing, such as a low melting temperature (T_m), low degree of crystallinity (X_c), high molecular mass up to the MDa range, and low polydispersity ($Đ_i$), hence a high uniformity of polyester chains in one and the same sample [41].

As described, it was only during the last decade, when profound information about the enzymatic and genomic particularities of *Hfx. mediterranei*, with special emphasis dedicated to the mechanisms involved in PHA biosynthesis, was elaborated [26]. This covers studies on the special *Hfx. mediterranei* PHA synthase enzymes [18], haloarchaeal phasins as enzymes essential for PHA granule formation [42], the identification and mapping of the phaB genes encoding PHA biosynthetic enzymes in *Hfx. mediterranei* [25], the multiple pathways generating the 3HV-precursor propionyl-CoA [39,43], or patatin, the first haloarchaeal enzyme identified to serve the in vivo mobilization (depolymerization) of native *Hfx. mediterranei* granules [43].

4. Process Parameters for Optimized *Hfx. mediterranei*-Mediated PHA Production

In 2017, Ferre-Güell and Winterburn investigated the impact of the nitrogen sources NH_4^+ and NO_3^- on biomass formation and PHA production by *Hfx. mediterranei*. In a N-rich medium based on glucose, yeast extract and 156 g/L NaCl, CDM and PHA content in CDM reached 10.7 g/L and 4.6 wt.%, respectively when using NH_4^+; with NO_3^-, only 5.6 g/L CDM, but 9.3 wt.% PHA in CDM were produced. Astonishingly, the type of N-source affected the composition of PHBHV copolyesters. While 16.9 mol.% 3HV were present in PHBHV when using NH_4^+, the 3HV fraction dropped to 12.5 mol.% when using NO_3^-. With NH_4^+, a low C/N-ratio of 42/1 resulted in reduced formation of active biomass, but increased the PHBHV share in CDM to 6.6 wt.%; the effect of the C/N-ratio was less pronounced when using NO_3^-. Remarkably, a lower C/N-ratio increased the 3HV share in PHBHV, which suggests an effect of the C/N-ratio on the activity of the propionyl-CoA generating pathways. Interestingly, no 3HV was detected in PHA before the polyester concentration reached 0.45 g/L. Hence, detailed understanding of the effect of type and concentration of different N-sources can contribute to the enhanced production of PHBHV copolyesters by haloarchaea with pre-defined composition and characteristics [44]. As a follow-up study, Melanie and colleagues examined the impact of the initial phosphate concentration on PHA production by *Hfx. mediterranei*; 0.95 g/L PHBHV (15.6% in CDM) with unexpectedly high 3HV content (22.36 mol.%) was produced after seven days of cultivation with 156 g/L NaCl and 0.5 g/L KH_2PO_4 as P-source in 500 mL shaking flasks. Lower initial KH_2PO_4 concentrations (0.25 or 0.00375 g/L) caused lower PHA productivity and lower 3HV fractions in PHBHV copolyesters. Thermal characterization of the products revealed data typical for PHA produced by *Hfx.*

mediterranei [45]. Moreover, Cui et al. studied the temperature effect on biomass and PHA formation by *Hfx. mediterranei*. This was done by developing, calibrating, and validating a mathematical model for growth and PHA production kinetics at 15, 20, 25, and 35 °C. The kinetic coefficients implemented into the model were obtained by experimental results from cultivations carried out in stirred and aerated flasks in a medium of similar composition to molasses wastewater. As a result, it was shown that the cultivation temperature considerably effects PHA production by *Hfx. mediterranei*; at 15 °C, the volumetric PHA productivity amounted to only 390 mg/(L·h), while 620 mg/(L·h) were obtained in cultivations at 35 °C. An Arrhenius equation plot was drawn that revealed the maximum specific growth rate (μ_{max}; 0.009 1/h at 15 °C, 0.033 1/h at 35 °C), maximum specific substrate uptake rate (q_{Smax}; 0.018 g/(g·h) at 15°C, 0.037 g/(g·h) at 35 °C), and specific decay rate (k_d; 0.0048 1/h at 15 °C, 0.0089 1/h at 35 °C) were higher at increased temperature. The calculated activation energy for biomass growth, decay, and substrate uptake were 58.31 kJ/mol, 22.38 kJ/mol, and 25.59 kJ/mol, respectively. For all investigated temperatures, the developed model was of high predictive power. Even with sufficient supply with nitrogen source, the elevated temperature level of 35 °C significantly improved PHA productivity; this temperature was therefore recommended as the optimal cultivation temperature to be used for this strain. Furthermore, the 3HV fraction in PHBHV turned out to be independent from temperature; under all temperature conditions, the PHBHV copolyesters contained 16.7 mol.% 3HV. These data are of high importance, because information in older literature for the temperature optimum of *Hfx. mediterranei* were inconsistent and even contradictory [46]. Unfortunately, the authors did not study higher temperatures, which were reported in the basic publications for *Hfx. mediterranei* as optimum for growth and PHA-biosynthesis (50 °C and 45 °C, respectively) [32].

5. Use of Different Feedstocks for PHA Biosynthesis by *Hfx. mediterranei*

A range of diverse inexpensive carbon-rich food and agro-industrial waste and side products have already been tested as feedstocks for PHA production by *Hfx. mediterranei*. This encompasses surplus whey from cheese and the dairy industry [41,47,48], crude glycerol phase (CGP) as the main by-product of biodiesel production [33], extruded corn starch [49], extruded rice bran [50], stillage from bioethanol manufacturing [51,52], molasses wastewater form sugar industry [48], olive mills wastewater [53], vinasse from molasses-based ethanol production [54], or macroalgae (seaweeds) hydrolyzed by advanced techniques [55].

5.1. Hfx. mediterranei on Hydrolyzed Whey Permeate

Indeed, *Hfx. mediterranei* is considered one of the most auspicious organisms for whey-based PHA production on a large scale because of its high production rates, high robustness and the stability of fermentation batches, as well as convenient product recovery via hypo-osmotic cell disintegration [41]. The strain grows excellently on both acidic or enzymatically hydrolyzed whey permeate (equimolar mixtures of glucose and galactose as main carbon source; permeate generally separated from whey retentate via ultrafiltration) but does not utilize intact lactose [56]. In bioreactor cultivation setups, high maximum specific growth rates (μ_{max}) of 0.11 1/h were reported when using hydrolyzed whey permeate as a substrate; this is substantially higher than specific growth values reported for other haloarchaea. Maximum values for specific PHA production (q_P) amounted to 0.08 g/(g·h). When optimizing the cultivation conditions (inoculum preparation, medium composition), these values were even enhanced to μ_{max} = 0.09 g/(L·h) and q_P = 0.15 g/(g·h), respectively. Biomass concentration and PHA fractions in biomass reached 16.8 g/L and 73 wt.%, respectively [57]. Importantly, when cultivated on hydrolyzed whey lactose (equimolar mixture of glucose and galactose), *Hfx. mediterranei* has a clear preference for glucose, which results in the accumulation of galactose in the fermentation medium, thus drastically increasing the biochemical oxygen demand of spent fermentation broth and causing the loss of a substantial part of the substrate. Suggested solutions to solve this ecological and economic shortcoming involve separating galactose from spent fermentation broth for further use, e.g., as a sweetener or nutritional and pharmaceutical additive; yet, this approach is economically rather doubtful. In this context, Pais et al. discovered that the activity of the strain's enzymes involved in galactose conversion can be increased by adaptation of the trace elements supply to the cultivation medium; this way, a more complete substrate conversion was achieved. Remarkably, the PHBHV copolyesters obtained in this study had a very low 3HV fraction of only 2 mol.% 3HV [48].

Attempts to further increase the material features of whey-based PHA produced by *Hfx. mediterranei*, the precursor substrates valeric acid and γ-butyrolactone (GBL) were supplied in bioreactor cultivations with 200 g/L NaCl and hydrolyzed whey permeate as main carbon source. These precursors were added in order to achieve higher 3HV fraction in PHA, and to introduce 4HB as an additional PHA building block. This way, a poly(3HB-*co*-21.8%-3HV-*co*-5.1%-4HB) terpolyester with encouraging material properties (low melting points, high molecular mass and low crystallinity) was produced, and suggested for further use in the medical field [47].

It is obvious that the highly saline waste streams of *Hfx. mediterranei* cultivations need appropriate handling, hence sustainable disposal or re-utilization in order to reduce process costs and to minimize the risk of environmental pollution. Importantly, disposing salt-rich materials after cell harvest and PHA recovery constitutes a real environmental threat, especially for large cultivation setups; the concentration of total dissolved solids (TDS) in disposed wastewater is limited with 2 g/L according to valid environmental norms. Therefore, the possibility of re-using saline cell debris, which remains after PHA recovery, as well as recycling the salt-rich spent fermentation broth by using it as mineral source in new cultivation setups was studied. Experiments with spent fermentation broth and saline cell debris were carried out; the results underlined the viability of recycling these waste streams. It was demonstrated that re-using spent fermentation broth for the preparation of new saline mineral medium drastically reduces the need for fresh salts. Furthermore, substituting up to 29% of yeast extract, typically a costly component used in *Hfx. mediterranei* cultivation media, for saline PHA-free cell debris gave growth rates similar to those obtained in the original cultivations [58].

Data from fed-batch cultivations on 200 L pilot scale, which are as yet the only results for large-scale PHA production using haloarchaea and inexpensive feedstock, were used for the cost assessment of PHBHV production on hydrolyzed whey by *Hfx. mediterranei*. This assessment encompassed the profits of solvent-free PHA recovery in distilled water, inexpensive acidic whey hydrolysis by mineral acids, abandoning any sterility provisions, copolyester production without the need for 3HV precursors, and the re-use of saline cell debris and spent fermentation broth in subsequent fermentation batches. A price of less than € 3 was estimated for the production of 1 kg PHA, which is significantly less than is typically reported PHA production prices of 5–10 €/kg. This pilot scale calculation delivered 7.2 g/L PHA and a volumetric productivity of 0.11 g/(L·h). The study also compared value creation for converting whey to, on the one hand, PHA, with, on the other hand, to whey powder, the currently most common application. In addition to cost assessment, a life cycle assessment (LCA) using the "sustainable process index" (SPI) as indicator for sustainability was carried out. A significant result of this process suggested that the estimated ecological footprint of whey-based PHA produced by *Hfx. mediterranei* is superior to fossil-based plastics if all process side streams are recycled [59]. This is in accordance with more recent considerations published by Narodoslawsky and colleagues, who concluded that the ecological footprint of "bioplastics" outperforms established plastics only when considering and optimizing the entire life cycle of the polymer [60].

5.2. Hfx. mediterranei on Crude Glycerol Phase from Biodiesel Industry

Beside hydrolyzed whey, strain *Hfx. mediterranei* accumulates PHA heteropolyesters (co- and terpolyesters) also when being fed with crude glycerol phase (CGP). CGP constitutes the main side-product of biodiesel production, which is steadily emerging in many global areas. In this context, CGP was used as feedstock for fed-batch bioreactor cultivations of *Hfx. mediterranei* in media containing 150 g/L NaCl; in these experiments, a volumetric PHBHV productivity of 0.12 g/(L·h) and a product fraction of 75 wt.% PHBHV (10 mol.% 3HV) in CDM were reached. Co-feeding the 4HB-precursor GBL together with the main substrate CGP, a PHA terpolyester containing 3HB (83 mol.%), 3HV (12 mol.%), and 4HB (5 mol.%) was synthesized [32]. Here, it should be added that the utilization of glycerol for PHA biosynthesis is not a typical feature for other haloarchaea; many of them use this substrate for the production of non-PHA biomass and maintenance energy, but not for PHA biosynthesis [61].

5.3. Hfx. mediterranei on Processed Starchy Materials

Extruded rice bran (ERB) and extruded cornstarch (ECS) were applied by Huang et al. as additional inexpensive substrates for PHA production by *Hfx. mediterranei*. For this purpose, 5 L scale bioreactor cultivations were performed in repeated fed-batch mode under pH-stat conditions with a medium containing 234 g/L NaCl and all other compounds required by the strain. Due to the insufficient utilization of non-processed ERB and ESC, these feedstocks were extruded before being supplied as substrate as ERB/ECS mixtures in a ratio of 1/8 (g/g). High values for CDM, PHA, and PHA content in biomass were reported: 140 g/L, 77.8 g/L, and 56 wt.%, respectively [51]. In a similar way, Chen and colleagues used cornstarch treated by an enzymatic (α-amylase) reactive extrusion process for PHA production by *Hfx. mediterranei*. The cultivation was performed in a 6 L bioreactor under pH-stat fed-batch cultivation conditions and a salinity of 234 g/L NaCl. Carbon and nitrogen concentration in the cultivation broth was kept constant by feeding a stream containing a 1/1.7 (g/g) mixture of extruded ECS (carbon source) and yeast extract (nitrogen source). After 70 h, the PHA concentration and PHA content in CDM reached 20 g/L and 51 wt.%, respectively. Similar to other PHA production setups carried out with *Hfx. mediterranei*, a PHBHV copolyester with 10.4 mol.% 3HV was produced by the strain without supply with 3HV precursor compounds. This process reached the as yet highest volumetric PHA productivity with *Hfx. mediterranei* with around 0.28 g/(L·h) [49].

5.4. Hfx. mediterranei on Waste Streams of Bioethanol Production

Vinasse constitutes a recalcitrant waste of ethanol production based on molasses. On shaking the flask scale, this waste product was studied as a potential substrate for PHA production by *Hfx. mediterranei*. Pre-treatment by adsorption on charcoal was carried out to remove inhibiting compounds from vinasse, mainly phenolic compounds. Using 25–50% (v/v) pre-treated vinasse delivered a maximum PHA content in biomass of 70 wt.%, a maximum PHA concentration of 19.7 g/L, a volumetric productivity for PHA of 0.21 g/(L·h), and a substrate conversion yield of 0.87 g/g. By this process, about 80% of the (bio)chemical oxygen demand of pre-treated vinasse were removed. Further, in these experiments PHA was recovered from biomass by cell lysis in a hypotonic medium and further purified by treatment with sodium hypochlorite and organic solvent. Again, the product was characterized as a PHBHV copolyester. Using 25% pre-treated vinasse, the 3HV fraction in PHBHV amounted to 12.4 mol.% and increased to 14.1 mol.% when using 50% pre-treated vinasse. The high medium salinity allowed the performance of the cultivation without prior sterilization of the bioreactor and medium, which contributes to production cost reduction. The authors underlined that the simple use of charcoal for vinasse detoxification was more economical than the above described processes using other waste materials as substrates; ultrafiltration to concentrate whey, or extrusion and enzymatic treatment of starchy materials contribute more to the production cost than charcoal pre-treatment. Analogous to the processes based on whey, which can be integrated into the production lines of dairies and cheese factories, where whey directly accrues as waste stream, the vinasse-based process can easily be integrated into distilleries, where it even contributes to the treatment of process wastewater [54].

As the main waste material stemming from ethanol manufacture based on rice, raw stillage was applied without pre-treatment by the same team of researchers as another inexpensive substrate for PHA biosynthesis by *Hfx. mediterranei*. These experiments strongly focused at closing material cycles within the process, and to reduce its environmental impact; for this purpose, medium salts from previous cultivation batches were directly recycled. 16.4 g/L PHA, about 70 wt.% PHA in biomass, a substrate conversion yield of 0.35 g/g, and a volumetric PHA productivity of 0.17 g/(L·h) were achieved in shaking flask experiments. In analogy to above described experiments performed using whey, a PHBHV copolyester with 15.3 mol.% 3HV was produced. A reduction of the (bio)chemical oxygen demand in feedstock stillage by about 85%, and a decrease of total dissolved solids (TDS) in spent fermentation broth to only 0.67 g/L were reached [50].

As follow up study, Bhattacharyya et al. performed a techno-economic assessment of *Hfx. mediterranei*-mediated PHA production on unsterile waste stillage from rice-based ethanol manufacturing. This process was performed in a plug-flow bioreactor made of plastic, which is normally used to study and optimize activated sludge processes. *Hfx. mediterranei* successfully utilized stillage, and produced 63 wt.% PHA in biomass, while PHA concentration, product yield, and volumetric productivity amounted to 13 g/L, 0.27 g/g, and 0.14 g/(L·h), respectively. A significant reduction of the (bio)chemical oxygen demand of stillage by 82% was reached. The accumulated PHA was identified as PHBHV copolyester with a molar 3HV fraction of 18%. An innovative desalination process of the supernatant of spent fermentation broth, consisting of two steps, was developed; this process involved stirring and heating the spent supernatant with decanoic acid. After cooling and settling, the mixture separated into three phases: salts precipitated and became available for subsequent fermentation batches, an organic phase of lower density (decanoic acid; to be applied for subsequent desalination cycles), and the heavier water phase. By this simple approach, it was possible to recover 99.3% of the medium salts and to re-use them in next PHA production batches. An assessment of cost for PHBHV produced via this process, which was suggested by the authors as the basis for the design of a pilot plant, estimated US $ 2.05 per kg of product; this calculation refers to a production plant with a production capacity of 1890 annual tons. It is important to note that desalination particularly contributed considerably to this low-cost estimate. Further, this techno-economic analysis holds promise for the realization of PHA production integrated in existing industrial production plants, in the case of rice-based stillage especially in emerging countries like India [52].

5.5. Hfx. mediterranei on Wastewater of Olive Oil Production

Using olive oil wastewater (OMW), a highly contaminated side stream of the olive processing industry, *Hfx. mediterranei* was cultivated in a one-stage cultivation process aiming at PHBHV production. In this process presented by Alsafadi and Al-Mashaqbeh, the inexpensive feedstock OMW was supplied to the culture without pre-treatment, which saved costly steps, e.g., for dephenolization. When the entire cultivation medium contained up to 25% OMW, the present phenolic compounds did not inhibit grow of the strain. The cultivation conditions were optimized to achieve maximum polymer yield and PHA fraction in biomass; this encompassed fine-tuning salinity, temperature, and oxygen supply. A salt concentration of 220 g/L NaCl and a temperature of 37 °C turned out to be best values for optimum PHA productivity. The accumulated biopolyester was recovered from biomass by hypotonic cell lysis assisted by SDS and vortexing, and further purified with sodium hypochlorite, thus using only minor amounts of organic solvents. The relative content of 3HV in the generated PHBHV copolyester amounted to 6%, which is significantly lower than 3HV fraction for *Hfx. mediterranei* PHBHV produced on other substrates as described above for whey, stillage, starchy materials, or vinasse. This process was suggested by the authors to enhance OMW valorization, and to reduce production cost of desired bioproducts. However, taking into account that the cultivations were performed only on a small shaking flask scale, upscaling to the bioreactor scale, to be carried out under controlled cultivation conditions, it will be necessary to assess industrial viability [53].

5.6. Hfx. mediterranei on Hydrolyzed Macroalgae

The green seaweed (microalga) *Ulva* sp., an organism typically producing unwanted algal blooms at coastal areas, was hydrolyzed by alkaline (4.8 mM KOH) and thermal (180 °C) batch treatment and used as substrate for shaking flask cultivations of *Hfx. mediterranei* at 42 °C and pH 7.2. PHA concentration, CDM, PHA content in biomass and 3HV fraction in PHA obtained when using 25% of Ulva hydrolysate reached 2.2 g/L, 3.8 g/L, 58 wt.%, and 0.08 mol/mol, respectively [55].

6. Microstructure of *Hfx. mediterranei* PHA Copolyesters

Hfx. mediterranei produces PHBHV copolyesters, which are not homogenous materials, but consist of diverse fractions of varying molecular mass and monomeric composition. PHBHV produced by *Hfx. mediterranei* using glucose and yeast extract was separated into two fractions with different 3HV contents by using a mixture of chloroform and acetone. The predominant fraction, which amounts to about 93% of the entire polymer had a 3HV fraction of 10.7 mol.% and a molecular mass of about 570 kDa, while the minor fraction contained considerably lower amounts of 3HV (12.3 mol.%) and had a significantly lower molecular mass of 78.2 kDa. This low-molecular mass fraction was soluble even in acetone, which typically is reported as an "anti-solvent" for short-chain-length PHA like PHB or PHBHV at temperatures below the boiling point of the solvent. T_m and T_g of both fractions were similar, and both had rather low Đᵢ values. By DSC characterization at heating rates below 20 °C/min, two overlapping melting peaks became visible in the DSC traces, with varying relative peak intensities when changing the heating rate; the authors supposed that this effect might originate from melt-and-recrystallization phenomena in PHA [62]. In another study, a low-molecular mass (209 kDa) fraction of a *Hfx. mediterranei* poly(3-hydroxybutyrate-*co*-3-hydroxyvalerate-*co*-4-hydroxybutyrate) terpolyester was extracted by acetone under reflux conditions in a Soxleth apparatus, while the major part of the product amounting to about 99%, which had a considerably higher molecular mass exceeding 1 MDa, was soluble in acetone only at a temperature exceeding acetone´s boiling point [63]. Both studies confirmed the presence of intracellular PHA blends in *Hfx. mediterranei*. More detailed insights into the microstructure of PHBHV produced by *Hfx. mediterranei* were disclosed by Han et al., who described the complex blocky structure of the biopolyester (*b*-PHA), which consists of alternating PHB and poly(3-hydroxyvalerate) (PHV) blocks, which are linked to blocks of randomly distributed PHBHV copolyesters. These researchers also demonstrated that *b*-PHA production by *Hfx. mediterranei* could be fine-tuned via the co-feeding of glucose and valerate. Because of this "blocky" structure and its high 3HV content, *Hfx. mediterranei* *b*-PHA displays exciting material features such as low degree of crystallinity and improved Young's modulus. Films of this polyester showed unique foveolar cluster-like surface morphology with high roughness. This enables a possible biomedical application of this *b*-PHA, as revealed by its better blood platelet adhesion and faster blood clotting behavior in comparison to randomly distributed PHBHV [64].

A fed-batch process using mixtures of butyric and valeric acid as substrate for *Hfx. mediterranei* was described by Ferre-Güell and Winterburn; this process was designed to synthesize PHBHV copolyesters with pre-defined composition in a reproducible way. Tween 80 added as emulsifier at a temperature of 37 °C improved the bioavailability of the substrates; the highest PHBV contents in biomass (59 wt.%) and volumetric productivity (10.2 mg/(L·h)) were reported for a butyric/valeric acid mix of 56/44. The biopolyester had a pre-defined 3HV fraction of 43 mol.%. This triggering of the PHBHV composition by adapting the composition of the substrate mix was realized both on shaking flask and bioreactor scale under different temperatures and emulsifier concentrations. Only insignificant variances in PHBHV product quality (molecular mass, thermo-mechanical properties) were observed for different production scales (bioreactor or shaking flask, respectively), which demonstrated the convenient scalability of this process [65]. A similar study aimed also at manufacturing *Hfx. mediterranei* PHBHV copolyesters of controlled composition and microstructure. Both *b*-PHBHV and PHBHV of random distribution were produced by supplying cultures with different fatty acids with an even (acetic, butyric, hexanoic, octatonic, and decanoic acid) or an odd number (propionic, valeric, heptanoic, nonanoic, undecanoic acid) of carbon atoms. Only those fatty acids with less than seven carbon atoms were accepted by the strain as a substrate for growth and PHA production. When feeding acetic acid, a PHBHV copolyester with about 10 mol.% 3HV was produced, which is in accordance to the 3HV fraction typically obtained when using glucose, glycerol, etc., while in the case of butyric acid, almost no 3HV was found in the copolyesters. Valeric acid used as sole carbon source resulted in an exceptionally high 3HV content of more than 90 mol.%. When using propionic acid, the 3HV content was lower than in the case of valeric because of the partial oxidative propionyl-CoA decarboxylation, which converts propionyl-CoA to CO_2 and acetyl-CoA, which acts as 3HB precursor. Applying different feeding strategies for butyric acid, valeric acid and mixtures of these acids, it was shown that sequential feeding creates b-PHBHV containing alternating PHB and PHV blocks, while random distribution of 3HB and 3HV occurs when co-feeding the substrates. Furthermore, higher 3HV fractions in randomly distributed PHBHV resulted in higher PHA chain mobility in the amorphous phase of the polyesters, which was evidenced by lower T_g values. In general, higher 3HV fractions in random PHBHV resulted in decreased polyester crystallinity, lower T_m, improved ductility, and higher elasticity, which consequently enhances processibility of the polymers [66].

7. Further Haloarchaeal Genera Encompassing PHA Producers

7.1. Haloarcula sp.

Beside the broadly described strain Hfx. mediterranei, other haloarchaea with more or less pronounced PHA production capacity were isolated from different saline environments. In this context, Altekar and Rajagopalan exposed the interrelation between PHA accumulation by the haloarchaea Hfx. mediterranei, Hfx. volcanii and Har. marismortui, and CO_2-fixation activity catalyzed by ribulose bisphosphate carboxylase (RuBisCo) present in the cell extracts of these strains [67]. Later, Nicolaus and colleagues isolated three previously unknown organisms from Tunesian marine salterns. These isolates grew well under extremely saline conditions (3.5 M NaCl, ~200 g/L). All three strains displayed parallel PHB production and EPS excretion when supplied with different carbon sources. By analyzing the strains´ lipid patterns, it was concluded that all of them belong to the genus Haloarcula (Har.). All of them grew on starch, and one (isolate "T5") grew expediently on the inexpensive substrate molasses. After growing about ten days on starch or glucose, this strain T5 accumulated 0.5 wt.% PHB in CDM, and 1 wt.% when growing on molasses. DNA-DNA hybridization tests and other biochemical studies identified strain T5 as a new Har. japonica ssp. [11]. Legat and colleagues later revealed by means of Nile Blue and Sudan Black staining and ^{1}H-NMR investigation of freeze-dried cells that also Har. hispanica strain DSM 4426[T] constitutes a producer of PHA, more precisely of PHBHV copolyesters [68].

Haloarcula sp. IRU1 was isolated from the hypersaline Iranian Urmia lake. In shaking flask cultivations, PHB production was optimized by using varying carbon, nitrogen, and phosphate source concentrations and by studying the effect of temperature in between 37 °C and 55 °C. Highest PHB contents in biomass of 63 wt.% were reported for 42 °C when using 2 g/L glucose, 0.2 g/L NH_4Cl, and 0.004 g/L KH_2PO_4 [69]. Later, glucose, fructose, sucrose, starch, acetate, and palmitic acid were tested as substrates for Har. sp. IRU1. Using glucose, CDM and PHB concentration substantially increased compared to all other carbon sources, while lowest CDM and PHB concentration were reported for acetate and palmitic acid [70]. The same organism, Har. sp. IRU1, was also cultivated on a medium containing petrochemical wastewater as carbon source; as determined by a Taguchi experimental design, highest PHB fractions in biomass (47 wt.%) were obtained with 2% wastewater, 0.8% tryptone, 0.001% KH_2PO_4 and a temperature of 47 °C [71]. In addition, Har. sp. IRU1 was also cultivated in minimal media containing crude oil as sole carbon source; axenic cultivations for five days with 2% crude oil, 0.4% yeast extract, and 0.016% NaH_2PO_4 at 47 °C were carried out. Unfortunately, only the highest PHB fraction in CDM (41 wt.%) was reported in this article, but no data for productivity or PHB concentration were reported. Still, Har. sp. IRU1 was proposed as promising organism for bioremediation of petrochemically polluted environments, combined with value-added PHB biosynthesis [72]. Finally, even textile wastewater was investigated as a substrate for this strain to thrive; in this study, PHA biosynthesis was not monitored [73].

Vinasse, a side-product of molasses-based ethanol manufacturing, comprises non-volatile phenolic compounds, which remain in the residue after distillative ethanol removal; these phenolic compounds are known inhibitors of microbial growth. In 2012, vinasse was studied by Pramanik and colleagues as a substrate to cultivate the extremely halophile *Har. marismortui* [74], the first haloarchaeon unambiguously shown to produce PHB by Kirk and Ginzburg already in the early 1970ies [10]. Using a highly saline (200 g/L NaCl) cultivation medium containing 10% raw vinasse, *Har. marismortui* accumulated 26 wt.% PHB in CDM and reached a volumetric PHB productivity of 0.015 g/(L·h) in shaking flask experiments. These values became considerably better after removing the phenolic compounds by well-established absorption on charcoal; in a medium consisting of 100% dephenolized vinasse, 30 wt.% PHB in CDM and a volumetric productivity of 0.02 g/(L·h) were reached [66]. However, it should be noticed that PHA biosynthesis is not observed in all *Har.* sp.; e.g., Oren and colleagues did not detect any PHA inclusions when investigating the red, square-shaped Egyptian brine-pool isolate *Har. quadrata* in details [75].

7.2. *Halogeometricum* sp.

In 2013, Salgaonkar et al. screened seven extremely halophilic Archaea isolated from brine and sediments of solar salterns in India. Defined saline cultivation media with 200 g/L NaCl turned out to be suitable for the growth of all seven microbial isolates; all of them also accumulated PHA. Based on phenotypic and genotypic tests, six strains out of them were grouped into the genus *Haloferax*, and named as strains TN4, TN5, TN6, TN7, TN10, and BBK2, while isolate TN9 was described as the new taxonomic species *Halogeometricum (Hgm.) borinquense*. This new organism performed most auspiciously among the seven isolates, and was investigated in more detail regarding its growth and PHA accumulation kinetics. It was revealed that highest PHA accumulation rates for strain *Hgm. borinquense* TN9 already take place during the exponential phase of growth, hence prior to depletion of growth-essential nutrients. This "growth-associated PHA-production" characteristic differs from most other reported PHA production strains, which typically display maximum PHA productivity not before nutrient deprivation. In biomass, 14 wt.% PHB homopolyester was produced after a cultivation period of five days [76]. Later, the same researchers isolated another haloarchaeon from the Marakkanam solar salterns in Tamil Nadu, India. This new organism was labeled *Hgm. borinquense* E3; the strain produced PHBHV copolyesters when growing in a highly saline medium on glucose as the sole carbon substrate. This copolyester production is similar to the findings discussed above for *Hfx. mediterranei*, but in contrast to the strain´s close relative *Hgm. borinquense* TN9, a strain which produced only PHB homopolyester on glucose. Shaking flask cultivation experiments lasting four days resulted in a high intracellular polymer content of 74 wt.% PHBHV (with 22 mol.% 3HV) in biomass [77]. Additionally, the same research team cultivated four wildtype haloarchaea on hydrolyzed sugarcane bagasse (hSCB), which constitutes an amply available by-product of sugar manufacturing mainly consisting of lignocelluloses. Among these organisms, *Hgm. borinquense* E3 exhibited the highest PHA productivity according to fluorescence measurements after Nile Red staining. The organisms were identified as *Haloferax volcanii* BBK2 (one of the strains isolated in [70]), *Haloarcula japonica* BS2, and *Halococcus salifodinae* BK6. As also described for *Hfx. mediterranei*, strain *Hgm. borinquense* E3 forms slightly pink colored colonies with a slimy appearance, which demonstrated pigment and EPS biosynthesis. In a medium containing 200 g/L NaCl at 37°C and 25% or 50% hSCB, *Hgm. borinquense* E3 was cultivated for six days in shaking flasks. The PHA fractions in biomass amounted to 50 wt.% (25% SCB) and 46 wt.% (50% hSBC), respectively, while specific production rates (q$_p$) were reported with 3.0 mg/(g·h) for 25% hSCB, and with 2.7 mg/(g·h) for 50% hSBC. A PHBHV copolyester with 13.3 mol.% 3HV was isolated from biomass [78]. Subsequent studies with *Hgm. borinquense* E3 resorted to starch-based waste materials used as substrates for PHA production. In this context, pure starch and acid-hydrolyzed cassava waste were used in parallel shaking flask cultivation experiments at a salinity of 200 g/L NaCl. After ten days, 4.6 g/L PHBHV (13.1 mol.% 3HV) were produced on pure starch, while the use of cassava waste delivered 1.5 g/L PHBHV (19.7 mol.% 3HV) [79]. It is noteworthy that, unfortunately, all cultivations with *Hgm. borinquense* were carried out on a shaking flask scale; scale up experiments under controlled conditions in bioreactors are still missing in the literature.

7.3. Halopiger sp.

A corrosion-resistant bioreactor consisting of polyether ether ketone (PEEK), tech glass and silicium nitrite ceramics was constructed by Hezayen and colleagues [80]. This new composite bioreactor was used for the cultivation of two new extremely halophilic isolates. One of them, "strain 56", today known as *Halopiger (Hpg.) aswanensis* DSM 13151, was studied for PHB production in a medium containing more than 200 g/L NaCl. The other strain, *Natrialba (Nab.)* sp., was used to synthesize poly(γ-glutamic acid) as target product in a medium of the same salinity. Both organisms were isolated from hypersaline samples taken from the soil of the Egyptian city Aswan. PHB production by "strain 56" (*Hpg. aswanensis*) on acetate and n-butyric acid as mixed substrate amounted to 4.6 g/L, and the PHB content in CDM to 53 wt.% after 12 days cultivation in batch-mode. It was determined that 40 °C was the optimal temperature to cultivate this strain. The isolated biopolyester had a M_w of 230,000 g/mol and a $Đ_i$ of about 1.4 [80]. In a follow-up study, Hezayen et al. reported for the first time on a PHA synthase of haloarchaea; here, the authors investigated crude extracts of "strain 56" in environments supporting PHA biosynthesis. A protocol for release of PHA granules by cell lysis in hypotonic medium and separation of granules by differential centrifugation was developed, and the granule-associated PHA synthase was studied and characterized [14]. Later, this strain, which forms Gram-negative, motile, pleomorphic pink rods, was biochemically and taxonomically categorized, and is nowadays known as *Hpg. aswanensis* DSM 13151. The organism was reported to produce large amounts of PHB; it also excretes an EPS, which causes high viscosity of the cultivation broth. High salinity of 220–250 g/L NaCl, pH-value 7.5 (range: 6–9.2) and a temperature of 40 °C (maximum accepted temperature: 55 °C) were determined as the optimum condition for this extreme halophilic species to thrive [81].

7.4. Halobiforma sp.

In the study published by Hezayen and colleagues [14], another red pigmented (carotenoid-rich) aerobic organism was isolated from hypersaline Egyptian soil in Aswan. When cultivated for eight days in shaking flasks on butyric acid, this "strain 135^T" accumulated up to 40 wt.% PHB in biomass; on complex substrates like casamino acids, peptone, or yeast extract, even 15 wt.% PHB in biomass were accumulated. This organism requires at least 130 g/L NaCl for biomass growth, and a temperature of 42 °C revealed best growth. The authors classified the new isolate as species *Halobiforma (Hbf.) haloterrestris* sp. nov. (DSM 13078^T) [82]. *Hbf. lacisalsi* sp. nov., a close microbial relative from the genus *Halobiforma*, was later isolated by Xu and associates from a salt lake in China. This organism was shown to grow optimally at 100 g/L NaCl; unfortunately, no tests were reported that refer to PHA biosynthesis [83].

7.5. Natrinema sp.

Danis et al. investigated five extremely halophilic archaeal isolates in order to identify new extremophilic strains with high capacity for PHA biosynthesis; the conversion of different inexpensive raw materials such as cornstarch, melon, apple, and tomato processing waste, sucrose, and whey. Among these materials, cornstarch appeared as the most encouraging substrate for PHA biosynthesis, while among the five isolated haloarchaea, strain 1KYS1 showed highest PHA production capacity. Via comparative 16S rRNA gene sequence analysis, it was revealed that strain 1KYS1 was closely related to the extremely halophilic genus *Natrinema (Nnm.)*, and, within this genus, to the strain *Nnm. pallidum* JCM 8980. When cultivated on starch as single carbon source and a salinity of 250 g/L NaCl, strain 1KYS1 accumulated 0.53 g PHA per g of its biomass. Transmission electron microscopy (TEM) revealed that the accumulated material, a PHBHV copolyester, forms large, uniform granules ("carbonosomes"), which, after cell lysis, is a considerable benefit for the convenient separation of PHA granules via floatation or centrifugation. In addition, this biopolyester was blended with low molar mass poly(ethylene glycol), which resulted in the preparation of a new type of biocompatible polymer film, which has been applied for drug release studies using the antibiotic Rifampicin [84]. In 2018, the haloarchaeon *Natrinema ajinwuensis* RM-G10 (synonym: *Natrinema altunense* strain RM-G10) was isolated from salt production pans in India. *Nnm. ajinwuensis* accumulated about 61 wt.% PHA in biomass and showed high volumetric PHA productivity of 0.21 g/(L·h) when cultivated for 72 h in repeated batch shaking flask cultivation setups on glucose. Using glycerol instead of glucose resulted in biomass formation, but not in PHA biosynthesis. The product based on glucose turned out to be a PHBHV copolyester with a 3HV fraction in PHBHV of 0.14 mol/mol, which is a value similar to those reported for other haloarchaeal strains (*vide supra*). When analyzed by DSC, the biopolyesters showed two separated melting endotherms (T_m 143 °C and 157.5 °C), T_g of −12.3 °C, an onset of decomposition temperature (T_d) of 284 °C, and a degree of crystallinity (X_c) of 35.45%. 200 g/L NaCl were reported as optimal salinity for both biomass growth and PHA production by this organism [61].

7.6. *Haloquadratum* sp.

Unusual organisms, originally isolated at the Egyptian Sinai Peninsula, were described in 1980 by Walsby, who was interested in the highly refractive gas vesicles produced by the microbes; this researcher described his isolates as "ultra-thin square bacteria" [85]. A quarter of a century later, Walsby reported that cells of this strain resemble "thin, square or rectangular sheets with sharp corners", and reported their dimensions being 2–5 µm wide but not even 0.2 µm thick. The outstanding low thickness of sheets makes them bulge slightly, with gas vesicles visible along their edges; he also noted that this organism thrives "at the edge of water activity". Importantly, Walsh also observed "poly-β-hydroxybutyrate granules in the corners" [86]. The organism was for a long time believed to be not culturable in monoseptic cultures, and its genome was deciphered not before 2006 by Bolhuis et al.; these researchers revealed that the strain´s genome encodes photoactive retinal proteins of the membrane and S-layer glycoproteins of the cell wall. In this study, the species name *Haloquadratum (Hqr.) walsbyi* was used [16]. Later, Burns et al. investigated two closely related novel square-shaped aerobic, extremely halophilic members of the haloarchaea, isolated from saltern crystallizers in Australia and Spain, and classified both of them as members of the new species *Hqr. walsbyi*. In this study, the authors described that growth of this occurs at pH 6.0–8.5, 25–45 °C and 14–360 g/L NaCl. The extremely halophilic cells lyse immediately in distilled water and a minimum of ~140 g/L salts is required for growth. Optimal growth occurs under neutral to alkaline conditions, above 180 g/L NaCl. By electron cryomicroscopy, PHA inclusions were reported by the authors, but not further studied or quantified [87]. Nile Blue A and Sudan Black staining of *Hqr. walsbyi* DSM 16790 grown in complex medium further substantiated PHA accumulation by this strain, which was confirmed by ^{1}H-NMR studies of fresh cells, evidencing the accumulation of PHB homopolyester, which, however, did not exceed 0.1% of CDM [69]. In 2011, the strain was grown aerobically with illumination on a medium containing 195 g/L NaCl and 0.5 g glycerol, 0.1 g yeast extract and 1 g sodium pyruvate as carbon sources; atomic force microscopy (AFM) was used for a detailed study of the cellular morphology. Importantly, these AFM studies showed corrugation of the cellular surface due to the presence of PHA granules, which were of almost uniform size within a single cell, and were packaged in tight bags. It was assumed that the primary function of these PHA granules was to reduce the cytosol volume, thus reducing the cellular energy demand for osmotic homeostasis; hence, they play a pivotal role for the strain to cope with the high salinity. In the supplementary material, the authors provided also impressing fluorescence microscope pictures of the cells with PHA granules visible as stained inclusions [88].

7.7. *Halococcus* sp.

A total of 20 haloarchaeal strains from strain collections were screened by Legat et al. via different PHA-staining techniques (Sudan Black B, Nile Blue A, and Nile Red). Both complex and defined cultivation media were used for the experiments. Further, PHA granules were visualized via TEM, while ^{1}H-NMR spectroscopy was applied to determine PHA composition. Beside strains known before as PHA producers like *Har. hispanica* DSM 4426[T] or *Hgr. walsbyi* DSM 16790, other organism like *Hbt.*

noricense DSM 9758[T], *Halococcus (Hcc.) dombrowskii* DSM 14522[T], *Hcc. hamelinensis* JCM 12892[T], *Hcc. morrhuae* DSM 1307[T], *Hcc. qingdaonensis* JCM 13587[T], *Hcc. saccharolyticus* DSM 5350[T], *Hcc. salifodinae* DSM 8989[T], *Hfx. volcanii* DSM 3757[T], *Halorubrum (Hrr.) chaoviator* DSM 19316[T], *Hrr. coriense* DSM 10284[T], *Natronococcus (Ncc.) occultus* DSM 3396[T], and *Natronobacterium (Nbt.) gregoryi* NCMB 2189[T] showed for the very first time accumulation of PHA when cultured in defined media with 200 g/L NaCl. By these tests, *Halococcus (Hcc.)* was identified as a new genus of PHA-producing microbes. While *Hcc. saccharolyticus* produced PHB homopolyester, all other strains produced PHBHV copolyesters without a supply of 3HV precursors. In this study, TEM pictures were produced for *Hcc. morrhue* and *Hcc. salifodine*, which showed the presence of at least one PHA carbonosome per cell, each about 0.05 to 0.3 μm in diameter [68].

7.8. *Halogranum* sp.

The haloarchaeon *Halogranum (Hgn.) amylolyticum* TNN58 was isolated in 2015 by Zhao and colleagues from marine solar salterns near Lianyungang in PR China. This organism was described to be a proficient producer of PHBHV copolyesters from simple, structurally unrelated substrates without being supplied with 3HV-related precursor compounds. Observed by TEM, a high number of PHA granules were visible inside the cells. High 3HV fractions in PHBHV exceeding 0.2 mol/mol are the up to now highest 3HV content in PHBHV reported for PHBHV copolyester production by wild-type organisms from unrelated carbon sources. Nitrogen limitation turned out to support PHBHV production by the strain *Hgn. amylolyticum* TNN58, though PHBHV accumulation occurred in an at least partially growth-associated way. Among the substrates acetate, benzoic acid, butyric acid, casamino acids, glucose, glycerol, lauric acid, and starch, the use of glucose allowed best biomass growth and highest PHA productivity. Fed-batch cultivations under controlled conditions in 7.5 L bioreactors were performed to investigate PHBHV production by *Hgn. amylolyticum* in more details. After 188 h of cultivation, CDM, PHBHV concentration, PHBHV fraction in biomass, and volumetric PHBHV productivity amounted to 29 g/L, 14 g/L, 48 wt.%, and 0.074 g/(L·h), respectively [89].

7.9. *Haloterrigena* sp.

The haloarchaeon *Haloterrigena (Htg.) hispanica* DSM 18328[T] was originally isolated as strain "FP1", and was the dominant organism thriving in a saltern crystallizer pond at Fuente dePiedra in the south of Spain. Romano and colleagues were the first who described this strain. The strain needs a minimum salinity of 150 g/L NaCl to grow optimally; growth occurs in a salinity range of 130–230 g/L NaCl, at pH-values between 6.5 and 8.5, and at temperatures between 37 °C and 60 °C. These authors also mentioned accumulation of "PHB" in this organism under nutritionally optimal cultivation conditions; corresponding to their publication, this postulation was made merely based on observation of PHA inclusions in the phase contrast microscope without characterizing the composition of the material at the level of monomers [90]. Later, *Htg. hispanica* DSM 18328[T] was cultivated by Di Donato and colleagues in a highly saline medium containing 200 g/L NaCl using carrot- or tomato waste, which accrues at enormous quantities in many countries like Italy, as sole carbon sources. This study confirmed that this thermophilic strain grows optimally at 50 °C; using this temperature, the organism was cultivated in batch bioreactor fermentation setups, which lasted five days, and also in dialysis fermentations, where bioreactors were equipped with a dialysis tube. Using a complex cultivation medium, the PHB homopolyester was produced at a quantity of 0.135 wt.% PHB in biomass; product composition was determined by ^{1}H-NMR analysis. When using carrot waste as substrate, 0.125 wt.% PHB in biomass were accumulated by *Htg. hispanica*, which is a quantity comparable to results obtained by cultivations on expensive media based on casamino acids and yeast extract. Astonishingly, ^{1}H-NMR analysis of this biopolyesters produced from carrot waste medium disclosed that the homopolyester poly(4-hydroxybutyrate), a highly flexible material with broad use in the surgical field, was produced instead of expected PHB [91].

7.10. Halorhabdus sp.

In 2000, the aerobic organism AX-2T was isolated by Wainø and colleagues from sediments of the Great Salt Lake in Utah, USA. This thermophilic strain grew optimally at extremely high NaCl concentrations of 270 g/L, which that time constituted the highest salinity optimum at all reported for any living species. Further, 50 °C and a neutral pH-value were determined as optimum growth parameters. Only a limited number of carbohydrates, namely glucose, fructose, and xylose, were accepted by the strain for biomass formation, while neither fatty acids nor complex substrates like peptone or yeast extract enabled microbial growth of this strain.

Cells of this isolate lyse instantly when exposed to distilled water, and were tested positively for PHA biosynthesis ("PHB is produced"); however, neither quantitative data for PHA production nor PHA composition were reported. Based on the outcomes of 16S rRNA analysis, the strain was classified as member of the Halobacteriaceae, but showed only limited similarity to other described species of this family.

The new taxon name *Halorhabdus (Hrd.) utahensis* was selected for this new strain, which is now deposited as DSM 12940^T [92]. *Hrd. tiamatea* is another representative of this genus. This extremely halophilic, non-pigmented archaeon was isolated in 2008 by Antunes and colleagues from a hypersaline, anoxic deep-sea brine-sediment interface of the Northern Red Sea, an unusual athalassohaline environment associated with tectonic activity. Also *Hrd. tiamatea* revealed optimal growth at a salinity of 270 g/L NaCl, neutral pH-value, a temperature of 45 °C, and the conversion of starch for biomass formation.

In contrast to *Hrd. utahensis*, which can be cultivated under both aerobic and anaerobic conditions, *Hrd. tiamatea* shows a clear preference for microaerophilic environments. However, the fact that *Hrd. tiamatea* accumulates PHA was revealed merely as a short annotation in this publication ("Poly-β-hydroxybutyrate is produced"); for this PHA-production test, based only on observation in phase-contrast microscope, cells were cultivated in HBM minimal medium supplemented with 0.005% (*w/v*) N H $_4$Cl and 0.5 to 1% (*w/v*) maltose [93]. Later, the same group of authors deciphered the complete genome of *Hrd. tiamatea*, which disclosed significant differences to the genome of *Hrd. utahensis*; for example, it was revealed that *Hrd. tiamatea* possesses putative trehalose and lactate dehydrogenase synthase genes, which are not found in *Hrd. utahensis* [94]. Finally, the facultative anaerobic strain *Hrd. rudnickae*, isolated from a borehole sample taken at a Polish salt mine, is the third member of the genus *Halorhabdus*, which was described to accumulate PHA.

This organism forms non-motile Gram-negative cocci, is red pigmented, und thrives best at a salinity of 200 g/L NaCl, a temperature of 40 °C and a neutral pH-range. Again, PHA inclusions in cells were spotted by TEM, but neither quantified nor characterized ("Poly-β-hydroxybutyrate is produced") [95]. Hence, production of PHA biopolyesters by the *Halorhabdus* genus is still awaiting its kinetic analysis and characterization at the monomeric level.

8. Conclusions

As detailed in the present review, a two-digit number of different haloarchaeal species were already described as potential PHA producers. However, most of these studies were restricted to modest cultivation scales, often merely reporting on microscopic observation and fluorescence staining of PHA granules. To the best of the author´s knowledge gained from the open literature and discussions with other scientists active in this field, there are not more than four haloarchaeal species (*Hfx. mediterranei*, *Hpg. aswanesnsis*, *Hgn. amylolyticum*, and *Htg. hispanica*), for which PHA accumulation was studied in cultivations performed under controlled conditions in bioreactors.

However, such bioreactor cultivation setups are the conditio sine qua non to get reliable kinetic data, and reasonable amounts of product for in-depth characterization. Most of all, sufficient amounts of product are needed for processing it to marketable prototype specimens; such processing is completely lacking in the case of haloarchael PHA. Moreover, techno-economic assessment of PHA production by haloarchaea, based on solid experimental data and holistic consideration of the entire production cycle, is only available for *Hfx. mediterranei*, for which economic and life cycle considerations were carried out based on the surplus substrates whey and waste stillage. Nevertheless, exactly these early techno-economic assessments already indicate the high potential of the extremely halophilic members of the Archaea domain for bio-economic biopolyester production of the future.

Taking advantage of the broad substrate spectrum, the formation of PHA heteropolyesters of tunable composition and microstructure in dependence on the cultivation strategy, the accessibility of haloarchaea towards inexpensive and convenient product recovery from biomass, the recyclability of process side-streams (spent fermentation broth and cell debris), the detailed knowledge about the complete genome of an increasing number of haloarchaea, and the expedient robustness of such cultivation batches sets haloarchaea at the forefront of efforts dedicated to finally make PHA economically competitive polymers with plastic-like properties, which also match the end-consumer´s expectations.

What is needed now is upscaling those processes at a promising lab-scale, and to tap the wealth of haloarchaea reported to produce PHA merely on a qualitative basis, or which have not yet been studied for PHA biosynthesis. In addition, one should be aware of parallel R&D activities with halophilic eubacteria as PHA production strains; here, especially the seminal works with *Halomonas bluephagenensis* TD01 should be mentioned, a proficient PHA production strain which can be cultivated in open bioreactor facilities [96], and which is well studied in terms of genetic manipulation [97,98]. Other examples for promising halophilic eubacteria as PHA producers encompass *Halomonas halophila* [99], or *Halomonas campaniensis* [100]. However, these organisms thrive best under salinities of about 60–70 g/L, which is drastically below the optimum salinity of haloarchaea, which makes the long-term stability of fermentation batches with *Halomonas* sp. uncertain compared with their "competitors" from the realm of haloarchaea.

To summarize the aforementioned, Tables 1 and 2 provide an overview of the PHA production processes by the individual haloarchaea discussed in the review, indicating the productivities, type of biopolyester produced, and studied production scale. While Table 1 collects the setups on smaller scale, Table 2 refers to the rather scarce number of setups carried out under controlled conditions in laboratory and pilot scale bioreactors.

Table 1. PHA production by haloarchaea on shaking flask and stirred flask scale-collected data from literature.

Species	Strain Isolation	Salinity in Medium, Substrates, T	Product	Production Scale/Productivity	Ref.
Hfx. mediterranei	Salt pond at the coast near Alicante, Spain	150 g/L NaCl; Molasses wastewater; T = 15, 20, 25, and 35 °C	PHBHV (16.7 mol.% 3HV)	2.5 L aerated and stirred flasks 0.62 (g/L·h), $q_{Pmax.}$ = 0.037 1/h (35 °C)	[46]
,,		200 g/L NaCl; T = 37 °C; 25–50% pre-treated vinasse	PHBHV (12.4 mol.% 3HV using 25% vinasse) (14.1 mol.% 3HV using 50% vinasse)	Shaking flask scale; 19.7 g/L PHA, 70 wt.% PHA in CDM, 0.21 g/(L·h)	[54]
,,		200 g/L NaCl; T = 37 °C; Rice-based stillage	PHBHV (15.3 mol.% 3HV)	Shaking flask scale; 16.4 g/L PHA, 70 wt.% PHA in CDM, 0.17 g/(L·h)	[51]
,,		190 g/L total salts; 144 g/L NaCl; Alkaline hydrolyzed *Ulva* sp. (macroalgae) as substrate; T = 42 °C	PHBHV (3 mol.% 3HV)	Shaking flask scale; batch cultivation; 2.2 g/L, 58% PHA in CDM, 0.035 g/(L·h)	[55]
,,		220 g/L NaCl; Dephenolized and native olive mill waste water (OMW); T = 37 °C	PHBHV (6.5 mol.% 3HV)	Shaking flask scale, batch cultivation; 43 wt.% PHA in CDM (concentration and productivity data inconsistent in publication)	[53]
,,		156 g/L NaCl; Glucose; nitrate or ammonia as N-source; T = 37 °C	PHBHV (12.5 mol.% 3HV using nitrate) (16.9 mol.% 3HV using ammonia)	Shaking flask scale, batch cultivation; 0.63 g/L, 4.6% PHA in CDM, 0.035 g/(L·h) with ammonia (C/N = 8) 0.80 g/L, 9.3% PHA in CDM, 0.035 g/(L·h) with ammonia (C/N =8)	[44]
,,		156 g/L NaCl; T = 37 °C; Glucose; varying phosphate concentrations	PHBHV (22.4 mol.% 3HV)	500 mL shaking flasks, batch 0.95 g/L PHA, 15.6 % PHA in CDM; 0.007 g/(L·h) with optimum phosphate concentration 0.5 g/L KH_2PO_4	[45]
,,		156 g/L NaCl; Different even- or add-numbered fatty acids; T = 37 °C	PHBHV (random or *b*-PHBHV) (<10 mol.% 3HV using even-numbered acids) (>87 mol.% 3HV using odd-numbered acids)	Shaking flask scale; batch and fed-batch 0.4–1.5 g/L PHA, 10.3–27.1 wt.% PHA in CDM, 0.003–0.010 g/(L·h) (fed-batch, dependent on C-source)	[66]
Hfx. volcanii	Dead Sea	200 g/L NaCl; T = 37 °C; Glucose	?	Shaking flask scale; Below detection limit	[68]

Table 1. *Cont.*

Species	Strain Isolation	Salinity in Medium, Substrates, T	Product	Production Scale/Productivity	Ref.
ʺ	ʺ	250 g/L NaCl; T = 37 °C; Glucose + yeast extract	"PHB"	Shaking flask scale; 7 wt.% PHA in CDM	[29]
(strain BBK2)	Solar salterns of Ribandar in Goa, India	200 g/L NaCl; T = 37 °C; Sugarcane bagasse hydrolysate	? (not identified)	Shaking flask scale; Not quantified	[78]
Hfx. gibbonsii	Salt pond at the coast near Alicante, Spain		"PHB"	Shaking flask scale; 1.2 wt.% PHA in CDM	[29]
Har. marismortui	Dead Sea	200 g/L NaCl; Raw and charcoal-pretreated vinasse from bioethanol production	PHB	Shaking flak scale; PHA content in CDM between 23 wt.% (10% non-detoxified vinasse) and 30 wt.% (100% charcoal-detoxified vinasse); 0.015 (non-detoxified) and 0.02 (detoxified) g/(L·h) PHB (2.8 and 4.5 g/L PHB, respectively)	[74]
Har. hispanica	Salt pond at the coast near Alicante, Spain	250 g/L NaCl; T = 37 °C; Glucose + yeast extract	"PHB"	Shaking flask scale; PHA content: 2.6 wt.% PHA in CDM	[29]
ʺ	ʺ	200 g/L NaCl; T = 37 °C; Glucose	PHBHV	0.09 wt.% PHA in CDM	[68]
Har. sp. IRU1	Hypersaline Urmia lake, Iran	250 g/L NaCl; 42 °C (other T tested); Glucose (other substrates tested)	PHB	Shaking flask scale; 66 wt.% PHB in CDM	[69]
ʺ		250 g/L NaCl; 42 °C (other T tested); Glucose (other substrates tested)	PHB	Shaking flask scale; 62 (glucose), 57 (starch), 56 (sucrose), 55 (fructose), 40 (acetate), 39 (palmitic acid) wt.% PHB in CDM; Max. PHA concentration and productivity: 0.98 g/L, 0.016 g/(L·h) (glucose)	[70]
ʺ		250 g/L NaCl; 47 °C (other T tested); Petrochemical wastewater, tryptone	PHB	Shaking flask scale; Max. 46.6 wt.% PHB in CDM (2% petrochemical wastewater, yeast extract, 47 °C)	[71]
ʺ		250 g/L NaCl; 47 °C (other T tested); Crude oil, yeast extract (other N-sources tested)	PHB	Shaking flask scale; Max. 41.3 wt.% PHB in CDM (2% crude oil, yeast extract, 47 °C)	[72]

Table 1. *Cont.*

Species	Strain Isolation	Salinity in Medium, Substrates, T	Product	Production Scale/Productivity	Ref.
Har. Japonica (strain BS2)	Solar salterns of Ribandar in Goa, India	200 g/L NaCl; T = 37 °C; Sugarcane bagasse hydrolysate	? (not identified)	Shaking flask scale; Not quantified	[78]
Hgm. borinquense (strain TN9)	Solar salterns of Marakkanam in Tamil Nadu, India	200 g/L NaCl; T = 37 °C; Glucose	PHB	Shaking flask scale; PHA content in CDM 14 wt.%; ca. 3 mg/(L·h) PHA	[76]
Hgm. Borinquense (strain E3)	Solar salterns of Marakkanam in Tamil Nadu, India	200 g/L NaCl; T = 37 °C; Glucose	PHBHV (21.5 mol.% 3HV)	Shaking flask scale; PHA content in CDM 74 wt.%; 0.21 g/(L·h) PHA	[77]
''	''	200 g/L NaCl; T = 37 °C; 25% and 50% hydrolyzed sugarcane bagasse	PHBHV (13.3 mol.% 3HV)	Shaking flak scale; PHA content in CDM between 45 and 50 wt.%; 0.0113 g/(L·h) PHBHV on 25%	[78]
''	''	200 g/L NaCl; T = 37 °C; Starch and carbon-rich fibrous waste (cassava bagasse)	PHBHV (13.1% 3HV with starch, 19.7% 3HV with cassava waste)	Shaking flask cultivations in batch mode; Starch: 4.6 g/L PHA, 0.02 g/(L·h), 74.2% PHA in CDM, Cassava bagasse: 1.52 g/L, 0.006 g/(L·h), 44.7% PHA in CDM	[79]
Hbt. noricense	Bore core of an Austrian Permian salt deposit	200 g/L NaCl; T = 37 °C; Glucose	PHBHV	Shaking flask scale; 0.11 wt.% PHA in CDM	[68]
Hcc. dombrowskii	Dry rock salt from Austrian alpine salt mine	Complex saline medium; T = 37 °C	PHBHV	Shaking flask scale; 0.16 wt.% PHA in CDM	[68]
Hcc. hamelinensis	Stromatolites from the Hamelin pool in the Australian Shark Bay	Complex saline medium; T = 37 °C	PHBHV	Shaking flask scale; Not quantified	[68]
Hcc. morrhuae	Dead Sea	Complex saline medium; T = 37 °C	PHBHV	Shaking flask scale; Not quantified	[68]
Hcc. qingdaonensis	Crude sea-salt sample collected near Qingdao, PR China	Complex saline medium; T = 37 °C	PHBHV	Shaking flask scale; Not quantified	[69]
Hcc. saccharolyticus	Salt; Cadiz, Spain	Complex saline medium; T = 37 °C	PHB	Shaking flask scale; 1.2 wt.% PHA in CDM	[68]
Hcc. salifodinae	Austrian alpine rock salt	Complex saline medium; T = 37 °C	PHBHV	Shaking flask scale; 0.06 wt.% PHA in CDM	[68]

Handbook of Polyhydroxyalkanoates (Bioengineering Essentials)

Table 1. *Cont.*

Species	Strain Isolation	Salinity in Medium, Substrates, T	Product	Production Scale/Productivity	Ref.
'' (strain BK6)	Solar salterns of Ribandar in Goa, India	200 g/L NaCl; T = 37 °C Sugarcane bagasse hydrolysate	n.d.	Shaking flask scale; Below detection limit	[78]
Hrr. chaviator	Sea salt in Baja California, Mexico, Western Australia and Greece	200 g/L NaCl; T = 37 °C Glucose	PHBHV	Shaking flask scale; Not quantified	[68]
Hrr. coriense	Dead Sea	200 g/L NaCl; T = 37 °C Glucose	PHBHV	Shaking flask scale; Not quantified	[68]
Hbf. Haloterrestris ("strain 135(T)")	Samples collected from surface of hypersaline soil collected in Aswan, Egypt	220 g/L NaCl; T = 42 °C (other T tested) Acetate + butyric acid or complex media	PHB	Shaking flask scale; 40 wt.% PHB in CDM on butyric acid, 15 wt.% PHB in CDM on complex medium	[83]
Nnm. ajinwuensis (=altunense)	Indian salt production pans	200 g/L NaCl (other salinities tested); T = 37 °C Glucose	PHBHV (13.9 mol.% 3HV)	Repeated batch cultivations in shaking flaks PHA content in CDM 61 wt.%; ca. 15 g/L PHA; 0.21 g/(L·h) PHA	[61]
Nnm. Palladium (strain JCM 8980, =isolate 1KYS1)	Kayacik saltern, Turkey	250 g/L NaCl; Starch	PHBHV (25 mol.% 3HV)	Shaking flak cultivations; PHA content in CDM 53 wt.%; 0.3 mg/(L·h) PHA	[84]
Nbt. gregoryi	Soda slat lake liquors from the East African Magadi soda lake	200 g/L NaCl; T = 37 °C alkaliphile; Carbohydrates	PHB	Shaking flask scale; 0.62 wt.% PHB	[68]
Ncc. occultus	Magadi Lake, Kenia	200 g/L NaCl; T = 37 °C alkaliphile; Glucose	PHBHV	Shaking flask scale; 3.1 wt.% PHB	[68]
Hrd. utahensis	Sediments of the Great Salt Lake in Utah	270 g/L NaCl (maximum described salinity optimum for living beings!); T = 50 °C Limited number of carbohydrates	Not specified ("PHB is produced")	Shaking flask scale; No quantitative data	[93]
Hrd. tiamatea	Hypersaline, anoxic deep-sea brine-sediment interface of the Red Sea	270 g/L NaCl (maximum described salinity optimum for living beings!); T = 45 °C Starch	Not specified ("PHB is produced")	Shaking flask scale; No quantitative data	[94]
Hrd. rudnickae	Borehole at Polish salt mine	200 g/L NaCl; T = 40 °C	Not specified ("PHB is produced")	Shaking flask scale; No quantitative data	[96]
Hqr. walsbyi	Sinai peninsula and saltern crystallizers in Australia and Spain	140–360 g/L NaCl for growth (optimum: >180 g/L); T = 25–45 °C	PHB	Shaking flaks scale; <1 wt.% PHA in CDM	[68,88, 89]

Table 2. PHA production by haloarchaea on bioreactor scale—collected data from literature.

Species	Strain Isolation	Salinity in Medium, Substrates, T	Product	Production Scale/Productivity	Ref.
Hfx. mediterranei	Salt pond at the coast near Alicante, Spain	250 g/L marine salts; Starch (20 g/L); Glucose (10 g/L); T = 38 °C (other T tested)	PHBHV (in publication: "PHB")	Stable (monoseptic) continuous cultivation over 3 months in 1.5 L bioreactor; 6.5 g/L PHA on starch 3.5 g/L on glucose	[32]
"	"	150 g/L NaCl; T = 37 °C; Glucose plus yeast extract	PHBHV (10 mol.% 3HV)	10 L bioreactor; fed-batch feeding; 0.21 g/(L·h), 13 g/L PHA, 0.7 g PHA in CDM	[37]
"	"	200 g/L NaCl; T = 37 °C; Hydrolyzed whey permeate; Hydrolyzed whey permeate plus GBL	PHBHV (6 mol.% 3HV); P(3HB-co-3HV-co-4HB) (21.8 mol.% 3HV, 5.1 mol.% 4HB)	42 L bioreactor fed-batch process; 0.09 g/(L·h), 12.2 g/L PHBHV 0.14 g/(L·h), 14.7 g/L poly(3HB-co-3HV-co-4HB)	[47]
"	"	150 g/L NaCl; T = 37 °C; Hydrolyzed whey permeate	PHBHV (10 mol.% 3HV)	200 L fed-batch **pilot process** (300 L bioreactor); **techno-economic assessment** 7.2 g/L PHA, 66 wt.% PHA in CDM, 0.11 g/(L·h)	[58,59]
"	"	200 g/L NaCl; T = 37 °C; Hydrolyzed whey permeate, spent fermentation broth and saline cell debris from previous whey-based processes	PHBHV (10 mol.% 3HV)	10 L bioreactor batch process 0.04 g/(L·h), 2.28 g/L PHA	[58]
"	"	156 g/L NaCl; T = 37 °C; Hydrolyzed whey permeate, elevated trace element concentration	PHBHV (<2 mol.% 3HV)	2 L bioreactor batch process 8 g/L PHBHV, 0.17 g/(L·h), 53 wt.% PHA in CDM	[48]
"	"	150 g/L NaCl; T = 37 °C; CGP; CGP plus GBL	PHBHV (10 mol.% 3HV); P(3HB-co-3HV-co-4HB) (11 mol.% 3HV, 5 mol.% 4HB)	42 L/10 L bioreactor fed-batch process; 0.12 g/(L·h), 16.2 g/L PHA 0.10 g/(L·h), 11.1 g/L PHA	[33]
"	"	200–230 g/L NaCl; T = 37 °C; Native cornstarch treated via enzymatic reactive extrusion	PHBHV (10.4 mol.% 3HV)	6 L bioreactor pH-stat fed-batch process; 0.28 g/(L·h), 0.508 g PHA in CDM; 20 g/L PHA	[49]

Table 2. *Cont.*

Species	Strain Isolation	Salinity in Medium, Substrates, T	Product	Production Scale/Productivity	Ref.
,,	,,	234 g/L NaCl; T = 37°C Mixtures of extruded rice bran plus extruded cornstarch	PHBHV (about 11 mol.% 3HV)	5 L bioreactor; pH-stat feeding strategy; 77.8 g/L PHA	[50]
,,	,,	200 g/L NaCl; Rice-based stillage T: n.r,	PHBHV (17.9 mol.% 3HV)	Unsterile 50 L plug-flow PMMA bioreactor; **techno-economic assessment** 13 g/L PHA, 63 wt.% PHA in CDM, 0.14 g/(L·h)	[52]
,,	,,	156 g/L NaCl; Mixes of butyric & valeric acid; Tween80 T = 37°C	PHBHV (43 mol.% 3HV at butyric/valeric acid = 56/44)	Fed-batch bioreactor cultivation 4.01 g/L PHA, 59 wt.% PHA in CDM; 0.01 g/(L·h)	[65]
,, (EPS-negative mutant; strain "ES1")	,,	140 g/L total salts (110 g/L NaCl) Glucose and valerate T = 37°C	b-PHBHV (up to 50 mol.% 3HV at end of fermentation)	7 L fed-batch bioreactor cultivation Results only reported for shaking flask experiments: max. ca. 5 g/L PHA, 50 wt.% PHA in CDM; 0.17 g/(L·h)	[64]
Hgr. amylolyticum	Tainan marine solar saltern near Lianyungang, PR China	200 g/L NaCl; T = 37°C Glucose	PHBHV (>20 mol.% 3HV)	7.5 L bioreactor; fed-batch feeding strategy; 0.074 g/(L·h), 14 g/L PHBHV, 48 wt.% PHA in CDM	[89]
Hpg. Aswanensis ("strain 56")	Samples collected from surface of hypersaline soil collected in Aswan, Egypt	250 g/L NaCl; T = 40°C Sodium acetate and butyric acid	PHB	Corrosion-resistant 8 L composite bioreactor; batch feeding; 0.0045 g/(L·h), 53 wt.% PHB in CDM, 4.6 g/L PHB, 0.018 g/(L·h)	[80]
Htg. hispanica	Saltern crystallizer pond at Fuente de Piedra saline lake, Malaga, Spain	200 g/L NaCl; T = 37°C Complex medium Carrot waste	PHB (complex medium) P(3HB-*co*-3HV-*co*-4HB) (carrot waste)	Bioreactor; batch setups and bioreactor equipped with ultrafiltration unit 0.135 wt.% PHA in CDM (complex medium); 0.125 wt.% PHA in CDM (carrot waste)	[91]

References

1. Lemoigne, M. Produits de Deshydration et de Polymerisation de L'acide β = Oxybutyrique. *Bull. Soc. Chim. Biol.* **1926**, *8*, 770–782.

2. Jendrossek, D. Polyhydroxyalkanoate granules are complex subcellular organelles (carbonosomes). *J. Bacteriol.* **2009**, *191*, 3195–3202. [CrossRef]

3. Kourmentza, C.; Plácido, J.; Venetsaneas, N.; Burniol-Figols, A.; Varrone, C.; Gavala, H.N.; Reis, M.A. Recent advances and challenges towards sustainable polyhydroxyalkanoate (PHA) production. *Bioengineering* **2017**, *4*, 55. [CrossRef]

4. Koller, M.; Maršálek, L.; Miranda de Sousa Dias, M.; Braunegg, G. Producing microbial polyhydroxyalkanoate (PHA) biopolyesters in a sustainable manner. *New Biotechnol.* **2017**, *37*, 24–38. [CrossRef]

5. Koller, M. Chemical and biochemical engineering approaches in manufacturing Polyhydroxyalkanoate (PHA) biopolyesters of tailored structure with focus on the diversity of building blocks. *Chem. Biochem. Eng. Q.* **2018**, *32*, 413–438. [CrossRef]

6. Koller, M. A review on established and emerging fermentation schemes for microbial production of Polyhydroxyalkanoate (PHA) biopolyesters. *Fermentation* **2018**, *4*, 30. [CrossRef]

7. Oren, A.; Ventosa, A. International Committee on Systematics of Prokaryotes Subcommittee on the taxonomy of Halobacteriaceae and subcommittee on the taxonomy of Halomonadaceae. Minutes of the joint open meeting, 23 May 2016, San Juan, Puerto Rico. *Int. J. Syst. Evol. Microbiol.* **2016**, *66*, 4291. [CrossRef]

8. Gupta, R.S.; Naushad, S.; Baker, S. Phylogenomic analyses and molecular signatures for the class *Halobacteria* and its two major clades: A proposal for division of the class *Halobacteria* into an emended order *Halobacteriales* and two new orders, *Haloferacales* ord. nov. and *Natrialbales* ord. nov., containing the novel families *Haloferacaceae* fam. nov. and *Natrialbaceae* fam. nov. *Int. J. Syst. Evol. Microbiol.* **2015**, *65*, 1050–1069. [CrossRef] [PubMed]

9. Gupta, R.S.; Naushad, S.; Fabros, R.; Adeolu, M. A phylogenomic reappraisal of family-level divisions within the class *Halobacteria*: Proposal to divide the order *Halobacteriales* into the families *Halobacteriaceae*, *Haloarculaceae* fam. nov., and *Halococcaceae* fam. nov., and the order *Haloferacales* into the families, *Haloferacaceae* and *Halorubraceae* fam nov. *Antonie van Leeuwenhoek* **2016**, *109*, 565–587. [CrossRef] [PubMed]

10. Kirk, R.G.; Ginzburg, M. Ultrastructure of two species of halobacterium. *J. Ultrastruct. Res.* **1972**, *41*, 80–94. [CrossRef]

11. Nicolaus, B.; Lama, L.; Esposito, E.; Manca, M.C.; Improta, R.; Bellitti, M.R.; Duckworth, A.W.; Grant, W.D.; Gambacorta, A. *Haloarcula* spp able to biosynthesize exo-and endopolymers. *J. Ind. Microbiol. Biotechnol.* **1999**, *23*, 489–496. [CrossRef]

12. Quillaguamán, J.; Guzmán, H.; Van-Thuoc, D.; Hatti-Kaul, R. Synthesis and production of polyhydroxyalkanoates by halophiles: Current potential and future prospects. *Appl. Microbiol. Biotechnol.* **2010**, *85*, 1687–1696. [CrossRef] [PubMed]

13. Mezzolla, V.; D'Urso, O.; Poltronieri, P. Role of PhaC type I and Type II enzymes during PHA biosynthesis. *Polymers* **2018**, *10*, 910. [CrossRef]

14. Hezayen, F.F.; Steinbüchel, A.; Rehm, B.H. Biochemical and enzymological properties of the polyhydroxybutyrate synthase from the extremely halophilic archaeon strain 56. *Arch. Biochem. Biophys.* **2002**, *403*, 284–291. [CrossRef]

15. Baliga, N.S.; Bonneau, R.; Facciotti, M.T.; Pan, M.; Glusman, G.; Deutsch, E.W.; Shannon, P.; Chiu, Y.; Weng, R.S.; Gan, R.R.; et al. Genome sequence of *Haloarcula marismortui*: A halophilic archaeon from the Dead Sea. *Genome Res.* **2004**, *14*, 2221–2234. [CrossRef]

16. Bolhuis, H.; Palm, P.; Wende, A.; Falb, M.; Rampp, M.; Rodriguez-Valera, F.; Pfeiffer, F.; Oesterhelt, D. The genome of the square archaeon *Haloquadratum walsbyi*: Life at the limits of water activity. *BMC Genom.* **2006**, *7*, 169–180. [CrossRef] [PubMed]

17. Han, J.; Lu, Q.; Zhou, L.; Zhou, J.; Xiang, H. Molecular characterization of the *phaEC*$_{Hm}$ genes, required for biosynthesis of poly(3-hydroxybutyrate) in the extremely halophilic archaeon *Haloarcula marismortui*. *Appl. Environ. Microbiol.* **2007**, *73*, 6058–6065. [CrossRef] [PubMed]

18. Lu, Q.; Han, J.; Zhou, L.; Zhou, J.; Xiang, H. Genetic and biochemical characterization of the poly(3-hydroxybutyrate-co-3-hydroxyvalerate) synthase in *Haloferax mediterranei*. *J. Bacteriol.* **2008**, *190*,

        4173–4180. [CrossRef]
19.   Han, J.; Li, M.; Hou, J.; Wu, L.; Zhou, J.; Xiang, H. Comparison of four *phaC* genes from *Haloferax mediterranei* and their function in different PHBV copolymer biosyntheses in *Haloarcula hispanica*. *Saline Syst.* **2010**, *6*, 9. [CrossRef]
20.   Han, J.; Zhang, F.; Hou, J.; Liu, X.; Li, M.; Liu, H.; Cai, L.; Zhang, B.; Chen, Y.; Zhou, J.; et al. Complete genome sequence of the metabolically versatile halophilic archaeon *Haloferax mediterranei*, a poly(3-hydroxybutyrate-*co*-3-hydroxyvalerate) producer. *J. Bacteriol.* **2012**, *194*, 4463–4464. [CrossRef]
21.   Ding, J.Y.; Chiang, P.W.; Hong, M.J.; Dyall-Smith, M.; Tang, S.L. Complete genome sequence of the extremely halophilic archaeon *Haloarcula hispanica* strain N601. *Genome Announc.* **2014**, *2*, e00178-14. [CrossRef] [PubMed]
22.   Hou, J.; Feng, B.; Han, J.; Liu, H.; Zhao, D.; Zhou, J.; Xiang, H. Haloarchaeal-type beta-ketothiolases involved in Poly(3-hydroxybutyrate-co-3-hydroxyvalerate) synthesis in *Haloferax mediterranei*. *Appl. Environ. Microbiol.* **2013**, *79*, 5104–5111. [CrossRef]
23.   Han, J.; Hou, J.; Liu, H.; Cai, S.; Feng, B.; Zhou, J.; Xiang, H. Wide distribution among halophilic archaea of a novel polyhydroxyalkanoate synthase subtype with homology to bacterial type III synthases. *Appl. Environ. Microbiol.* **2010**, *76*, 7811–7819. [CrossRef] [PubMed]
24.   Han, J.; Lu, Q.; Zhou, L.; Liu, H.; Xiang, H. Identification of the polyhydroxyalkanoate (PHA)-specific acetoacetyl coenzyme A reductase among multiple FabG paralogs in *Haloarcula hispanica* and reconstruction of the PHA biosynthetic pathway in *Haloferax volcanii*. *Appl. Environ. Microbiol.* **2009**, *75*, 6168–6175. [CrossRef] [PubMed]
25.   Feng, B.; Cai, S.; Han, J.; Liu, H.; Zhou, J.; Xiang, H. Identification of the *phaB* genes and analysis of the PHBHV precursor supplying pathway in *Haloferax mediterranei*. *Acta Microbiol. Sin.* **2010**, *50*, 1305–1312.
26.   Xiang, H. PHBHV Biosynthesis by *Haloferax mediterranei*: From Genetics, Metabolism, and Engineering to Economical Production. In *Microbial Biopolyester Production, Performance and Processing. Microbiology, Feedsstocks, and Metabolism*; Koller, M., Ed.; Book Series: Recent Advances in Biotechnology; Bentham Science Publishers: Sharjah, United Arab Emirates, 2016; Volume 1, pp. 348–379.
27.   Rodríguez-Valera, F.; Ruiz-Berraquero, F.; Ramos-Cormenzana, A. Behaviour of mixed populations of halophilic bacteria in continuous cultures. *Can. J. Microbiol.* **1980**, *26*, 1259–1263. [CrossRef]
28.   Rodriguez-Valera, F.; Juez, G.; Kushner, D.J. *Halobacterium mediterranei* spec, nov., a new carbohydrate-utilizing extreme halophile. *Syst. Appl. Microbiol.* **1983**, *4*, 369–381. [CrossRef]
29.   Fernandez-Castillo, R.; Rodriguez-Valera, F.; Gonzalez-Ramos, J.; Ruiz-Berraquero, F. Accumulation of poly(β-hydroxybutyrate) by halobacteria. *Appl. Environ. Microbiol.* **1986**, *51*, 214–216.
30.   Torreblanca, M.; Rodriguez-Valera, F.; Juez, G.; Ventosa, A.; Kamekura, M.; Kates, M. Classification of non-alkaliphilic halobacteria based on numerical taxonomy and polar lipid composition, and description of *Haloarcula* gen. nov. and *Haloferax* gen. nov. *Syst. Appl. Microbiol.* **1986**, *8*, 89–99. [CrossRef]
31.   Rodriguez-Valera, F.; Lillo, J. Halobacteria as producers of polyhydroxyalkanoates. *FEMS Microbiol. Lett.* **1982**, *103*, 181–186. [CrossRef]
32.   Lillo, J.G.; Rodriguez-Valera, F. Effects of culture conditions on poly(β-hydroxybutyric acid) production by *Haloferax mediterranei*. *Appl. Environ. Microbiol.* **1990**, *56*, 2517–2521.
33.   Antón, J.; Meseguer, I.; Rodriguez-Valera, F. Production of an extracellular polysaccharide by *Haloferax mediterranei*. *Appl. Environ. Microbiol.* **1988**, *54*, 2381–2386.
34.   Hermann-Krauss, C.; Koller, M.; Muhr, A.; Fasl, H.; Stelzer, F.; Braunegg, G. Archaeal production of polyhydroxyalkanoate (PHA) co-and terpolyesters from biodiesel industry-derived by-products. *Archaea* **2013**, *2013*, 129268. [CrossRef]
35.   Oren, A. Microbial life at high salt concentrations: Phylogenetic and metabolic diversity. *Saline Syst.* **2008**, *4*, 1–13. [CrossRef]
36.   Parolis, H.; Parolis, L.A.; Boán, I.F.; Rodríguez-Valera, F.; Widmalm, G.; Manca, M.C.; Jansson, P.-E.; Sutherland, I.W. The structure of the exopolysaccharide produced by the halophilic Archaeon *Haloferax mediterranei* strain R4 (ATCC 33500). *Carbohydr. Res.* **1996**, *295*, 147–156. [CrossRef]
37.   Koller, M.; Chiellini, E.; Braunegg, G. Study on the production and re-use of poly(3-hydroxybutyrate-co-3-hydroxyvalerate) and extracellular polysaccharide by the archaeon *Haloferax mediterranei* strain DSM 1411.

Chem. Biochem. Eng. Q. **2015**, *29*, 87–98. [CrossRef]

38. Cui, Y.W.; Gong, X.Y.; Shi, Y.P.; Wang, Z.D. Salinity effect on production of PHA and EPS by *Haloferax mediterranei*. *RSC Adv.* **2017**, *7*, 53587–53595. [CrossRef]

39. Han, J.; Hou, J.; Zhang, F.; Ai, G.; Li, M.; Cai, S.; Liu, H.; Wang, L.; Wang, Z.; Zhang, S.; Cai, L.; Zhao, D.; Zhou, J.; Xiang, H. Multiple propionyl coenzyme A-supplying pathways for production of the bioplastic poly(3-hydroxybutyrate-*co*-3-hydroxyvalerate) in *Haloferax mediterranei*. *Appl. Environ. Microbiol.* **2013**, *79*, 2922–2931. [CrossRef]

40. Koller, M. Switching from petro-plastics to microbial polyhydroxyalkanoates (PHA): The biotechnological escape route of choice out of the plastic predicament? *The EuroBiotech J.* **2019**, *3*, 32–44. [CrossRef]

41. Koller, M.; Hesse, P.; Bona, R.; Kutschera, C.; Atlić, A.; Braunegg, G. Potential of various archae-and eubacterial strains as industrial polyhydroxyalkanoate producers from whey. *Macromol. Biosci.* **2007**, *7*, 218–226. [CrossRef]

42. Cai, S.; Cai, L.; Liu, H.; Liu, X.; Han, J.; Zhou, J.; Xiang, H. Identification of the haloarchaeal phasin (PhaP) that functions in polyhydroxyalkanoate accumulation and granule formation in *Haloferax mediterranei*. *Appl. Environ. Microbiol.* **2012**, *78*, 1946–1952. [CrossRef]

43. Liu, G.; Hou, J.; Cai, S.; Zhao, D.; Cai, L.; Han, J.; Zhou, J.; Xiang, H. A patatin-like protein associated with the polyhydroxyalkanoate (PHA) granules of *Haloferax mediterranei* acts as an efficient depolymerase in the degradation of native PHA. *Appl. Environ. Microbiol.* **2015**, *81*, 3029–3038. [CrossRef]

44. Ferre-Güell, A.; Winterburn, J. Production of the copolymer poly(3-hydroxybutyrate-*co*-3-hydroxyvalerate) with varied composition using different nitrogen sources with *Haloferax mediterranei*. *Extremophiles* **2017**, *21*, 1037–1047. [CrossRef]

45. Melanie, S.; Winterburn, J.B.; Devianto, H. Production of biopolymer Polyhydroxyalkanoates (PHA) by extreme halophilic marine Archaea *Haloferax mediterranei* in medium with varying phosphorus concentration. *J. Eng. Technol. Sci.* **2018**, *50*, 255–271. [CrossRef]

46. Cui, Y.W.; Zhang, H.Y.; Ji, S.Y.; Wang, Z.W. Kinetic analysis of the temperature effect on polyhydroxyalkanoate production by *Haloferax mediterranei* in synthetic molasses wastewater. *J. Polym. Environ.* **2017**, *25*, 277–285. [CrossRef]

47. Koller, M.; Hesse, P.; Bona, R.; Kutschera, C.; Atlić, A.; Braunegg, G. Biosynthesis of high quality polyhydroxyalkanoate co-and terpolyesters for potential medical application by the archaeon *Haloferax mediterranei*. *Macromol. Symp.* **2007**, *253*, 33–39. [CrossRef]

48. Pais, J.; Serafim, L.S.; Freitas, F.; Reis, M.A. Conversion of cheese whey into poly(3-hydroxybutyrate-*co*-3-hydroxyvalerate) by *Haloferax mediterranei*. *New Biotechnol.* **2016**, *33*, 224–230. [CrossRef]

49. Chen, C.W.; Don, T.M.; Yen, H.F. Enzymatic extruded starch as a carbon source for the production of poly(3-hydroxybutyrate-*co*-3-hydroxyvalerate) by *Haloferax mediterranei*. *Process Biochem.* **2006**, *41*, 2289–2296. [CrossRef]

50. Huang, T.Y.; Duan, K.J.; Huang, S.Y.; Chen, C.W. Production of polyhydroxyalkanoates from inexpensive extruded rice bran and starch by *Haloferax mediterranei*. *J. Ind. Microbiol. Biotechnol.* **2006**, *33*, 701–706. [CrossRef]

51. Bhattacharyya, A.; Saha, J.; Haldar, S.; Bhowmic, A.; Mukhopadhyay, U.K.; Mukherjee, J. Production of poly-3-(hydroxybutyrate-*co*-hydroxyvalerate) by *Haloferax mediterranei* using rice-based ethanol stillage with simultaneous recovery and re-use of medium salts. *Extremophiles* **2014**, *18*, 463–470. [CrossRef]

52. Bhattacharyya, A.; Jana, K.; Haldar, S.; Bhowmic, A.; Mukhopadhyay, U.K.; De, S.; Mukherjee, J. Integration of poly-3-(hydroxybutyrate-*co*-hydroxyvalerate) production by *Haloferax mediterranei* through utilization of stillage from rice-based ethanol manufacture in India and its techno-economic analysis. *World J. Microbiol. Biotechnol.* **2015**, *31*, 717–727. [CrossRef]

53. Alsafadi, D.; Al-Mashaqbeh, O. A one-stage cultivation process for the production of poly-3-(hydroxybutyrate-*co*-hydroxyvalerate) from olive mill wastewater by *Haloferax mediterranei*. *New Biotechnol.* **2017**, *34*, 47–53. [CrossRef]

54. Bhattacharyya, A.; Pramanik, A.; Maji, S.K.; Haldar, S.; Mukhopadhyay, U.K.; Mukherjee, J. Utilization of vinasse for production of poly-3-(hydroxybutyrate-*co*-hydroxyvalerate) by *Haloferax mediterranei*. *AMB Express* **2012**, *2*, 34. [CrossRef]

55. Ghosh, S.; Gnaim, R.; Greiserman, S.; Fadeev, L.; Gozin, M.; Golberg, A. Macroalgal biomass subcritical hydrolysates for the production of polyhydroxyalkanoate (PHA) by *Haloferax mediterranei*. *Bioresour. Technol.* **2019**, *271*, 166–173. [CrossRef]

56. Koller, M.; Puppi, D.; Chiellini, F.; Braunegg, G. Comparing chemical and enzymatic Hydrolysis of whey lactose to generate feedstocks for haloarchaeal poly(3-hydroxybutyrate-*co*-3-hydroxyvalerate) biosynthesis. *Int. J. Pharm. Sci. Res.* **2016**, *3*, 112. [CrossRef]

57. Koller, M.; Atlić, A.; Gonzalez-Garcia, Y.; Kutschera, C.; Braunegg, G. Polyhydroxyalkanoate (PHA) biosynthesis from whey lactose. *Macromol. Symp.* **2008**, *27*, 287–292. [CrossRef]

58. Koller, M. Recycling of waste streams of the biotechnological poly(hydroxyalkanoate) production by *Haloferax mediterranei* on whey. *Int. J. Polym. Sci.* **2015**, *2015*, 370164. [CrossRef]

59. Koller, M.; Sandholzer, D.; Salerno, A.; Braunegg, G.; Narodoslawsky, M. Biopolymer from industrial residues: Life cycle assessment of poly(hydroxyalkanoates) from whey. *Resour. Conserv. Recycl.* **2013**, *73*, 64–71. [CrossRef]

60. Narodoslawsky, M.; Shazad, K.; Kollmann, R.; Schnitzer, H. LCA of PHA production–Identifying the ecological potential of bio-plastic. *Chem. Biochem. Eng. Q.* **2015**, *29*, 299–305. [CrossRef]

61. Mahansaria, R.; Dhara, A.; Saha, A.; Haldar, S.; Mukherjee, J. Production enhancement and characterization of the polyhydroxyalkanoate produced by *Natrinema ajinwuensis* (as synonym)≡ *Natrinema altunense* strain RM-G10. *Int. J. Biol. Macromol.* **2018**, *107*, 1480–1490. [CrossRef]

62. Don, T.M.; Chen, C.W.; Chan, T.H. Preparation and characterization of poly(hydroxyalkanoate) from the fermentation of *Haloferax mediterranei*. *J. Biomater. Sci. Polym. E* **2016**, *17*, 1425–1438. [CrossRef]

63. Koller, M.; Bona, R.; Chiellini, E.; Braunegg, G. Extraction of short-chain-length poly-[(*R*)-hydroxyalkanoates] (*scl*-PHA) by the "anti-solvent" acetone under elevated temperature and pressure. *Biotechnol. Lett.* **2013**, *35*, 1023–1028. [CrossRef]

64. Han, J.; Wu, L.P.; Hou, J.; Zhao, D.; Xiang, H. Biosynthesis, characterization, and hemostasis potential of tailor-made poly(3-hydroxybutyrate-*co*-3-hydroxyvalerate) produced by *Haloferax mediterranei*. *Biomacromolecules* **2015**, *16*, 578–588. [CrossRef]

65. Ferre-Güell, A.; Winterburn, J. Increased production of polyhydroxyalkanoates with controllable composition and consistent material properties by fed-batch fermentation. *Biochem. Eng. J.* **2019**, *141*, 35–42. [CrossRef]

66. Ferre-Güell, A.; Winterburn, J. Biosynthesis and characterization of Polyhydroxyalkanoates with controlled composition and microstructure. *Biomacromolecules* **2018**, *19*, 996–1005. [CrossRef] [PubMed]

67. Altekar, W.; Rajagopalan, R. Ribulose bisphosphate carboxylase activity in halophilic Archaebacteria. *Arch. Microbiol.* **1990**, *153*, 169–174. [CrossRef]

68. Legat, A.; Gruber, C.; Zangger, K.; Wanner, G.; Stan-Lotter, H. Identification of polyhydroxyalkanoates in *Halococcus* and other haloarchaeal species. *Appl. Microbiol. Biotechnol.* **2010**, *87*, 1119–1127. [CrossRef] [PubMed]

69. Taran, M.; Amirkhani, H. Strategies of poly (3-hydroxybutyrate) synthesis by *Haloarcula* sp. IRU1 utilizing glucose as carbon source: Optimization of culture conditions by Taguchi methodology. *Int. J. Biol. Macromol.* **2010**, *47*, 632–634. [CrossRef]

70. Taran, M. Synthesis of poly(3-hydroxybutyrate) from different carbon sources by *Haloarcula* sp. IRU1. *Polym.-Plast. Technol.* **2011**, *50*, 530–532. [CrossRef]

71. Taran, M. Utilization of petrochemical wastewater for the production of poly(3-hydroxybutyrate) by *Haloarcula* sp. IRU1. *J. Hazard. Mater.* **2011**, *188*, 26–28. [CrossRef]

72. Taran, M. Poly(3-hydroxybutyrate) production from crude oil by *Haloarcula* sp. IRU1: Optimization of culture conditions by Taguchi method. *Pet. Sci. Technol.* **2011**, *29*, 1264–1269. [CrossRef]

73. Taran, M.; Sharifi, M.; Bagheri, S. Utilization of textile wastewater as carbon source by newly isolated *Haloarcula* sp. IRU1: Optimization of conditions by Taguchi methodology. *Clean Technol. Environ.* **2011**, *13*, 535–538. [CrossRef]

74. Pramanik, A.; Mitra, A.; Arumugam, M.; Bhattacharyya, A.; Sadhukhan, S.; Ray, A.; Haldar, S.; Mukhopadhyay, U.K.; Mukherjee, J. Utilization of vinasse for the production of polyhydroxybutyrate by *Haloarcula marismortui*. *Folia Microbiol.* **2012**, *57*, 71–79. [CrossRef]

75. Oren, A.; Ventosa, A.; Gutiérrez, M.C.; Kamekura, M. *Haloarcula quadrata* sp. nov., a square, motile archaeon isolated from a brine pool in Sinai (Egypt). *Int. J. Syst. Evol. Microbiol.* **1999**, *49*, 1149–1155. [CrossRef] [PubMed]

76. Salgaonkar, B.B.; Mani, K.; Bragança, J.M. Accumulation of polyhydroxyalkanoates by halophilic archaea isolated from traditional solar salterns of India. *Extremophiles* **2013**, *17*, 787–795. [CrossRef]

77. Salgaonkar, B.B.; Bragança, J.M. Biosynthesis of poly(3-hydroxybutyrate-*co*-3-hydroxyvalerate) by *Halogeometricum borinquense* strain E3. *Int. J. Biol. Macromol.* **2015**, *78*, 339–346. [CrossRef]

78. Salgaonkar, B.B.; Bragança, J.M. Utilization of sugarcane bagasse by *Halogeometricum borinquense* strain E3 for biosynthesis of poly(3-hydroxybutyrate-*co*-3-hydroxyvalerate). *Bioengineering* **2017**, *4*, 50. [CrossRef] [PubMed]

79. Salgaonkar, B.B.; Mani, K.; Bragança, J.M. Sustainable bioconversion of cassava waste to Poly(3-hydroxybutyrate-*co*-3-hydroxyvalerate) by *Halogeometricum borinquense* strain E3. *J. Polym. Environ.* **2019**, *27*, 299–308. [CrossRef]

80. Hezayen, F.F.; Rehm, B.H.A.; Eberhardt, R.; Steinbüchel, A. Polymer production by two newly isolated extremely halophilic archaea: Application of a novel corrosion-resistant bioreactor. *Appl. Microbiol. Biotechnol.* **2000**, *54*, 319–325. [CrossRef] [PubMed]

81. Hezayen, F.F.; Gutiérrez, M.C.; Steinbüchel, A.; Tindall, B.J.; Rehm, B.H.A. *Halopiger aswanensis* sp. nov., a polymer-producing and extremely halophilic archaeon isolated from hypersaline soil. *Int. J. Syst. Evol. Microbiol.* **2010**, *60*, 633–637. [CrossRef]

82. Hezayen, F.F.; Tindall, B.J.; Steinbüchel, A.; Rehm, B.H.A. Characterization of a novel halophilic archaeon, *Halobiforma haloterrestris* gen. nov., sp. nov., and transfer of *Natronobacterium nitratireducens* to *Halobiforma nitratireducens* comb. nov. *Int. J. Syst. Evol. Microbiol.* **2002**, *52*, 2271–2280. [CrossRef]

83. Xu, X.W.; Wu, M.; Zhou, P.J.; Liu, S.J. *Halobiforma lacisalsi* sp. nov., isolated from a salt lake in China. *Int. J. Syst. Evol. Microbiol.* **2005**, *55*, 1949–1952. [CrossRef]

84. Danis, O.; Ogan, A.; Tatlican, P.; Attar, A.; Cakmakci, E.; Mertoglu, B.; Birbir, M. Preparation of poly(3-hydroxybutyrate-*co*-hydroxyvalerate) films from halophilic archaea and their potential use in drug delivery. *Extremophiles* **2015**, *19*, 515–524. [CrossRef]

85. Walsby, A.E. A square bacterium. *Nature* **1980**, *283*, 69. [CrossRef]

86. Walsby, A.E. Archaea with square cells. *Trends Microbiol.* **2005**, *13*, 193–195. [CrossRef] [PubMed]

87. Burns, D.G.; Janssen, P.H.; Itoh, T.; Kamekura, M.; Li, Z.; Jensen, G.; Rodríguez-Valera, F.; Bolhuis, H.; Dyall-Smith, M.L. *Haloquadratum walsbyi* gen. nov., sp. nov., the square haloarchaeon of Walsby, isolated from saltern crystallizers in Australia and Spain. *Int. J. Syst. Evol. Microbiol.* **2007**, *57*, 387–392. [CrossRef]

88. Saponetti, M.S.; Bobba, F.; Salerno, G.; Scarfato, A.; Corcelli, A.; Cucolo, A. Morphological and structural aspects of the extremely halophilic archaeon *Haloquadratum walsbyi*. *PLoS ONE* **2011**, *6*, e18653. [CrossRef]

89. Zhao, Y.X.; Rao, Z.M.; Xue, Y.F.; Gong, P.; Ji, Y.Z.; Ma, Y.H. Poly(3-hydroxybutyrate-*co*-3-hydroxyvalerate) production by Haloarchaeon *Halogranum amylolyticum*. *Appl. Microbiol. Biotechnol.* **2015**, *99*, 7639–7649. [CrossRef] [PubMed]

90. Romano, I.; Poli, A.; Finore, I.; Huertas, F.J.; Gambacorta, A.; Pelliccione, S.; Nicolaus, G.; Lama, L.; Nicolaus, B. *Haloterrigena hispanica* sp. nov., an extremely halophilic archaeon from Fuente de Piedra, southern Spain. *Int. J. Syst. Evol. Microbiol.* **2007**, *57*, 1499–1503. [CrossRef] [PubMed]

91. Di Donato, P.; Fiorentino, G.; Anzelmo, G.; Tommonaro, G.; Nicolaus, B.; Poli, A. Re-use of vegetable wastes as cheap substrates for extremophile biomass production. *Waste Biomass Valoriz.* **2011**, *2*, 103–111. [CrossRef]

92. Wainø, M.; Tindall, B.J.; Ingvorsen, K. *Halorhabdus utahensis* gen. nov., sp. nov., an aerobic, extremely halophilic member of the Archaea from Great Salt Lake, Utah. *Int. J. Syst. Evol. Microbiol.* **2000**, *50*, 183–190. [CrossRef] [PubMed]

93. Antunes, A.; Taborda, M.; Huber, R.; Moissl, C.; Nobre, M.F.; da Costa, M.S. *Halorhabdus tiamatea* sp. nov., a non-pigmented, extremely halophilic archaeon from a deep-sea, hypersaline anoxic basin of the Red Sea, and emended description of the genus *Halorhabdus*. *Int. J. Syst. Evol. Microbiol.* **2008**, *58*, 215–220. [CrossRef]

94. Antunes, A.; Alam, I.; Bajic, V.B.; Stingl, U. Genome sequence of *Halorhabdus tiamatea*, the first archaeon isolated from a deep-sea anoxic brine lake. *J. Bacteriol.* **2011**, *193*, 4553–4554. [CrossRef]

95. Albuquerque, L.; Kowalewicz-Kulbat, M.; Drzewiecka, D.; Stączek, P.; d'Auria, G.; Rosselló-Móra, R.; da Costa, M.S. *Halorhabdus rudnickae* sp. nov., a halophilic archaeon isolated from a salt mine borehole in Poland. *Syst. Appl. Microbiol.* **2016**, *39*, 100–105. [CrossRef] [PubMed]

96. Ye, J.; Huang, W.; Wang, D.; Chen, F.; Yin, J.; Li, T.; Zhang, H.; Chen, G.Q. Pilot Scale-up of Poly(3-hydroxybutyrate-*co*-4-hydroxybutyrate) Production by *Halomonas bluephagenesis* via Cell Growth Adapted Optimization Process. *Biotechnol. J.* **2018**, *13*, 1800074. [CrossRef] [PubMed]

97. Chen, X.; Yin, J.; Ye, J.; Zhang, H.; Che, X.; Ma, Y.; Li, M.; Wu, L.-P.; Chen, G.Q. Engineering *Halomonas bluephagenesis* TD01 for non-sterile production of poly(3-hydroxybutyrate-*co*-4-hydroxybutyrate). *Bioresour. Technol.* **2017**, *244*, 534–541. [CrossRef] [PubMed]

98. Ye, J.; Hu, D.; Che, X.; Jiang, X.; Li, T.; Chen, J.; Zhang, H.M.; Chen, G.Q. Engineering of *Halomonas bluephagenesis* for low cost production of poly (3-hydroxybutyrate-*co*-4-hydroxybutyrate) from glucose. *Metab. Eng.* **2018**, *47*, 143–152. [CrossRef] [PubMed]

99. Kucera, D.; Pernicová, I.; Kovalcik, A.; Koller, M.; Mullerova, L.; Sedlacek, P.; Mravec, F.; Nebesarova, J.; Kalina, M.; Marova, I.; et al. Characterization of the promising poly(3-hydroxybutyrate) producing halophilic bacterium *Halomonas halophila*. *Bioresour. Technol.* **2018**, *256*, 552–556. [CrossRef]

100. Yue, H.; Ling, C.; Yang, T.; Chen, X.; Chen, Y.; Deng, H.; Wu, Q.; Chen, J.; Chen, G.Q. A seawater-based open and continuous process for polyhydroxyalkanoates production by recombinant *Halomonas campaniensis* LS21 grown in mixed substrates. *Biotechnol. Biofuels* **2014**, *7*, 108. [CrossRef]

13

Polyhydroxyalkanoate Production on Waste Water Treatment Plants: Process Scheme, Operating Conditions and Potential Analysis for German and European Municipal Waste Water Treatment Plants

Timo Pittmann [1],* and Heidrun Steinmetz [2]

[1]   TBF + Partner AG, Herrenberger Strasse 14, 71032 Boeblingen, Germany
[2]   Department of Resource Efficient Wastewater Technology, University of Kaiserslautern, Paul-Ehrlich-Str. 14, 67663 Kaiserslautern, Germany; heidrun.steinmetz@bauing.uni-kl.de
*   Correspondence: pit@tbf.ch

Academic Editor: Martin Koller

Abstract: This work describes the production of polyhydroxyalkanoates (PHA) as a side stream process on a municipal waste water treatment plant (WWTP) and a subsequent analysis of the production potential in Germany and the European Union (EU). Therefore, tests with different types of sludge from a WWTP were investigated regarding their volatile fatty acids (VFA) production-potential. Afterwards, primary sludge was used as substrate to test a series of operating conditions (temperature, pH, retention time (RT) and withdrawal (WD)) in order to find suitable settings for a high and stable VFA production. In a second step, various tests regarding a high PHA production and stable PHA composition to determine the influence of substrate concentration, temperature, pH and cycle time of an installed feast/famine-regime were conducted. Experiments with a semi-continuous reactor operation showed that a short RT of 4 days and a small WD of 25% at pH = 6 and around 30 °C is preferable for a high VFA production rate (PR) of 1913 mg$_{VFA}$/(L×d) and a stable VFA composition. A high PHA production up to 28.4% of cell dry weight (CDW) was reached at lower substrate concentration, 20 °C, neutral pH-value and a 24 h cycle time. A final step a potential analysis, based on the results and detailed data from German waste water treatment plants, showed that the theoretically possible production of biopolymers in Germany amounts to more than 19% of the 2016 worldwide biopolymer production. In addition, a profound estimation regarding the EU showed that in theory about 120% of the worldwide biopolymer production (in 2016) could be produced on European waste water treatment plants.

Keywords: biopolymer; municipal sewage plant; PHA; primary sludge; VFA

---

1. Introduction

Common plastic is derived from petrochemicals based on the limited natural resource petroleum. Besides the exploitation of natural resources, the use of plastic is responsible for major waste problems, as common plastic is non- or poor biodegradable [1].

Biopolymers present a possible alternative to common plastics. If they are fully biodegradable [2,3] their use not only allows the preservation of limited resources, but also suits the idea of sustainability.

The term "biopolymer" or "bioplastic" is not yet uniformly defined. Common definitions of the term "biopolymer" also include biodegradable plastics from fossil fuels and non-biodegradable plastics from renewable resources as seen in Figure 1. To eliminate the problems accompanied by polymer production from crude oil a more stringent definition is introduced by the authors:

"Biopolymers are made from renewable resources and/or biodegradable waste materials (e.g., waste water, sewage sludge, organic waste) and are fully biodegradable by naturally occurring microorganisms."

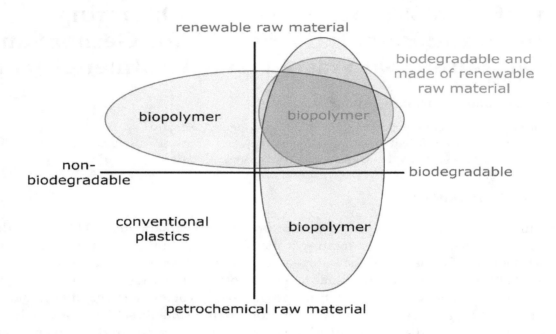

Figure 1. Definition for biopolymers, including the stringent definition on the upper right, modified after [4].

This definition ensures that polymers from fossil resources and non-biodegradable polymers, which cause at least one of the mentioned problems, are excluded and that the term biopolymer is just used for polymers, which allow the preservation of limited resources and also suit the idea of sustainability. This type of biopolymers is shown in the upper right of Figure 1.

Beside other polymers polyhydroxyalkanoates (PHA), which are biodegradable polyesters accumulated by bacteria under nutrient limited conditions [5] or under balanced growth, are a source for bioplastic production matching the above mentioned strict definition. More than 150 component parts of PHA have been identified so far [6]. The possibility for chemical modification of PHA provide a wide range of material properties and an even wider range of use [7,8]. However, so far the main raw material for the biopolymer production are starchy plants like maize [9], constituting the disadvantages of high land consumption, diminishing food resources as well as problems like leaching of nutrients, input of pesticide and soil erosion [10].

So far, municipal waste water treatment plants (WWTP) as alternative raw material and biomass source for the PHA production have not been widely investigated, although they offer the opportunity to compensate the disadvantages of the common PHA production using starchy plants.

PHA production in WWTP takes place in two steps, which composes the production of volatile fatty acids (VFA) in an anaerobic process and finally the PHA production in an aerobic process (see also Figure 2). In contrast to [11,12] the PHA production process described in this work is designed as a side stream process of a municipal WWTP and does not include the treatment of waste water. Therefore, the whole process can focus on polymer production only.

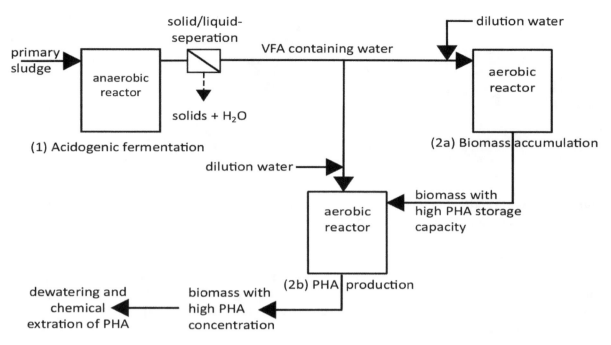

Figure 2. PHA production scheme.

The possibility to use ice-cream waste water as alternative source material for the VFA production was shown by [13] while [14] investigated the effect of pH, sludge-retention time (RT) and acetate concentration on the PHA production from municipal waste water. Diverse authors [15–18] stated that there is a general possibility to produce PHA from activated sludge.

In many of the research projects on PHA production, synthetic waste water was used to gain knowledge about one part of the PHA production or the production's operating conditions [19–25]. However, so far no research group has investigated the general possibility and all operating conditions of a biopolymer production using only material flows of a WWTP.

PHA production itself is based on a bacteria mixed culture selection from excess sludge via aerobic dynamic feeding. The installed feast/famine regime for enrichment of PHA producing bacteria is state of the art and tested by many authors [19,23,26,27]. The feast-phase is defined as a period of substrate availability and could be monitored via the reactors oxygen concentration. During the period of starvation (famine-phase) bacteria with the ability of polymer-storage gained a selection advantage as they are able to use the stored polymers as carbon and energy source.

The objective of this research project was to find the most suitable raw material and all operating conditions for the VFA and PHA production process using only material flows of a WWTP. At first the suitability of different raw materials of a municipal WWTP for VFA-production were investigated and afterwards the influence of operating conditions (temperature, pH, retention time (RT) and withdrawal (WD)) and reactor operation method. Another concern was, how the tested operating conditions or the diversity of the used material flows of a WWTP influence the VFA composition and the type of PHA produced. As there is a variation in the composition of the used material flows (different sludge) of a WWTP, it is of particular importance to observe their influence on VFA production and composition.

Then the possibility to produce PHA out of the VFA containing substrate was tested using a feast/famine regime (as shown in Figure 3). Subsequently, the influence of operating conditions (temperature, pH, cycle time (CT) and substrate concentration) on PHA production were investigated.

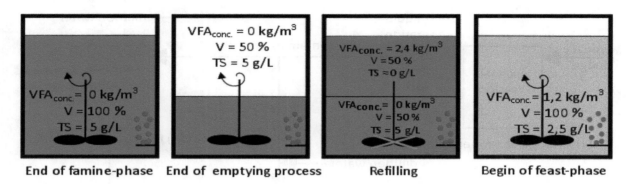

Figure 3. Emptying and refilling process during aerobic dynamic feeding (values shown are an example for investigated operation conditions), modified after [28].

The PHA potential on German WWTPs was calculated based on detailed data from operators of WWTPs [29] and the results of the PHA production experiments mentioned above. Finally, a profound estimation of the biopolymer potential of all WWTPs in the 28 member states of the European Union (EU) was made based on data provided by the EU [30] and the mentioned PHA production experiments.

2. Materials and Methods

2.1. VFA Production

2.1.1. Conception of the Experiments

For the VFA production ((1) Acidogenic fermentation in Figure 2) anaerobic reactors of different sizes (4 L, 15 L) were operated as batch reactors or as semi-continuous reactors. While the batch operation is defined as a one-time substrate filling at the beginning of the experiment with no withdrawal and refill during the test, a semi-continuously operation method allows to introduce and withdraw substrate to or from the reactor. Semi-continuously is defined as one-time substrate filling at the beginning of every cycle, e.g., daily within a test duration of one month.

All tests were conducted without sedimentation or biomass recirculation. Therefore, the hydraulic retention time equals the sludge age and both will be referred hereinafter as RT.

The raw material is the most important base for gaining high PHA production rates, so that the selection of suitable raw material has the number one priority. The improvement of the VFA production's operation conditions was examined afterwards with the most appropriate raw material found. A chronological test order was implemented as follows:

1. Selection of raw material
2. Investigation of the most suitable pH-level
3. Evaluation of a retention time (RT) range
4. Selection of a suitable combination of RT and withdrawal (WD)

2.1.2. Selection of Raw Material

For raw material selection continuously stirred batch reactors with a volume of 4 L were used. Four different types of sludge, namely primary sludge (average total solid $TS_a = 43$ g/L), excess sludge ($TS_a = 10$ g/L), a one to one mixture of primary-and digested sludge ($TS_a = 37.5$ g/L) and a one to one mixture of excess- and digested sludge ($TS_a = 21$ g/L) from a municipal WWTP were treated under anaerobic conditions. Thereby the digested sludge from the WWTP's digester was only used as inoculum for the anaerobic process in order to find out, if it could accelerate the process. All types of sludge were investigated under four different conditions: pH controlled at pH = 6, without pH-control and each at around 20 °C or around 30 °C reactor temperature. In summary 16 different tests were performed. The reactors were filled at the beginning of the experiments and samples of 50 mL were retrieved every day to determine the VFA concentration and composition. To achieve the selected temperature, the reactors were situated in temperature-controlled rooms. For pH-controlled tests, the pH-value was measured by a mobile pH meter (WTW pH340i) and adjusted with NaOH by hand twice a day. The test duration for all experiments was 18 days to 20 days. A sample of all tested types of sludge was taken before and after the tests to determine the chemical oxygen demand (COD), the total Kjeldahl nitrogen(TKN) and total P.

2.1.3. Evaluation of Operating Conditions

As the influence of fermentation temperature was already observed during the selection of the raw material, three additional operating conditions (pH, RT, WD) were investigated in continuous stirred tank reactors (CSTR) with a volume of 15 L. For all experiments primary sludge from a municipal WWTP was used as raw material and a sample was retrieved prior to the experiments to determine COD, TKN and total P. All reactors were placed in a temperature-controlled room at around 30 °C. pH was monitored at all times via a Metrom Profitrode pH probe and automatically adjusted with NaOH during all tests.

For the batch tests (pH, pre-RT) all reactors were filled with primary sludge at the beginning and samples of 100 mL were retrieved every day to determine the VFA concentration and composition. The batch test period was 18 days to 20 days long. Former studies show a high VFA production in a pH-range between 5 [11] and 9 [31] or 11 [32]. In consequence a range of pH-levels (pH = 6, 6.5, 7, 8, 10) were tested.

For the semi-continuously operated tests all reactors were filled with primary sludge at the beginning and operated under pH-controlled conditions at pH = 6. After a starting phase of 10 days, to accumulate VFAs, the semi-continuous operation phase began, for which a certain amount of the sludge in the reactor was exchanged. Samples of 100 mL were retrieved to analyse the VFA concentration and composition for about 40 days with a RT of 4 days, 6 days and 8 days, each with 25% and 50% WD. Additionally, a 75% WD was performed with a RT of 4 days. A RT of 2 days was also tested with a WD of 50%. RT and WD are related factors, e.g., a RT of 4 days was used when 25% of the sludge was exchanged every day, 50% every second day or 75% every third day.

2.1.4. Analytical Procedures

COD, TKN, total P and total solids (TS) were determined according to standard methods.

The concentration and composition of volatile fatty acids, namely formate (Fo), acetate (Ac), propionate (Pro) and butyrate (Bu), were detected by high performance liquid chromatography (HPLC). Therefore, the sample was acidified to pH = 2 and filtered through 0.45 μm membrane filter. Afterwards HPLC detection was performed using a HP1100 chromatographer equipped with an UV detector and a Varian Metacarb 87H column. Sulphuric acid (0.05 M) was used as eluent at a flow rate of 0.6 mL/min. The detection wavelength was 210 nm. Volatile fatty acid's concentration was calibrated using 4 nmol to 4000 nmol standards.

As the results of formic acid detection was below detection point for all except one test, formic acid is not shown in the VFA composition.

2.1.5. Conversion of Units

For a better comparability all results regarding VFA concentrations, COD, TKN and total P were converted into mg/L.

The concentration of VFA in terms of mg/L is defined as:

$$VFA = Ac + Pro + Bu \tag{1}$$

The degree of acidification (DA) was calculated according to [33] as shown in Equation (2). As VFA results are given in mg_{VFA}/L they have to be converted into COD units as shown in Equation (3)

$$DA = VFA/COD_S \text{ in } (mg_{COD}/L)/(mg_{COD}/L) \qquad (2)$$

with COD_S = COD of Substrate at the start.

$$COD_{VFA} = (\text{conc. } VFA_i/\text{molar mass } VFA_i) \text{ oxygen demand} \qquad (3)$$

with i = Ac, Pro, Bu.

For a better comparability regarding the different RT and WD, the average of the VFA concentration during the test period was calculated. In a second step, the average VFA concentration was used to calculate the average VFA production rate (PR_{VFA}). This step also eliminates the reactors size and can hence be considered as average VFA production rate per day and litre, which will be referred to production rate (PR) hereinafter (Equation (4)).

$$PR_{VFA} = \text{av. VFA conc.}/RT \text{ in } mg_{VFA}/(L \times day) \qquad (4)$$

This calculation helps to compare results from different reactor sizes and retention times.

2.2. PHA Production

2.2.1. Experimental Set-Up

The overall process describing the production of biopolymers from municipal waste water is displayed in Figure 2. For PHA production only "(2a): Biomass accumulation" and "(2b): PHA production" is considered. The first step, the volatile fatty acids (VFA) production, was already discussed in Chapter 2.1. The substrate produced in step 1 under anaerobic fermentation process at 30 °C, pH = 6, RT = 4 days and a withdrawal of 25% was frozen at −18 °C and defrosted about 24 h prior to its use as input material for phase 2 PHA production.

During all PHA production tests continuous stirred tank reactor (CSTR) with a volume of 15 L were used. All reactors were equipped with a pH- and oxygen-probe. If necessary, pH-value was adjusted via a dosing pump. pH levels were controlled by the pH probe and adjusted with NaOH or H_2SO_4. The reactor temperature was controlled via the pH-probe and manually adjusted using a heating bath (Haake DC30) and a heat exchanger, installed in the reactor. To maintain aerobic conditions an aerator was installed in the reactor. All reactors were operated in batch mode. The batch operation is defined as a one-time substrate filling at the beginning of the aerobic dynamic feeding-cycle with no withdrawal and refill during the cycle. There was no sedimentation or biomass recirculation in all tests. Therefore, the hydraulic retention time equals the sludge age and both will be referred hereinafter as RT. At the end of every cycle 7.5 L were withdrawn from reactor 2a and filled into Reactor 2b. Hence a RT of 2 days was implemented. Afterwards both reactors were filled with 7.5 L of substrate and fresh water to achieve the working volume of 15 L per reactor. Reactor 2b was emptied at the end of the feast-phase, and samples were taken to measure the PHA concentration and composition.

Both reactors were operated under similar conditions, just differing concerning nutrient availability. As the VFA enriched substrate showed nutrient limited conditions, with a Carbon:Nitrogen:Phosphorus (C:N:P)-ratio in a range of 100:2:0.5 to 100:3:0.8 [34] CH_4N_2O and KH_2PO_4 were added to Reactor 2a to create optimal conditions for the bacteria growth (C:N:P = 100:5:1) and selection process. Reactor 2b, however, was operated under the named nutrient limited conditions to reach a higher PHA concentration.

Samples to determine the PHA concentration were taken at the end of the feast-phase. The total solid (TS) concentration was measured at the end of each cycle and was about 5 ± 0.5 g/L in both reactors. This was done to figure out if the selection process in Reactor 2a was working correctly or if the biomass concentration was decreasing e.g., for lack of nutrients. As the biomass concentration was stable throughout all experiments no further tests regarding cell growth were conducted.

Experiments regarding the substrate concentration and temperature selection had the top priority. Afterwards the optimisation of all other operation conditions concerning the PHA production was examined with the most appropriate reactor temperature and substrate concentration found. A chronological test order was implemented as follows:

1.   Selection of a suitable substrate concentration
2.   Investigation of the reactor temperature
3.   Evaluation of a suitable pH-level
4.   Selection of a suitable cycle time

2.2.2. Investigation Concerning the Best Substrate Concentration

As there is a big variation in substrate concentration in literature, different substrate concentration of 1200 mg_{VFA}/L and 2000 mg_{VFA}/L were tested at 20 °C or 30 °C, pH = 7 or without pH control and a CT = 24 h. The named concentrations were chosen to avoid possible problems triggered by substrate inhibition.

2.2.3. Investigation Concerning the Best Temperature

The temperatures of material flows from a municipal WWTP are about 15 °C to 20 °C in temperate climates and might exceed 30 °C in hot climates or after mesophilic acidification. So tests with 15 °C, 20 °C and 30 °C were performed with a substrate concentration of 1200 mg_{VFA}/L, pH = 7 and a CT of 24 h or 48 h to find the best reactor temperature regarding PHA production. To avoid a substrate induced influence all tests were conducted with the same substrate batch.

2.2.4. Investigation Concerning the Best pH Level

As the best pH level to produce PHA depends on the used substrate, various pH levels (6, 7, 8, 9, without pH control) were tested in this study with a substrate concentration of 1200 mg_{VFA}/L, a temperature of 20 °C and a CT = 24 h.

2.2.5. Investigation Concerning the Best Cycle Time

To find the most suitable cycle time (CT) for the bacteria selection process, experiments with a substrate concentration of 1200 mg_{VFA}/L were conducted at a temperature of 20 °C and at pH = 7 or 8. As the feast/famine ratio is more important than the overall CT a constant substrate concentration should ensure that the feast phases of the cycles were constant and the variation in CT resulted in a different famine phase, only. [25] stated that the feast-phase should not last longer than 20% of the overall CT to create a selection pressure on non-PHA accumulating bacteria. Therefore, CTs of 24 h, 48 h and 72 h were tested.

2.2.6. Analytical Procedures

COD, TKN, total P and total solids (TS) were determined according to standard methods.

The concentration and composition of polyhydroxyalkanoates (PHA), namely polyhydroxybutyrate (PHB) and polyhydroxyvalerate (PHV), were detected by gas chromatography, according to [35] with some variations. Therefore, the biomass was separated via a centrifuge at 10,000 rpm for 20 min and dried at 105 °C. Afterwards the sample was pulverised with a ball mill and about 100 mg were digested and analysed. Detection was performed using a Perkin Elmer Autosystem XL chromatographer and a VF5ms 30 m × 0.25 column. Helium was used as carrier gas. PHA concentration was calibrated using 4 mL standards. The concentration and composition of volatile fatty acids, namely formate (Fo), acetate (Ac), propionate (Pro) and butyrate (Bu), were detected as described in Chapter 2.1.3.

2.2.7. Calculation of Parameters

For better comparability, all results regarding VFA concentrations, COD, TKN and total P were converted into mg/L. The concentration of PHA in terms of % cell dry weight (CDW) is defined as:

$$PHA = PHB + PHV \tag{5}$$

2.3. Potential Analysis

Calculations

Based on the PHA production results described in Chapter 3, a potential analysis was performed. The aim of the analysis was to determine the potential of biopolymer production (based on renewal resources and biodegradable, see definition in Chapter 1) on German and European waste water treatment plants (WWTP) by using sewage sludge as a substrate. All input data used for the calculations can be found in Table 1.

A plausibility analyses was performed to cross-check the most important input data like the amount of primary sludge (PS) per population equivalent (PE).

As detailed data about waste water and sewage sludge production are available in Germany, the first step of the potential analysis was calculated using these data together with results presented in Chapter 3. In a second calculation step, data provided by the European Union (EU) were used to create an in-depth estimation of the biopolymer potential considering all 28 member states.

Table 1. Input data used during the potential analysis.

Parameter	Unit	Value	Literature
Connected people equivalents (PE) on German WWTPs	Mio. PE	115.7	[29]
Proportion of PSP *-PEs regarding total PEs in Germany	%	92	[29]
PEs with PSP * in Germany	Mio. PE	106.56	
Amount of primary sludge per PE	L/(PE×d)	1.1	[36]
Total solid conc. of primary sludge/acidified material	g/L	35	[4]
VFA concentration	g_{VFA}/m^3	7,653	[4]
Retention time and withdrawal at the first production step	d; %/d	4; 25	[4]
Total solid concentration in the aerobic Reactors 2a/2b	g/L	5.0	[4]
Loading rate for PHA production	kg_{VFA}/m^3	1.2	[4]
Retention time and withdrawal at Reactor 2a	d and %/d	2 and 50	[4]
PHA proportion based on cell dry weight	CDW.-%	28.4	[36]
Yearly sewage sludge amount in the EU	t_{TS}/a	13,245,180	[30]
Yearly sewage sludge amount in Germany	t_{TS}/a	1,815,150	[30]

* PSP = German WWTPs with preliminary sedimentation potential (PSP = more than 10,000 PE).

3. Results and Discussion

3.1. VFA Production

3.1.1. Potential Analysis

Table 2 displays the results of the performed investigations ordered by degree of acidification. Primary sludge performed best under three out of four conditions and yielded by far the best degree of acidification (DA) with 31% at 30 °C under pH-controlled conditions [34]. The second best carbon source, a one to one mixture of primary and digested sludge at 20 °C under pH-uncontrolled conditions, achieved only a DA of 14%. In five out of eight experiments pH-uncontrolled conditions resulted in a higher DA. Therefore, a fermentation without pH control should be considered for all fermentation raw materials. However, primary sludge yielded better DAs with pH control at both investigated temperatures. At 30 °C the DA of primary sludge was twice as high than without pH control.

Table 2. Degree of acidification and VFA composition in dependence of substrate and operation conditions (batch-tests, 4 L).

Carbon Source	pH	Temperature	Max. Conc.	DA	Ac/Pro/Bu
		(°C)	(Day)	(%)	(%)
Primary sludge	6 *	30	9	31	52/48/0
Primary sludge	6 *	20	7	14	56/44/0
Primary sludge	4.6	30	10	14	41/59/0
Primary-/digested sludge	7	20	14	14	79/21/0
Primary sludge	4.5	20	15	13	42/58/0
Primary-/digested sludge	7.5	30	14	12	84/16/0
Excess sludge	7	30	5	10	59/20/20
Excess sludge	6.5	20	4	8	60/20/20
Primary-/digested sludge	6 *	30	5	7	75/25/0
Excess sludge	6 *	20	7	6	24/76/7
Excess sludge	6 *	30	5	6	67/33/0
Primary-/digested sludge	6 *	20	2	3	57/43/0
Excess-/digested sludge	6 *	30	4	3	100/0/0
Excess-/digested sludge	8	30	3	3	76/0/24
Excess-/digested sludge	7.5	20	7	2	100/0/0
Excess-/digested sludge	6 *	20	2	1	100/0/0

* Marks conditions pH-controlled.

Table 2 also shows the composition of the VFA. The results varied strongly between 24/76/7 (%Ac/%Pro/%Bu) and 100/0/0 depending on the used raw material. Primary sludge produced none butyric acid and acetic and propionic acid in nearly two equal sections. Excess sludge on the other hand produced up to 21% butyric acid, while the one to one mixture of primary and digested sludge produced the most acetic acid (up to 84%) of all tested raw materials. The results show that the raw material has a major influence on the VFA composition.

As the use of primary sludge resulted in highest DA and showed only small variations in VFA composition under the tested conditions, it was chosen as raw material and used in all further tests.

Beside the ability to produce VFAs, primary sludge has other advantages as raw material for the PHA production. As primary sludge is a mixture of organic material, water and fermenting microorganisms no longsome biological adaptation-phase or biomass recirculation for the fermentation process was necessary. During all experiments primary sludge showed nutrient limited conditions, as described in Chapter 2.2.1. This is of particular significance given that nutrient limited conditions are essential for the later PHA production [19,37].

3.1.2. Temperature

The aim of the investigations at two temperature levels was to ascertain if the VFA production at ambient temperature (20 °C) can reach the same VFA production compared with heating the sludge.

In six out of eight tested combinations a temperature increase from 20 to 30 °C caused a higher VFA production as shown in Table 2. The experiment confirmed the results of [32], who stated that the VFA concentration increases with higher fermentation temperature. Using primary sludge as raw material (under pH-controlled conditions) the temperature change from 20 to 30 °C caused a DA increase from 14% to 31%.

The general assumption that the acidification rate is higher at 30 °C than at 20 °C could not be confirmed. Only three out of eight tested combinations reached their VFA maximum at 30 °C in a shorter span of time than at 20 °C. Four out of them even reached their VFA maximum at 20 °C in a shorter span of time than at 30 °C. The results can be seen in Table 2. Primary sludge under pH-controlled conditions obtained its VFA maximum after 7 days at 20 °C and after 9 days at 30 °C. Nevertheless, the fact that the DA of primary sludge under pH-controlled conditions at 30 °C was twice as high as the DA at 20 °C is all the more important as the VFA production at 30 °C lasted only about 30% longer.

The variation of temperature has a wide range of influence on the VFA composition, depending on the used substrate. As primary sludge was already chosen as substrate for the optimisation tests, only its VFA composition was of interest for further tests. However, in the case of primary sludge the temperature change investigated resulted only in marginal changes in the VFA composition.

Consequently, a temperature around 30 °C for the further experiments was considered as reasonable.

3.1.3. pH

As illustrated in Figure 4, no big difference in the maximum VFA concentration between pH = 6 and pH = 8 was observed. A pH value of 7 yielded the highest result with 18,286 mg$_{VFA}$/L after a RT of 10 days. The fermentation at pH = 10 reached significantly worse results with a maximum of 10,050 mg$_{VFA}$/L only at 18 days retention time. This is in contrast to the results of [31] showing the best result at pH = 9 with excess sludge and food waste as source material and [32] yielding the highest result at pH = 11 with excess sludge as raw material.

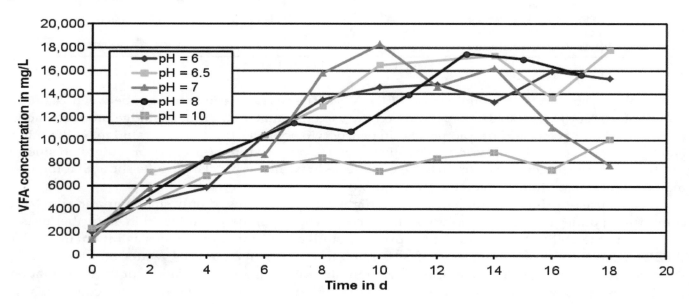

Figure 4. VFA concentrations during tests at different pH levels.

Although pH = 7 yielded the best result, methane production turned out to be an issue at this pH-value. After about 15 days the acetate concentration was falling rapidly and the overall VFA concentration after 18 days was less than 44% of the maximum (Figure 4). During this period more than 15 Vol.% methane was detected in the reactor via gas measurement. To prevent methanogenic conditions a pH-level of 6 has to be kept [38]. Consequently, further investigations were performed under pH = 6, although it produced about 12% less VFA within the batch experiments.

The variation of pH-level showed a strong influence on the VFA composition. Changing the pH from 6 to 7 within the batch experiments caused a constant decrease in the acetic acid ratio as shown in Table 3. In the same tests, the propionic acid ratio increased, while the butyric acid ratio decreased to zero. At pH = 8 conditions a reverse trend was observed. With 60% the maximum acetic acid ratio as well as the minimum propionic acid ratio (37%) was detected, while butyric acid was produced in small amounts (3%). In comparison to pH = 8, a reduction of acetic acid and propionic acid production was detected at the highest tested pH-level (pH = 10), while the butyric acid ratio increased to the highest level (8%) observed. In contrast to the other pH-values tested, formic acid was produced at pH = 10 with a ratio of 16.

Table 3. VFA composition and DA in dependence of pH or RT and WD.

Batch			Semi Continuous			
pH	Ac/Pro/Bu	DA	pH	RT and WD	Ac/Pro/Bu	DA
	(%)	(%)		(day and %)	(%)	(%)
6	45/51/4	29	6	4 and 25	49/38/13	15
6.5	37/61/2	29	6	4 and 50	46/47/7	14
7	28/72/0	39	6	4 and 75	48/45/7	14
8	60/37/3	29	6	6 and 25	51/38/11	22
10	45/31/8*	14	6	6 and 50	46/48/6	17
			6	8 and 25	46/41/13	20
			6	8 and 50	43/49/8	19

* Missing to 100% is formate.

3.1.4. RT and WD

To get an idea about how much adapted bacteria are needed in the reactor to produce the most VFAs a wide range of RT and WD was tested. RT = 2 days and WD = 50% yielded poor results (VFA$_{max}$ < 2000 mg$_{VFA}$/L) and after 10 days of semi-continuous operation the test was shut down. Therefore, these results are not shown. Further results ranged in a broad band between 5000 mg$_{VFA}$/L and 10,000 mg$_{VFA}$/L.

Obviously, the VFA production with short RTs and small WDs fluctuated less than using long RTs and large WDs, what can be explained by the changing composition of the introduced primary sludge. These changes of the used primary sludge are mostly due to weather events. A rainfall after a period of dry weather can transport a huge amount of organic matter to the WWTP. Daily changes in the waste water's composition or different contents during the week could be another reason for changing the primary sludge's composition. Smaller WD stabilises the fermentation process because only little material is turned over and the reactor is less sensitive to heterogeneous primary sludge input.

Figure 5 shows the average VFA concentration over a period of 40 days for all investigated combinations. Both, RT and WD influenced the VFA production. With higher WD the VFA concentration was decreasing at all tested RTs. The highest overall VFA concentration was reached at a RT of 6 days with a WD of 25%. Longer and shorter RTs (with a WD of 25%, too) resulted in lower VFA concentrations.

In order to have a high PHA-production in the second stage the VFA production rate (PR) is more important than the VFA concentration, which could, if it exceeds a certain value, lead to a substrate inhibition [23]. Therefore, the PR was calculated on Equation (4). A RT = 4 days and WD = 25% yielded the top production rate with PR = 1913 mg$_{VFA}$/(L×d) at a VFA concentration of 7653 mg$_{VFA}$/L on average [4].

The variation of RT influenced the VFA composition only slightly as shown in Table 3. Acetic acid and propionic acid were produced in a similar range, while butyric acid was always the smallest part. Nevertheless, the variation of WD had an effect. With a WD of 25% the fraction of propionic acid was about 20% smaller throughout all tests, compared to a WD of 50% and 75%, while the butyric acid ratio was nearly twice as much. The acetic acid ratio with a WD of 25% was slightly higher than for any other WDs.

As a stable VFA composition is necessary for high quality PHA production the possible fluctuation of the VFA composition during the whole test period is important. Figure 6 is exemplary for all semi-continuously operated tests and shows the concentrations of acetate, propionate and butyrate at RT = 4 days, WD = 50% during the whole test period of 44 days. After a starting phase of 10 days (not shown in the figure) the semi-continuous operation began. Due to the change in the operation method (from batch to semi-continuously) a transition phase with a decrease in VFA concentration was observed for the first six days of semi-continuous operation. Fluctuations in the VFA concentration between day six and day 44 were due to the changing concentration and composition of the introduced primary sludge. Although a fluctuation in VFA concentration after the fermentation step was observed, only small changes in the VFA composition were detected. Thus, it was possible to show that the variability of the raw material primary sludge did not affect the VFA composition significantly.

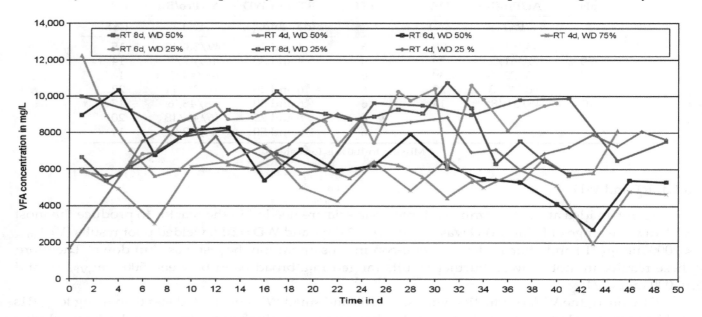

Figure 5. VFA concentration in dependence of RT and WD.

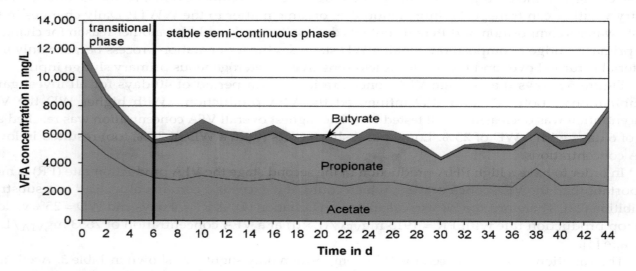

Figure 6. Development of the VFA concentration and composition at RT = 4 days, WD = 50%.

3.2. PHA Production

The acidified primary sludge from phase 1 was used for the production of PHA in step two where the influence of different operating conditions was investigated.

3.2.1. Substrate Concentration

The results of the conducted experiments are displayed in Table 4. It appears that higher PHA concentrations were gained at the lower substrate concentration tested. The maximum PHA concentration of 25.9% based on cell dry weight (CDW) was reached at a pH of 7 and a reactor temperature of 20 °C. With a VFA concentration of 2000 mg/L and a reactor temperature of 20 °C the highest PHA concentration achieved was 4.8% CDW, only and therefore much less than at a substrate concentration of 1200 mg/L at the same temperature. This confirms the observations of [8,23,39] that an increasing substrate concentration could result in a substrate inhibition.

Table 4. PHA production in dependence of two tested substrate concentrations.

Substrate Conc.	Temp.	pH	PHA	PHB/PHV
(mg$_{VFA}$/L)	(°C)		(% CDW)	(% CDW/% CDW)
1200	20	*	13.2	7.0/6.2
1200	20	7	25.9	13.2/12.7
1200	30	*	3.4	2.3/1.1
2000	20	*	4.8	3.3/1.5
2000	20	7	1.8	<2/1.8
2000	30	*	5.8	3.8/2.0

* Marks conditions without pH-control.

Furthermore, the experiment at the higher temperature and the lower substrate concentration (30 °C, 1200 mg/L) leads to a significant lower PHA production than at 20 °C and 1200 mg/L. This indicates that a reactor temperature of 20 °C may be preferable for a primary sludge based PHA production (see also Chapter 3.2.2).

PHA composition did not show any dependence regarding to substrate concentration. During all tests higher proportions of PHB than PHV were produced. About twice as much PHB (than PHV) was produced at all experiments at a substrate concentration of 2000 mg/L and at 30 °C with 1200 mg/L, while a nearly equal proportion of both PHAs was reached at 1200 mg/L and 20 °C.

3.2.2. Temperature

Table 5 displays the results of the PHA production at different temperatures. As the bacteria metabolism is slower at lower temperatures a very long feast-phase was observed at 15 °C. To ensure a sufficiently long famine-phase the cycle time of 24 h was doubled at all test with a reactor temperature of 15 °C. Figure 7 shows the length of the feast-phase based on the reactors oxygen concentrations. The very long feast-phase at 15 °C (around 22 h) is clearly visible. All other feast-phases at 20 °C or 30 °C were significantly shorter and not longer than 600 min. A somewhat surprising fact was that a faster bacteria metabolism at 30 °C did not result in a shorter feast-phase than at 20 °C. These observations are in contrast to [40], who stated that shorter fest-phases are observed at higher temperatures. At the same time, they confirm the results of [41], who gained the highest PHA concentration at reactor temperatures around 20 °C. This effect may origin in the fact, that the used sewage sludge was already adopted to temperature around 20 °C.

Table 5. PHA production in dependence of the reactor temperature.

Temp.	pH	CT	PHA	PHB/PHV
(°C)		(h)	(% CDW)	(% CDW/% CDW)
15	7	24/48	4.2	2.5/1.7
15	8	24/48	3.9	2.5/1.4
20	7	24	25.9	13.2/12.7
20	*	24	13.2	7.0/6.2
30	7	24	0.6	<2/0.6
30	*	24	3.4	2.3/1.1

* Marks conditions without pH-control.

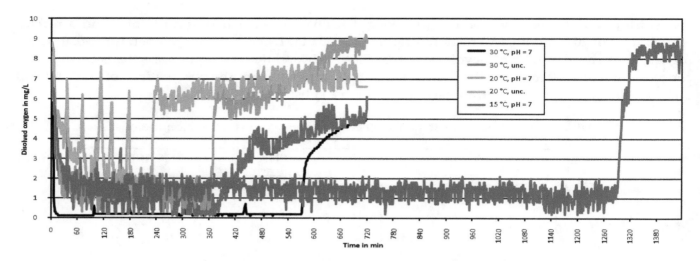

Figure 7. Reactor temperature influence on the feast/famine phase length.

There is no influence on PHA composition due to temperature changes. The results of Chapter 3.2.1 that the PHB/PHV ratio is about 2/1 at lower PHA concentrations produced, was confirmed (Table 5 at 15 °C and 30 °C), while both proportion are more or less equal at a higher PHA production (Table 5 at 20 °C).

As listed in Table 5 conditions of 20 °C yielded the highest PHA concentrations with 13.2% CDW and 25.9% CDW. At a reactor temperature of 15 °C or 30 °C a maximum PHA concentration of less than 4.5% CDW was reached, only. In consequence further experiments were conducted at 20 °C to achieve the best possible PHA concentration [42].

3.2.3. pH

Table 6 illustrates big differences in the maximum PHA concentration between the tested pH levels. All tests were operated in batch-mode with a cycle time of one day. The substrate's pH was adjusted before adding it into the reactor. An exception was the pH uncontrolled test. This experiment should clarify if a high PHA production is possible at fluctuating pH value. During the whole cycle the pH varied between 7.3 at the beginning of the feast-phase and 9.3 at the end of the famine-phase, with an average pH around 9.

Table 6. PHA production in dependence of the pH.

pH	PHA	PHB/PHV
	(% CDW)	(% CDW/% CDW)
unc. (av. 9)	13.2	7.0/6.2
6	—	—
7	25.9	13.2/12.7
8	28.4	14.7/13.7
9	4.4	3.0/1.4

The experiments with controlled pH values showed highly varying results. While there was no detectable PHA production at pH = 6, more than 25% CDW was produced between pH 7 and 8, with a maximum PHA production of 28.4% CDW at pH = 8. Then again, at the highest pH of 9 a low PHA production of 4.4% was reached only. The experiment without pH control produced around half as much PHA as the tests at pH 7 or 8. Nevertheless the results of the uncontrolled test were far better than at the same pH-level of 9 during the controlled test. Still a pH controlled PHA production is preferable, as the maximum PHA production was obtained at pH controlled operation [42].

pH changes did not influence the PHA composition. As described in Sections 3.2.2 and 3.2.3 the PHB/PHV ratio is 2/1 at low PHA concentrations and nearly 1/1 at a higher PHA production.

3.2.4. Cycle Time

When using a PHA production based on a bacteria mixed culture a feast/famine regime is crucial. This is the only way for PHA producing bacteria to gain a significant selection advantage over other microorganisms.

The produced amount of PHA in dependence of the length of the feast/famine-phase is displayed in Table 7. A correlation between the cycle time (CT) and the length of the feast-phase was observed, showing that a longer CT leads to a shorter feast-phase. This could be due to a longer and therefore harder famine-phase leading to a long period of starvation. It seems that after a bigger starvation, the bacteria's substrate uptake is higher and faster than at shorter cycle times. [39] highlights the cycle time must be such that the complete PHA is metabolised at the end of the famine-phase. This guideline was confirmed by testing samples, taken at the end of each famine-phase. No PHA was detected and therefore the required operating condition was kept during all tests. Thus, a cycle time of 24 h is sufficient.

Table 7. PHA production in dependence of the cycle time.

CT	Feast/Famine	Feast/Famine-Ratio	PHA	PHB/PHV
(h)	(min)	(%/%)	(% CDW)	(% CDW/% CDW)
24	524/916	36/64	28.4	14.7/13.7
48	500/2380	17/83	18.3	8.0/10.3
72	448/3872	10/90	21.4	8.3/13.1

A former study concluded that the proportion of the feast-phase should not exceed 20% of the CT, because a longer famine phase could lead to a lower selection pressure on non-PHA accumulating bacteria [26]. Table 7 shows a feast-phase proportion of 36% at a cycle time of 24 h and therefore nearly twice as high as suggested by [26]. Regardless, at this cycle time the highest PHA concentration (28.4% CDW) was reached. A feast-phase proportion of 17% at a CT of 48 h led to a PHA production of 18.3% CDW, while at the longest tested CT of 72 h a PHA production of 21.4% CDW was observed with a feast-phase proportion of 10%. This indicates that a fixed feast-phase proportion is unnecessary and the demanded operation condition by [39] is preferable. However, more experiments will be necessary to confirm these results.

An influence of the cycle time length on the PHA composition is shown in Table 7, also. While all other experiments reached a higher PHB than PHV proportion a cycle time of 48 h or longer led to a higher PHV production. This could be due to the fact that the PHA accumulation bacteria produced PHB at first while the PHV production started later. As no samples were taken at low oxygen levels during the feast-phase and no uptake rate was measured for the VFAs (acetate, butyrate, valerate) during the experiments this presumption could not be confirmed.

The preferable operating conditions for the VFA production described in Chapter 3.1 provide VFA every day. Regarding the pairing of both productions steps (VFA and PHA production) and having in mind that with a cycle time of one day the highest PHA concentration (28.4% CDW) was yielded, a CT = 24 h is favourable [42].

3.3. Potential Analysis

3.3.1. Calculation for German Waste Water Treatment Plants

Figure 8 shows the results for possible biopolymer production on German WWTPs. All material streams and reactor volumes in this figure are theoretical values showing the size of the flows, if the

best substrate for acidification [34] of all German WWTPs with preliminary sedimentation potential
(PSP = more than 10,000 PE) would be used for PHA production.

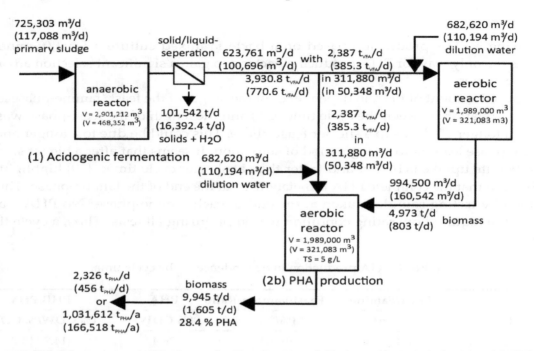

Figure 8. Biopolymer potential scheme for European and German (figures in brackets) WWTPs.

First of all, the calculation for the amount of primary sludge in Germany regarding the PEs
(Table 1) connected to German WWTPs is shown in Equation (6).

$$115,700,000 \text{ PE} \times 1.1 \frac{\text{L}_{PS}}{\text{PE} \times \text{d}} = 1,272,700,000 \frac{\text{L}_{PS}}{\text{d}} = 127,270 \frac{\text{m}^3_{PS}}{\text{d}} \qquad (6)$$

Around 92% of PEs are coming from WWTPs with preliminary sedimentation potential (Table 1),
on which a primary clarifier is installed or the construction of a primary clarifier would be preferable.
Thus, the actual amount of primary sludge is calculated as follows:

$$127,270 \frac{\text{m}^3_{PS}}{\text{d}} \times 92\% = 117,088.4 \frac{\text{m}^3_{PS}}{\text{d}} \qquad (7)$$

Using these data and the results of own experiments [34,42] and of Chapter 3.1 and 3.2 a calculation
of the possible PHA production through various steps can be performed.

Implementing the best reactor operation method, using a retention time of 4 days and a daily
withdrawal of 25% (Table 1) 117,088.4 $\frac{\text{m}^3}{\text{d}}$ acidified material could be used for PHA production every
day (Equation (8)).

$$117,088.4 \frac{\text{m}^3_{PS}}{\text{d}} \times 4 \text{ d} \times 25 \frac{\%}{\text{d}} = 117,088.4 \frac{\text{m}^3}{\text{d}} \qquad (8)$$

The total solid (TS) concentration of the acidified material of 35 $\frac{\text{kg}}{\text{m}^3}$ (Table 1) and the assumed
residual moisture after de-watering via centrifuge of 75% leads to a daily acidified liquid production
of 100,696 $\frac{\text{m}^3}{\text{d}}$ (Equations (9)–(11)).

$$117,088.4 \frac{\text{m}^3}{\text{d}} \times 35 \frac{\text{kg}_{TS}}{\text{m}^3} = 4,098,094 \frac{\text{kg}_{TS}}{\text{d}} = 4098.1 \frac{\text{t}_{TS}}{\text{d}} \qquad (9)$$

The assumed residual moisture of 75% means that the calculated 4098.1 $\frac{t_{TS}}{d}$ biomass is 25% of the total mass separated by the centrifuge. Accordingly, 75% of the separated total mass is water. Assuming that the solid phase is completely separated it follows:

$$4098.1\,\frac{t_{TS}}{d} + \frac{4098.1\,\frac{t_{TS}}{d}}{25\,\%} \times 75\%\,\frac{t_{H_2O}}{d} = 16{,}392.4\,\frac{t_{TS+H_2O}}{d} \tag{10}$$

With an assumed average density of $1\,\frac{t}{m^3}$ the amount of

$$117{,}088.4\,\frac{m^3}{d} - \frac{16{,}392.4\,\frac{t}{d}}{1}\,\frac{\frac{t}{d}}{\frac{t}{m^3}} = 100{,}696\,\frac{m^3}{d} \tag{11}$$

of acidified liquid is available for the PHA production step.

Regarding the average VFA concentration of 7653 $\frac{mg_{VFA}}{L}$ (Table 1) the amount of VFA in the acidified water can be calculated (Equation (12)).

$$7653\,\frac{g_{VFA}}{m^3} \times 100{,}696\,\frac{m^3}{d} = 770.6\,\frac{t_{VFA}}{d} \tag{12}$$

The substrate is divided into equal parts to both reactors (2a and 2b) of the second production step so that each reactor is supplied with 50,348 $\frac{m^3}{d}$ of acidified liquid (Equation (13)), containing 385.3 $\frac{t_{VFA}}{d}$ (Equation (14)).

$$\frac{100{,}696\,\frac{m^3}{d}}{2} = 50{,}348\,\frac{m^3}{d} \tag{13}$$

$$\frac{770.6\,\frac{t_{VFA}}{d}}{2} = 385.3\,\frac{t_{VFA}}{d} \tag{14}$$

In order to achieve the required loading rate of 1.2 $\frac{kg_{VFA}}{m^3 \times d}$ in both reactors of the second production step, the volume of Reactor 2a (sum over Germany) and 2b (sum over Germany) should be 321,083 m^3 each (Equation (15)).

$$\frac{385{,}300\,\frac{kg_{VFA}}{d}}{1.2\,\frac{kg_{VFA}}{m^3 \times d}} = 321{,}083.3\,m^3 \tag{15}$$

Reactor 2a is operated with a retention time of 2 days, a daily withdrawal of 50% of the reactors volume and a total solid concentration of TS = 5 $\frac{g_{TS}}{L}$ (Table 1). Due to the fact that there is no biomass sedimentation before the removal of the material, the withdrawn material can be considered as fully mixed. Therefore, the amount of bacteria must double every cycle to achieve a constant concentration of total solids. During the own experiments it was shown that this necessary condition was fulfilled [42].

The removal of 50% of the reactor volume leads to the amount of substrate, which has to be filled in the Reactor 2b every day (Equation (16)).

$$321{,}083.3\,\frac{m^3}{d} \times 50\% = 160{,}541.7\,\frac{m^3}{d} \tag{16}$$

The emptying and refilling process at the beginning of every cycle is represented in Figure 3. In order to reach a VFA concentration of 1.2 $\frac{kg_{VFA}}{m^3 \times d}$ and regarding to the fact that the VFA concentration at the end of a cycle is zero, a substrate with a VFA concentration of 2.4 $\frac{kg_{VFA}}{m^3 \times d}$ is needed if half of the reactors volume is exchanged.

Equation (16) shows the required amount of substrate for one day for one reactor regarding the needed VFA concentration. Hence the amount of VFA rich liquid of 50,348 $\frac{m^3}{d}$ (Equation (13)) has to be diluted with 110,193.7 $\frac{m^3}{d}$ fresh water (Equation (17)).

$$160,541.7 \, \frac{m^3}{d} - 50,348 \, \frac{m^3}{d} = 110,193.7 \, \frac{m^3}{d} \tag{17}$$

When calculating the amount of dilution water it should be noted that the large amount of water is due to the reactors comparatively low total solid concentration of TS = 5 $\frac{g_{TS}}{L}$, which was installed during the experiments in [42]. This concentration is used for the potential analysis as well, to keep the calculation as close as possible to the operation conditions of the carried out experiments. Of course, a much higher solid concentration could be installed, leading to significantly smaller reactor volumes as well as less dilution water. As the amount of dilution water does not affect the result of the potential analysis it was kept.

As described, the biomass concentration in Reactor 2a was 5 $\frac{g_{TS}}{L}$ at the end of a cycle. Regarding the withdrawal of 50% of the reactor's volume a total of

$$321,083.3 \, \frac{m^3}{d} \times 5 \, \frac{kg_{TS}}{m^3} \times 5\% = 802,708.3 \, \frac{kg_{TS}}{d} = 802.71 \, \frac{t_{TS}}{d} \tag{18}$$

biomass is transferred into Reactor 2b. This reactor also had a dry matter content of 5 $\frac{kg_{TS}}{m^3}$ after the PHA production step (Table 1). Considering a cycle time of one day and the volume of 321,083 m³, the amount of biomass in Reactor 2b sums up to (Equation (19)):

$$\frac{321,083.3 \, m^3}{1 \, d} \times 5 \, \frac{kg_{TS}}{m^3} = 1,605,416.5 \, \frac{kg_{TS}}{d} = 1605.4 \, \frac{t_{TS}}{d} \tag{19}$$

With the reached PHA concentration of 28.4% of the cell dry weight (CDW) [42] the daily amount of biopolymer is calculated in Equation (20).

$$1605.4 \, \frac{t_{TS}}{d} \times 28.4\% \, \frac{PHA}{TS} = 455.9 \, \frac{t_{PHA}}{d} \tag{20}$$

Finally, the possible annual amount of PHA production on German waste water treatment plants can be calculated in Equation (21).

$$455.9 \, \frac{t_{PHA}}{d} \times 365.25 \, \frac{d}{a} = 166,517.5 \, \frac{t_{PHA}}{a} \tag{21}$$

Dividing the reactor volume (sum of all production stages) of 1,110,518 m³ by the people's equivalent with preliminary sedimentation potential of 106.56 Mio. PE (Table 1) results in a reactor volume (sum of all production stages) per capita of around 10.4 L/PE. Each PE can contribute to the production of 1.6 kg$_{PHA}$/(PE×a). Keeping in mind that the aeration tank volume on a WWTP with 100,000 PE sums up to 10,000 m³–15,000 m³ an additional reactor volume of approximately 1050 m³ would be needed for the biopolymer production, only. Thus, the extra volume would not be disproportional.

3.3.2. Estimation for European Waste Water Treatment Plants

The Biopolymer potential for European WWTPs are calculated similar to Section 3.3.1 and also shown in Figure 8.

As there are missing data about the amount of connected persons (PEs) for many EU member states as well as about the amount of municipal waste water, it is impossible to calculate the EU-wide production of primary sludge analogous to Equations (6) and (7). However, there are data for all 28 EU member states regarding the production of sewage sludge. These data show the dry weight of sewage sludge in $\frac{t_{TS}}{a}$ (Table 1) and hence their unit must be transferred into $\frac{m^3}{a}$ (Equation (22)) to compare them with the data used for German WWTP in Equation (6). Therefore, an average total solid concentration of 15 $\frac{g_{TS}}{L}$ or 66.67 $\frac{m^3}{t_{TS}}$ for European sewage sludge (primary and secondary) is assumed.

$$13,245,180 \; \frac{t_{TS}}{a} \times 66.67 \; \frac{m^3}{t_{TS}} = 883,056,150 \; \frac{m^3}{a} \tag{22}$$

On the assumption that the proportion of primary sludge in the amount of sewage sludge is more or less constant in all member states, the percentage can be calculated (Equation (25)) using the theoretical yearly amount of primary sludge produced in Germany (Equations (6) and (23)) and the yearly amount of German sewage sludge (Table 1) (Equation (24)).

$$115,700,000 \; PE \times 1.1 \; \frac{L_{PS}}{PE \times d} \times 365.25 \; \frac{d}{a} = 46,485,367.5 \; \frac{m^3_{PS}}{a} \tag{23}$$

$$1,815,150 \; \frac{t_{TS}}{a} \times 66.67 \; \frac{m^3}{t_{TS}} = 121,016,051 \; \frac{m^3}{a} \tag{24}$$

$$\frac{46,485,367.5 \; \frac{m^3_{PS}}{a}}{121,016,051 \; \frac{m^3}{a}} \times 100\% = 38.4\% \tag{25}$$

Assuming that not all European waste water treatment plants are equipped with a primary clarifier, the proportion of primary sludge is rounded off to 30%, so that the yearly amount of European primary sludge sums up to 265 Mio. $\frac{m^3_{PS}}{a}$ (Equation (26)) or 725,303 $\frac{m^3_{PS}}{d}$ (Equation (27)).

$$883,056,151 \; \frac{m^3_{PS}}{a} \times 30\% = 264,916,845.3 \; \frac{m^3_{PS}}{a} \tag{26}$$

$$\frac{264,916,845.3 \; \frac{m^3_{PS}}{a}}{365.25 \; \frac{d}{a}} = 725,302.8 \; \frac{m^3_{PS}}{d} \tag{27}$$

By now, the European biopolymer potential can be calculated analogous to Equations (8)–(21). The amount of acidified material is:

$$725,303 \; \frac{m^3_{PS}}{d} \times 4 \; d \times 25 \; \frac{\%}{d} = 725,303 \; \frac{m^3}{d} \tag{28}$$

Using Equations (9)–(11) the amount of acidified liquid can be calculated (Equation (31)):

$$725,303 \; \frac{m^3}{d} \times 35 \; \frac{kg_{TS}}{m^3} = 25,385,605 \; \frac{kg_{TS}}{d} = 25,385.6 \; \frac{t_{TS}}{d} \tag{29}$$

$$25,385.6 \; \frac{t_{TS}}{d} + \frac{25,385.6 \; \frac{t_{TS}}{d}}{25\%} \times 75\% \; \frac{t_{H_2O}}{d} = 101,542.4 \; \frac{t_{TS+H2O}}{d} \tag{30}$$

With an average density of 1 $\frac{t}{m^3}$

$$725,303 \; \frac{m^3}{d} - \frac{101,542.4 \; \frac{t}{d}}{1 \; \frac{t}{m^3}} = 623,760.6 \; \frac{m^3}{d} \tag{31}$$

of acidified liquid can be used within the second PHA production step. This leads to the amount of VFAs in the acidified water (Equation (32)):

$$7653 \; \frac{g_{VFA}}{m^3} \times 623,760.6 \; \frac{m^3}{d} = 4773.6 \; \frac{t_{VFA}}{d} \tag{32}$$

Analogous to Equations (13) and (14) both reactors of the second production step are supplied with 311,880 $\frac{m^3}{d}$ (Equation (33)) of acidified liquid containing 2386.8 $\frac{t_{VFA}}{d}$ (Equation (34)).

$$\frac{623,760.6 \frac{m^3}{d}}{2} = 311,880.3 \frac{m^3}{d} \tag{33}$$

$$\frac{4773.6 \frac{t_{VFA}}{d}}{2} = 2386.8 \frac{t_{VFA}}{d} \tag{34}$$

The reactor volumes (Equation (35)) can be calculated analogous to Equation (15):

$$\frac{2,386,800 \frac{kg_{VFA}}{d}}{1.2 \frac{kg_{VFA}}{m^3 \times d}} = 1,989,000 \text{ m}^3 \tag{35}$$

The daily substrate amount for one reactor is (Equation (36)):

$$1,989,000 \frac{m^3}{d} \times 50\% = 994,500 \frac{m^3}{d} \tag{36}$$

Analogous to Equation (17) the amount of dilution water can be calculated (Equation (37)):

$$994,500 \frac{m^3}{d} - 311,880.3 \frac{m^3}{d} = 682,619.7 \frac{m^3}{d} \tag{37}$$

With a withdrawal of 50% a biomass transfer to Reactor 2b of 4094.58 $\frac{t_{TS}}{d}$ is necessary (Equation (38)).

$$1,989,000 \frac{m^3}{d} \times 5 \frac{kg_{TS}}{m^3} \times 50\% = 4,972,500 \frac{kg_{TS}}{d} = 4972.5 \frac{t_{TS}}{d} \tag{38}$$

With the described total solid concentration of 5 $\frac{kg_{TS}}{m^3}$ after the PHA production the amount of biomass in Reactor 2b is (Equation (39)):

$$1,989,000 \text{ m}^3 \times 5 \frac{kg_{TS}}{m^3} = 9,945,000 \frac{kg_{TS}}{d} = 9945 \frac{t_{TS}}{d} \tag{39}$$

With a cycle time of one day and the reached PHA concentration of 28.4% CDW (Table 1) the daily amount of PHA sums up to (Equation (40)):

$$9945 \frac{t_{TS}}{d} \times 28.4\% \frac{PHA}{TS} = 2824.4 \frac{t_{PHA}}{d} \tag{40}$$

Finally, the possible annual amount of PHA production on European waste water treatment plants can be calculated in Equation (41).

$$2824.4 \frac{t_{PHA}}{d} \times 365.25 \frac{d}{a} = 1,031,612.1 \frac{t_{PHA}}{a} \tag{41}$$

3.3.3. Summary of the Results and Optimization Potential

The market for biopolymers is predicted to grow continuously [43]. In 2016, a worldwide biopolymer production of 4.16 Mio. $\frac{t}{a}$, of which 861,120 $\frac{t}{a}$ or 20.7% [43] suit the criteria of the stringent definition for biopolymers, introduced in Chapter 1, were achieved. Taking Equation (21) into account approximately 4.0% of the worldwide biopolymer production (bio- and non-biodegradable) could be produced just by using primary sludge from German WWTPs. Around 19.3% of the worldwide biopolymers could be produced on WWTPs in Germany considering the stringent definition, only.

For the biopolymer production on European WWTPs (Equation (41)) approximately 24.7% of 2016's worldwide biopolymer production (bio- and non-biodegradable) or around 119.8% of 2016's worldwide biopolymer production due to the stringent definition could be produced.

Assuming an improved PHA production with an achievable PHA concentration of 0.5 $\frac{g_{PHA}}{g_{VSS}}$ [44] or even around 60% CDW [45,46] a total amount of 2,179,446.8 $\frac{t_{PHA}}{a}$ (Equation (42)) could be produced on European WWTPs by using primary sludge, only.

$$9945 \frac{t_{TS}}{d} \times 60\% \times 36,525 \frac{d}{a} = 2,179,446.8 \frac{t_{PHA}}{a} \tag{42}$$

Thus, approximately 52.4% of 2016's worldwide biopolymer production (bio- and non-biodegradable) or 253.1% of 2016's worldwide biopolymer production due to the stringent definition could be produced in an improved production on WWTPs in the EU.

A large proportion of polymers (biopolymers and those from synthetic production) is used for packing materials. The PHAs feature similar characteristics like polypropylene (PP), which is the mostly sold plastic in the EU with 18.8% (around 8.6 Mio. $\frac{t}{a}$) market share in 2012 [47]. The potential analysis for Germany equates to approximately 1.9% of the EUs PP production. Using the calculation for the EU for PHA production from primary sludge around 12.0% of the conventional PP sold in the EU could be substituted which is a significant potential.

3.3.4. Plausibility Analysis

As some input parameters do have a strong effect to the calculations, a plausibility analysis was carried out. All critical parameters, like total solid concentration of the primary sludge, the daily amount of primary sludge per PE, or the daily amount of primary sludge per PE were analysed and considered plausible. A more detailed description of the plausibility analysis can be found in [28].

4. Summary and Conclusions

From the results, it could be concluded that the production of high amounts of VFAs with a stable VFA composition on a WWTP is possible. Using different raw materials shows a strong influences on degree of acidification and VFA composition. The VFA production and composition is strongly influenced by a pH-level change in the reactor. A semi-continuous operation method of the reactor with a short RT and small WD is preferable. With primary sludge as raw material no biomass recirculation is needed during the fermentation process.

The results showed that the produces VFA are suitable for PHA production in a second stage. This amount of PHA produced is strongly influenced by the reactors operating conditions (temperature, pH-level and substrate concentration), while the PHA composition is influenced by cycle time changes. At preferred conditions, a stable PHB/PHV composition was reached and both PHAs were produced in nearly the same proportion.

Nevertheless, further research is needed to couple both processes for constant and long term PHA production and for upscaling.

The results of the presented potential analysis clearly indicate the possibility to produce large amounts of PHAs on German and European WWTPs. It has been shown that municipal WWTPs could be used as a significant source for biopolymers and waste water is an important substituent for plant-based raw materials in the PHA production.

More than twice the amount of 2016's worldwide biopolymer production could be produced on European WWTPs with an upgraded operation. Thus, the production of biopolymers on waste water treatment plants contribute to a recycling of the organic material contained in waste water.

Acknowledgments: We thank the WILLY-HAGER-STIFTUNG, Stuttgart for funding the research project.

Author Contributions: Timo Pittmann and Heidrun Steinmetz conceived and designed the experiments; Timo Pittmann performed the experiments; Timo Pittmann and Heidrun Steinmetz analyzed the data; Timo Pittmann contributed reagents/materials/analysis tools; Timo Pittmann and Heidrun Steinmetz wrote the paper.

References

1. United Nations Environment Programme (UNEP). *Marine Litter-A Global Challenge*; UNEP: Nairobi, Kenia, 2009.
2. Jendrossek, D.U.; Handrick, R. Microbial degradation of polyhydroxyalkanoates. *Annu. Rev. Microbiol.* **2002**, *56*, 403–432. [CrossRef] [PubMed]
3. Choi, G.G.; Kim, H.W.U.; Rhee, Y.H. Enzymatic and non-enzymatic degradation of poly(3-hydroxybutyrate-co-3-hydroxyvalerate) copolyesters produced by *Alcaligenes* sp. MT-16. *J. Microbiol.* **2004**, *42*, 346–352. [PubMed]
4. Pittmann, T. Herstellung von Biokunststoffen aus Stoffströmen einer kommunalen Kläranlage. Dissertation thesis, Institute for Sanitary Engineering, Water Quality and Solid Waste Management, Stuttgart, Germany, 2015.
5. Nikodinovic-Runic, J.U.; Guzik, M. Carbon-rich wastes as feedstocks for biodegradable polymer (polyhydroxyalkanoate) production using bacteria. *Adv. Appl. Microbiol.* **2013**, *84*, 139–200. [CrossRef] [PubMed]
6. Cavalheiro, J.M.B.T.; de Almeida, C.M.M.D.; Grandfils, C.; da Fonseca, M.M.R. Poly(3-hydroxybutyrate) production by *Cupriavidus necator* using waste glycerol. *Process Biochem.* **2009**, *44*, 509–515. [CrossRef]
7. Zinn, M.U.; Hany, R. Tailored material properties of polyhydroxyalkanoates through biosynthesis and chemical modification. *Adv. Eng. Mater.* **2005**, *7*, 408–411. [CrossRef]
8. Akaraonye, E.; Keshavarz, T.U.; Roy, I. Production of polyhydroxyalkanoates: The future green materials of choice. *J. Chem. Technol. Biotechnol.* **2010**, *85*, 732–743. [CrossRef]
9. Steinbuechel, A. *Angewandte Mikrobiologie, Biopolymere und Vorstufen*; Springer: Berlin/Heidelberg, Germany, 2005.
10. Faulstich, M.; Greiff, K.B. Klimaschutz durch biomasse, ergebnisse des SRU-sondergutachtens. *Umweltwissenschaften und Schadstoff-Forschung* **2007**. [CrossRef]
11. Bengtsson, S.; Werker, A.; Christensson, M.; Welander, T. Production of polyhydroxyalkanoates by activated sludge treating a paper mill wastewater. *Bioresour. Technol.* **2008**, *99*, 509–516. [CrossRef] [PubMed]
12. Morgan-Sagastume, F.; Valentino, F.; Hjort, M.; Cirne, D.; Karabegovic, L.; Geradin, F.; Dupont, O.; Johansson, P.; Karlsson, A.; Magnusson, P.; et al. Biopolymer production from sludge and municipal wastewater treatment. *Water Sci. Technol.* **2014**, *69*, 177–184. [CrossRef] [PubMed]
13. Chakravarty, P.; Mhaisalkar, V.; Chakrabarti, T. Study on poly-hydroxy-alkanoate (PHA) production in pilot scale continuous mode wastewater treatment system. *Bioresour. Technol.* **2010**, *101*, 2896–2899. [CrossRef] [PubMed]
14. Chua, A.; Takabatake, H.; Satoh, H.; Mino, T. Production of polyhydroxyalkanoates (PHA) by activated sludge treating municipal wastewater: Effect of pH, sludge retention time (SRT), and acetate concentration in influent. *Water Res.* **2003**, *37*, 3602–3611. [CrossRef]
15. Chua, H.; Yu, P. Production of biodegradable plastics from chemical wastewater-A novel method to reduce excess activated sludge generated from industrial wastewater treatment. *Water Sci. Technol.* **1999**, *39*, 273–280. [CrossRef]
16. Dionisi, D.; Majone, M.; Papa, V.; Beccari, M. Biodegradable polymers from organic acids by using activated sludge enriched by aerobic periodic feeding. *Biotechnol. Bioeng.* **2004**, *85*, 569–579. [CrossRef] [PubMed]
17. Reddy, S.V.; Thirumala, M.; Reddy, T.K.; Mahmood, S.K. Isolation of bacteria producing polyhydroxyalkanoates (PHA) from municipal sewage sludge. *World J. Microbiol. Biotechnol.* **2008**, *24*, 2949–2955. [CrossRef]
18. Lemos, P.C.; Serafim, L.S.; Reis, M. Synthesis of polyhydroxyalkanoates from different short-chain fatty acids by mixed cultures submitted to aerobic dynamic feeding. *J. Biotechnol.* **2006**, *122*, 226–238. [CrossRef] [PubMed]
19. Albuquerque, M.; Eiroa, M.; Torres, C.; Nunes, B.R.; Reis, M. Strategies for the development of a side stream process for polyhydroxyalkanoate (PHA) production from sugar cane molasses. *J. Biotechnol.* **2007**, *130*,

411–421. [CrossRef] [PubMed]

20. Albuquerque, M.G.E.; Martino, V.; Pollet, E.; Avérous, L.; Reis, M. Mixed culture polyhydroxyalkanoate (PHA) production from volatile fatty acid (VFA)-rich streams: Effect of substrate composition and feeding regime on PHA productivity, composition and properties. *J. Biotechnol.* **2011**, *151*, 66–76. [CrossRef] [PubMed]

21. Bengtsson, S. The utilization of glycogen accumulating organisms for mixed culture production of polyhydroxyalkanoates. *Biotechnol. Bioeng.* **2009**, *104*, 698–708. [CrossRef] [PubMed]

22. Bengtsson, S.; Pisco, A.R.; Reis, M.; Lemos, P.C. Production of polyhydroxyalkanoates from fermented sugar cane molasses by a mixed culture enriched in glycogen accumulating organisms. *J. Biotechnol.* **2010**, *145*, 253–263. [CrossRef] [PubMed]

23. Albuquerque, M.; Concas, S.; Bengtsoon, S.; Reis, M. Mixed culture polyhydroxyalkanoates production from sugar cane molasses: The use of a 2-stage CSTR system for culture selection. *Bioresour. Technol.* **2010**, *101*, 7112–7122. [CrossRef] [PubMed]

24. Choi, J.; Lee, S.Y. Process analysis and economic evaluation for Poly(3-hydroxybutyrate) production by fermentation. *Bioprocess Eng.* **1997**, *17*, 335. [CrossRef]

25. Dionisi, D.; Carucci, G.; Petrangeli Papini, M.; Riccardi, C.; Majone, M.; Carrasco, F. Olive oil mill effluents as a feedstock for production of biodegradable polymers. *Water Res.* **2005**, *39*, 2076–2084. [CrossRef] [PubMed]

26. Dionisi, D.; Majone, M.; Vallini, G.; Di Gregorio, S.; Beccari, M. Effect of the applied organic load rate on biodegradable polymer production by mixed microbial cultures in a sequencing batch reactor. *Biotechnol. Bioeng.* **2005**, *93*, 76–88. [CrossRef] [PubMed]

27. Johnson, K.; Jiang, Y.; Kleerebezem, R.; Muyzer, G.; van Loosdrecht, M. Enrichment of a mixed bacterial culture with a high polyhydroxyalkanoate storage capacity. *Biomacromolecules* **2009**, *10*, 670–676. [CrossRef] [PubMed]

28. Pittmann, T.; Steinmetz, H. Potential for polyhydroxyalkanoate production on German or European municipal waste water treatment plants. *Bioresour. Technol.* **2016**, *214*, 9–15. [CrossRef] [PubMed]

29. DWA: 28th Performance Comparison of Municipal Waste Water Treatment Plants. 2016. Available online: http://de.dwa.de/tl_files/_media/content/PDFs/1_Aktuelles/leistungsvergleich_2015.PDF (accessed on 2 June 2017).

30. Eurostat: Sewage Sludge Production and Disposal. Available online: http://ec.europa.eu/eurostat/data/database?nodecode=envwwspd (accessed on 2 June 2017).

31. Chen, H.; Meng, H.; Nie, Z.; Zhang, M. Polyhydroxyalkanoate production from fermented volatile fatty acids: Effect of pH and feeding regimes. *Bioresour. Technol.* **2013**, *128*, 533–538. [CrossRef] [PubMed]

32. Mengmeng, C.; Hong, C.; Qingliang, Z.; Shirley, S.N.; Jie, R. Optimal production of polyhydroxyalkanoates (PHA) in activated sludge fed by volatile fatty acids (VFAs) generated from alkaline excess sludge fermentation. *Bioresour. Technol.* **2009**, *100*, 1399–1405. [CrossRef] [PubMed]

33. Bengtsson, S.; Hallquist, J.; Werker, A.; Welander, T. Acidogenic fermentation of industrial wastewaters. Effects of chemostat retention time and pH on volatile fatty acids production. *Biochem. Eng. J.* **2008**, *40*, 492–499. [CrossRef]

34. Pittmann, T.; Steinmetz, H. Influence of operating conditions for volatile fatty acids enrichment as a first step for polyhydroxyalkanoate production on a municipal waste water treatment plant. *Bioresour. Technol.* **2013**, *148*, 270–276. [CrossRef] [PubMed]

35. Queirós, D.; Rossetti, S.; Serafim, L.S. PHA production by mixed cultures: A way to valorize wastes from pulp industry. *Bioresour. Technol.* **2014**, *157*, 197–205. [CrossRef] [PubMed]

36. *Dimensioning of a Single-Stage Biological Wastewater Tretment Plant*; Publishing Company of ATV-DVWK, Water, Wastewater, Waste: Hennef, Germany, 2000; Standard ATV-DVWK-A131. [CrossRef]

37. Serafim, L.S.; Lemos, P.C.; Oliveira, R.; Reis, M. Optimization of polyhydroxybutyrate production by mixed cultures submitted to aerobic dynamic feeding conditions. *Biotechnol. Bioeng.* **2004**, *87*, 145–160. [CrossRef] [PubMed]

38. Bischofsberger, W.; Dichtl, N.; Rosenwinkel, K.-H.; Seyfried, C.F.; Bohnke, B. *Anaerobtechnik*; Springer: Berlin/Heidelberg, Germany, 2015. [CrossRef]

39. Morgan-Sagastume, F.; Karlsson, A.; Johansson, P.; Pratt, S.; Boon, N.; Lant, P.; Werker, A. Production of polyhydroxyalkanoates in open, mixed cultures from a waste sludge stream containing high levels of soluble

organics, nitrogen and phosphorus. *Water Res.* **2010**, *44*, 5196–5211. [CrossRef] [PubMed]

40. Johnson, K.; Kleerebezem, R.; van Loosdrecht, M.C.M. Influence of ammonium on the accumulation of polyhydroxybutyrate (PHB) in aerobic open mixed cultures. *J. Biotechnol.* **2010**, *147*, 73–79. [CrossRef] [PubMed]

41. Wen, Q.; Chen, Z.; Tian, T.; Chen, W. Effects of phosphorus and nitrogen limitation on PHA production in activated sludge. *J. Environ. Sci.* **2010**, *22*, 1602–1607. [CrossRef]

42. Pittmann, T.; Steinmetz, H. Polyhydroxyalkanoate production as a side stream process on a municipal waste water treatment plant. *Bioresour. Technol.* **2014**, *167*, 297–302. [CrossRef] [PubMed]

43. European Bioplastik e.V. Bioplastics. Bioplastics-Facts and Figures. 2016. Available online: http://docs. european-bioplastics.org/publications/EUBP_Facts_and_figures.pdf (accessed on 19 April 2017).

44. Morgan-Sagastume, F.; Hjort, M. Integrated production of polyhydroxyalkanoates (PHAs) with municipal wastewater and sludge treatment at pilot scale. *Bioresour. Technol.* **2015**, *181*, 78–89. [CrossRef] [PubMed]

45. Jia, Q.; Wang, H.; Wang, X. Dynamic synthesis of polyhydroxyalkanoates by bacterial consortium from simulated excess sludge fermentation liquid. *Bioresour. Technol.* **2013**, *140*, 328–336. [CrossRef] [PubMed]

46. Jia, Q.; Xiong, H. Production of polyhydroxyalkanoates (PHA) by bacterial consortium from excess sludge fermentation liquid at laboratory and pilot scales. *Bioresour. Technol.* **2014**, *171*, 159–167. [CrossRef] [PubMed]

47. Bonten, C. *Kunststofftechnik. Einführung und Grundlagen*; Hanser Carl Verlag: München, Germany, 2014.

PERMISSIONS

LIST OF CONTRIBUTORS

Stéphane Bruzaud
Institut de Recherche Dupuy de Lôme (IRDL), Université de Bretagne Sud (UBS), EA 3884 Lorient, France

Anne Elain
Institut de Recherche Dupuy de Lôme (IRDL), Université de Bretagne Sud (UBS), 56300 Pontivy, France

Tatiana Thomas
Institut de Recherche Dupuy de Lôme (IRDL), Université de Bretagne Sud (UBS), EA 3884 Lorient, France
Institut de Recherche Dupuy de Lôme (IRDL), Université de Bretagne Sud (UBS), 56300 Pontivy, France

Kumar Sudesh, Hua Tiang Tan and Hui Lim
School of Biological Sciences, Universiti Sains Malaysia (USM), Penang 11800, Malaysia

Alexis Bazire
Laboratoire de Biotechnologie et Chimie Marines (LBCM), IUEM, Université de Bretagne-Sud (UBS), EA 3884 Lorient, France

Hadiqa Javaid, Ali Nawaz, Hamid Mukhtar, Ikram-Ul-Haq, Kanita Ahmed Shah, Syeda Michelle Naqvi and Sheeba Shakoor
Institute of Industrial Biotechnology (IIB), Government College University, Lahore 54000, Pakistan

Naveeda Riaz
Department of Biological Sciences, International Islamic University, Islamabad 45550, Pakistan

Hooria Khan and Imdad Kaleem
Department of Biosciences, COMSATS University Islamabad (CUI), Islamabad 45550, Pakistan

Aamir Rasool
Institute of Biochemistry, University of Balochistan, Quetta 87300, Pakistan

Kaleem Ullah
Department of Microbiology, University of Balochistan, Quetta 87300, Pakistan

Robina Manzoor
Faculty of Marine Sciences, Lasbella University of Agriculture, Water and Marine Sciences, Balochistan 90150, Pakistan

Ghulam Murtaza
Department of Zoology, University of Gujrat, Gujrat 50700, Pakistan

Ayaka Hokamura, Yuko Yunoue, Saki Goto and Hiromi Matsusaki
Department of Food and Health Sciences, Faculty of Environmental and Symbiotic Sciences, Prefectural University of Kumamoto, Kumamoto 862-8502, Japan

Christoph Herwig
Institute of Chemical, Environmental and Bioscience Engineering, Research Area Biochemical Engineering, Technische Universität Wien, 1060 Vienna, Austria

Donya Kamravamanesh
Institute of Chemical, Environmental and Bioscience Engineering, Research Area Biochemical Engineering, Technische Universität Wien, 1060 Vienna, Austria
Lackner Ventures and Consulting GmbH, Hofherr Schrantz Gasse 2, 1210 Vienna, Austria

Maximilian Lackner
Lackner Ventures and Consulting GmbH, Hofherr Schrantz Gasse 2, 1210 Vienna, Austria
Institute of Industrial Engineering, University of Applied Sciences FH Technikum Wien, Höchstädtplatz 6, 1200 Vienna, Austria

Bhakti B. Salgaonkar and Judith M. Bragança
Department of Biological Sciences, Birla Institute of Technology and Science Pilani, K K Birla, Goa Campus, NH-17B, Zuarinagar, Goa 403 726, India

Björn Gutschmann, Manon T. H. Weiske, Peter Neubauer and Sebastian L. Riedel
Bioprocess Engineering, Department of Biotechnology, Technische Universität Berlin, 13355 Berlin, Germany

Thomas Schiewe and Roland Hass
innoFSPEC, University of Potsdam, 14476 Potsdam, Germany

Rodrigo Yoji Uwamori Takahashi, Nathalia Aparecida Santos Castilho, Marcus Adonai Castro da Silva, Maria Cecilia Miotto and André Oliveira de Souza Lima
Centro de Ciências Tecnológicas da Terra e do Mar, Universidade do Vale do Itajaí, R. Uruguai 458, Itajaí-SC 88302-202, Brazil

Gianni Pecorini
Department of Chemistry and Industrial Chemistry, University of Pisa, UdR INSTM – Pisa, Via G. Moruzzi 13, 56124 Pisa, Italy

Dario Puppi, Andrea Morelli and Federica Chiellini
BIOLab Research Group, Department of Chemistry and Industrial Chemistry, University of Pisa, UdR INSTM Pisa, via Moruzzi 13, 56124 Pisa, Italy

Ivana Novackova
Faculty of Chemistry, Brno University of Technology, Purkynova 118, 612 00 Brno, Czech Republic

Petr Sedlacek
Material Research Centre, Faculty of Chemistry, Brno University of Technology, Purkynova 118, 612 00 Brno, Czech Republic

Dan Kucera, Iva Pernicova and Stanislav Obruca
Faculty of Chemistry, Brno University of Technology, Purkynova 118, 612 00 Brno, Czech Republic
Material Research Centre, Faculty of Chemistry, Brno University of Technology, Purkynova 118, 612 00 Brno, Czech Republic

Eduarda Morgana da Silva Montenegro and Gabriela Scholante Delabary
Centro de Ciências Tecnológicas da Terra e do Mar, Universidade do Vale do Itajaí, R. Uruguai 458, 88302-202 Itajaí-SC, Brazil

Fernando Dini Andreote
Department of Soil Science, "Luiz de Queiroz" College of Agriculture, University of São Paulo, Piracicaba-SP 13418-260, Brazil

Martin Koller
Office of Research Management and Service, c/o Institute of Chemistry, University of Graz, NAWI Graz, Heinrichstrasse 28/III, 8010 Graz, Austria
ARENA—Association for Resource Efficient and Sustainable Technologies, Inffeldgasse 21b, 8010 Graz, Austria

Timo Pittmann
TBF + Partner AG, Herrenberger Strasse 14, 71032 Boeblingen, Germany

Heidrun Steinmetz
Department of Resource Efficient Wastewater Technology, University of Kaiserslautern, Paul-Ehrlich-Str. 14, 67663 Kaiserslautern, Germany

Index

9 781647 403980